QH   2503
361   Glass, Bentley
G 55 Forerunners Of Darwin

DATE

JAN 6 '69
MAY 18 '79

Waubonsee Community College

FORERUNNERS OF DARWIN: 1745-1859

FORERUNNERS OF ARWIN: 1745–1859

EDITED BY *BENTLEY GLASS*

OWSEI TEMKIN

WILLIAM L. STRAUS, JR.

under the auspices of
The Johns Hopkins History of Ideas Club

BALTIMORE: *The Johns Hopkins Press*

© 1959 by The Johns Hopkins Press, Baltimore 18, Md.

Distributed in Great Britain by Oxford University Press, London

Printed in the United States of America by J. H. Furst Co., Baltimore

This book has been brought to publication with the assistance of a grant from The Ford Foundation.

The Library of Congress has cataloged this book as follows:

Glass, Hiram Bentley, 1906–   *ed.*

   Forerunners of Darwin: 1745–1859. Edited by Bentley Glass, Owsei Temkin ₍and₎ William L. Straus, Jr., under the auspices of the Johns Hopkins History of Ideas Club. Baltimore, Johns Hopkins Press ₍1959₎

   471 pp. illus. 24 cm.

   Includes bibliography.

   1. Evolution—Hist.   I. Title.

QH361.G55   575.016   59–9978 ‡

Library of Congress

PREFACE

This book, which is issued on the centennial of the publication of Charles Darwin's monumental achievement, is in no way intended to detract from his proper glory, but rather to provide a setting whereby to judge it aright. All scientific discoveries and developments have their antecedents, and those of us who labor in science know full well that the connections between antecedent and current work oft-times become so blurred that few workers can really say how and where they obtained their own viewpoints and conceived their lines of attack. Darwin may indeed attribute his own evolutionary inspiration to Malthus and to Lyell, but how fully was that true? And whence did Malthus and Lyell derive their own conceptions? Alfred Russel Wallace likewise arrived at the theory of natural selection in 1858, in complete ignorance of Darwin's long effort to write *The Origin of Species*—and from whence did he receive the conception? Sharp observation of natural phenomena unquestionably plays its part, but the thought-pattern present in the observing mind is the ground, fertile or stony, upon which the seed falls.

Wisps of evolutionary thought thread the philosophy of the Greeks and the religious thought of church fathers and scholastics. The rise of modern science, during the fifteenth, sixteenth, and seventeenth centuries, forced the observing mind to doubt traditional explanations of origins and to wonder about fossils and living species, about time and transmutation, about the dissemination of life in New and Old Worlds, about Providence and Plan. By the middle of the eighteenth century, theories of evolution were being heatedly debated. As the evidence in favor of organic evolution grew, the struggle between two seeming alternatives, the one, Divine Plan and Providence in an Original Creation and the other, a godless mechanism of chance and of blind cause and effect, became not so much a debate that divided scientists and philosophers into two camps as a cleavage in the heart of each individual man. Thus, as the individual often resolves an implacable conflict by repressing it into the subconscious, so human thought in the first half of the nineteenth century stubbornly and blindly repressed the implications of the growing evidence in regard to the origin of species, including his own. Darwin was the outburst of those repressed conclusions, the victory of that submerged scientific conviction. His was the magnificent synthesis of evidence, all known before, and of theory, adumbrated in every postulate by his forerunners—a synthesis so compelling in honesty and comprehensiveness that it forced such men as Thomas Henry Huxley to say: How stupid not to have realized that before!

This volume is devoted to the forerunners who made Charles Darwin's achievement possible. Many of them he had hardly heard of, some of them perhaps not at all. Certain of them were hardly evolutionists; others, in their own eyes, not evolutionists at all. Some, who lived into the period after 1859, even hated the Darwinian teaching and fought it vehemently. Yet one and all they formed the great, steadily enlarging current of biological thought which eventuated in Charles Darwin. The number of such forerunners who are treated in the present volume is by no means complete. Malthus and many another who might legitimately be added are missing from the gallery of precursors. Perhaps another centennial will produce a more ambitious and definitive work. What this one attempts to do is merely to establish the existence and to signify the importance of the descent of ideas for the doctrine of the descent of species.

The history of science is also a segment of the history of ideas, and yields us a glimpse of the enlarging mind of man. It is thus much more than a coincidence that the person who, more than fifty years ago, wrote the first penetrating essays on the early evolutionists later became the progenitor of the study of the history of ideas, and founder of The History of Ideas Club at Johns Hopkins. Some of those early essays of Arthur O. Lovejoy are here revised and republished, together with a new contribution made by him to this subject. Other members of the History of Ideas Club have contributed studies to the project, and guests of other institutions of learning have presented papers to the History of Ideas Club which are likewise included. This is the way the book has grown, a somewhat haphazard way governed by individual predilections, and therefore explaining its obvious omissions and lapses. If there be countervailing merit, perchance it lies in the enthusiasms with which the several authors have treated their chosen topics.

Bentley Glass

November 12, 1958

CONTENTS

Preface   v

I. *Introductory Background*

1. Fossils and Early Cosmology   3
   FRANCIS C. HABER
2. The Germination of the Idea of Biological Species   30
   BENTLEY GLASS

II. *The Eighteenth Century*

3. Maupertuis, Pioneer of Genetics and Evolution   51
   BENTLEY GLASS
4. Buffon and the Problem of Species   84
   ARTHUR O. LOVEJOY
5. Diderot and Eighteenth Century French Transformism   114
   LESTER G. CROCKER
6. Heredity and Variation in the Eighteenth Century Concept of the Species   144
   BENTLEY GLASS
7. Kant and Evolution   173
   ARTHUR O. LOVEJOY
8. Herder: Progressionism without Transformism   207
   ARTHUR O. LOVEJOY
9. Fossils and the Idea of a Process of Time in Natural History   222
   FRANCIS C. HABER

III. *The Nineteenth Century*

10. Lamarck and Darwin in the History of Science   265
    CHARLES COULSTON GILLISPIE

11. An Embryological Enigma in the *Origin of Species*   292
    JANE OPPENHEIMER

12. The Idea of Descent in Post-Romantic German Biology: 1848–1858   323
    OWSEI TEMKIN

13. The Argument for Organic Evolution before the *Origin of Species*, 1830–1858   356
    ARTHUR O. LOVEJOY

14. Schopenhauer as an Evolutionist   415
    ARTHUR O. LOVEJOY

15. Recent Criticism of the Darwinian Theory of Recapitulation: Its Grounds and Its Initiator   438
    ARTHUR O. LOVEJOY

Index   459

List of Illustrations

         Ray, FACING 34

         Maupertuis, FACING 54

         Buffon, FACING 86

         Bonnet, FACING 164

         Lamarck, FACING 266

ONE

OSSILS AND EARLY COSMOLOGY

FRANCIS C. HABER

> The leading idea which is present in all our researches, and which accompanies every fresh observation, the sound which to the ear of the student of Nature seems continually echoed from every part of her works, is—
> Time!—Time!—Time!
>
> (George Poulett Scrope, 1858)

I

The growth of knowledge about fossil forms of life, out of which a genealogy of species could be drawn, and the emergence of an appreciation of the immense time required by natural processes to fulfill the history of life indicated by the fossil record, were among the fundamental developments in the background of Charles Darwin's *Origin of Species*. The genealogy of fossil species was an innovation of but a few decades before Darwin turned his attention to evolution, and it was Darwin's theory which transformed it from a chart of special creations to a pattern of evolving life. That he was able to make this transformation was, in part, because he had come to a full recognition of the time process represented by the fossil record, a singular achievement before 1858.

At the beginning of the nineteenth century, the prevailing view on the age of the earth was based on Biblical chronology, which placed the date of creation at about four thousand years before

Christ. The exact date of creation had often been disputed by scholars because of variant texts of the old Testament, but the range of time in question did not materially affect the idea of a short duration for the earth. The chronology determined by Archbishop Ussher in 1650 was added to the Authorized Version of the English Bible, and at the popular level its users accepted the chronology as a part of the Old Testament itself, including Darwin until 1861, when he expressed surprise at learning that such was not the case.[1]

The survival of Reformation bibliolatry into the nineteenth century, with a resulting narrow, literal interpretation of Genesis as revealed history, gave a powerful theological sanction to the concept of a short duration of temporal existence. But even among critical rationalists who questioned the authorship of Genesis by Moses, it was not unusual to find an acceptance of the time scale of Biblical chronology. Infinite time and indeterminate eons had long been familiar ideas in physics, mathematics, philosophy, and literature, particularly ancient and Oriental literature, yet, in notions of actual, elapsed historical time, the Holy Scriptures were generally accepted in the entire Western World as the most ancient of literary sources with a record of historical events. While it was mandatory for the bibliolatrists to concur with the chronology of Genesis, there was little reason to think that the duration of the earth was greatly in excess of the span given by Biblical chronology as long as the only evidence on the age of the earth came from literary sources. In the first decade of the nineteenth century, there was still no strong challenge to the Biblical age of the earth from archeology, Egyptology, Assyriology, cultural anthropology, linguistic history, or astronomy, although sporadic questionings of Genesis had been ventured on some points in these fields. The philosophers of the Enlightenment often disregarded Biblical chronology, adopting vague eons of time in its stead, but the first violent scientific onslaught against the traditional Christian concept of the age of the world came from geology at about the beginning of the nineteenth century, though it had been long in preparation.

Concessions to geo-chronology were made piecemeal by the orthodox, but by the 1840's the age of the earth had been extended from six thousand years to millions. The evidences of antiquity bore in upon geologists and paleontologists rapidly. Still, there was a

[1] Francis Darwin and A. C. Seward, eds., *More Letters of Charles Darwin* (New York, 1903), II, 30-31.

reluctance to go beyond positive evidence on the part of scientists, and when the *Origin of Species* was written, Darwin anticipated a strong objection to his theory of evolution because of an absence of fossil evidence of mutations between species and a prejudice regarding time which had deterred geologists from interpolating the necessary unrecorded operations in the fossil series. Darwin turned to the geological record and pictured it as a history of the world, imperfectly kept, written in a changing dialect, of which the last volume alone had been preserved, and of that, only short, broken chapters here and there, with a few lines left on a page. When the record was so viewed, the magnitude of time in the course of nature could be fully appreciated, and, he reasoned, it was miraculous, not that links between the species were missing, but that any fossils should have survived the long course of destruction in nature. The importance of a conception of adequate time for the acceptance of his theory of evolution, Darwin expressed as follows:

> The belief that species were immutable productions was almost unavoidable as long as the history of the world was thought to be of short duration; and now that we have acquired some idea of the lapse of time, we are too apt to assume, without proof, that the geological record is so perfect that it would have afforded us plain evidence of the mutation of species, if they had undergone mutation.[2]

There were aspects of the Christian view of history which held a potential for a genetic view of development, but until Darwin's theory of evolution gained acceptance, this potential was scarcely brought to light, and orthodoxy supported a fixed scheme of things in a finished Creation of a few thousand years' existence. Such a view was early challenged by evidence of a geological nature, and, although there were a host of other important influences involved, the study of fossil beds were of primary significance in raising a knowledge of nature's past beyond the historic period. Not only did the fossil strata eventually furnish the content and chronology of prehistory, but they also stirred man to the realization that nature had indeed undergone a series of extensive changes in the course of time. The earth has the appearance of stability and permanence, despite the occurrence of earthquakes, volcanoes, floods,

---

[2] Charles Darwin, *On the Origin of Species* . . . , 6th ed. (New York, 1873), p. 266; or Modern Library edition, p. 368.

and small changes here and there. In the face of the assumption that the features of earth are permanent, the presence of masses of marine shells, sometimes hundreds of feet in depth, in the inland mountains raised the question, "how did they get there?" For the sake of simplicity, I shall call this problem the fossil enigma.

This question had stimulated man's curiosity from time immemorial. It appeared in antiquity, in the Middle Ages, in the Renaissance, and with ever greater intensity in the modern period. Systems of cosmogony had to take the question into account, and interest in the question was a strong motivating force in the exploration of the earth's surface. Georges Cuvier, who did so much in demonstrating that the fossil beds were archives of a prehistoric world, wrote in 1811 that it was to these fossil remains alone that we owe even the commencement of a theory of the earth, and that without them, we should perhaps have never suspected that any successive epochs or series of different operations in the formation of the world had existed.[3]

Cuvier stated the role of fossils in cosmogony rather strongly, but there is no question of its importance. Without attempting to give a connected view of the development of geology or paleontology, a brief sketch of the long history of fossils in cosmogony and their influence in giving rise to a dynamic, yet concrete, view of the process of time in nature may prove helpful in showing how an awareness of a history of nature was introduced into natural history.

II

The first person known to have explicitly recognized fossils as memorials of geologic change and the succession of life on the earth appears to have been Xenophanes of Colophon, although it is possible that fossils had a part in the thought of Thales and Anaximander on cosmogony. Xenophanes was reported to have been of the opinion

> that there had been a mixture of the earth with the sea, and that in process of time it was disengaged from the moisture, alleging

---

[3] Georges Cuvier, *Essay on the Theory of the Earth*, tr. with notes by Robert Jameson, 4th ed. (Edinburgh, 1822), p. 54. By a "Theory of the Earth," Cuvier meant the "true" theory of his own day.

that he could produce such proofs as the following: that in the midst of earth, and in mountains, shells are discovered; and also in Syracuse he affirms was found in the quarries the print of a fish and of seals, and in Paros an image of a laurel in the bottom of a stone, and in Melita parts of all sorts of marine animals. And he says that these were generated when all things originally were embedded in mud, and that an impression of them was dried in the mud, but that all men had perished when the earth, being precipitated into the sea, was converted into mud; then, again, that it originated generation, and that this overthrow occurred to all worlds.[4]

This report, made by Hippolytus about the year 200, would indicate that his view of the connection between fossils and geologic change had been kept alive for some six centuries of ancient intellectual life.

Xanthus of Lydia also drew attention to shells in the lands of Armenia, Phrygia, and Lydia, and concluded that these places had once been the bed of the sea.[5] He further suggested that the land and sea areas must be constantly undergoing a change of positions. Herodotus noted fossil shells in the hills of Egypt and deduced that those lands were once submerged.[6] Eratosthenes, in the third century before Christ, trying to explain the changes taking place on

---

[4] Hippolytus, *The Refutation of All Heresies*, Bk. I, ch. XII, in *The Ante-Nicene Fathers*, ed., A. Cleveland Coxe (New York, 1893-6), v, 17. A different translation, and the most widely accepted, is given by Hermann Diels, *Fragmente der Vorsokratiker* (Berlin, 1906), 1^5, 122, englished by Morris R. Cohen and I. E. Drabkin, *A Source Book in Greek Science* (McGraw-Hill, New York, 1948), p. 378. See also Kathleen Freeman, *The Pre-Socratic Philosophers, A Companion to Diels, Fragmente der Vorsokratiker* (Blackwell, Oxford, 1946), pp. 89, 102. A translation is also available in John Burnet, *Early Greek Philosophy*, 4th ed. (London, 1930), pp. 123-4. The scientific importance of this passage was made known by Alexander von Humboldt in *Kosmos*. Karl Gustav Fiedler, *Reise durch alle Theile von Griechenland*, 1834-7 (Leipzig, 1840) pointed out that Xenophanes could not have seen fossils in Paros, but could have seen them in Pharos (see Ernst von Lasaulx, "Die Geologie der Greichen und Römer. Ein Beitrag zur Philosophie der Geschichte," *Abhandlungen der Philosophisch-Philologischen Classe der Koeniglich Bayerischen Akademie der Wissenschaften*, Bd. XXVII, Abth. 3 [München], 1852, S. 519n). Gronovius had early faced the problem of finding the proper rendering of "seals" and "laurel," and this has continued to draw attention, e. g., Diels; Freeman; Burnet; Theodor Gompertz, *Greek Thinkers* (London, 1901-12), I, 162, 551; and *Isis*, XXXIII (1942), 689-90.

Burnet says of this passage, "This is, of course, the theory of Anaximander, and we may perhaps credit him rather than Xenophanes with the observation of fossils" (p. 124).

[5] *The Geography of Strabo*, Intro., Bk. I, Ch. III, par. 4.

[6] Herodotus, *History*, II, 10.

the earth's surface, observed "that this question in particular has presented a problem: how does it come about that large quantities of mussel-shells, oyster-shells, scallop-shells and also salt-marshes are found in the interior at a distance of two or three thousand stadia from the sea." [7]

Even at this early period, the two opposing outlooks on geologic process, catastrophism and uniformitarianism, or actualism, became manifest. Popular in Greek thought was the view that the past consisted of a succession of recurring cycles called Great Years. In the cycle there was a Great Summer when the earth slowly dessicated and finally burst into flames, followed by a Great Winter when the earth was subjected to violent inundations. Some such view seems to have been at the heart of the system of Xenophanes. It was also expounded by Seneca in *Quaestiones Naturales*. Between the two, the early and late in ancient science, the cataclysmic Great Year worked its spell, even in the cosmology of Plato, after whom one version of the Great Year has been called the Platonic Year.[8]

Among those Greek philosophers who believed in a created world, some claimed that the waters of the earth were diminishing, as evidenced by marine fossils in the mountains, and as there had been a beginning to the process of evaporation, so, too, there must be an end when the world would be extinguished by a great conflagration.[9] Perhaps one phase of the Great Year was understood in this creationist philosophy, instead of a unique existence for the earth, but in any case, the pattern of events in both the creationists philosophy and that of periodic regeneration was marked by catastrophism.

Aristotle directed the arguments of his *Meteorologica* against the creationists.[10] The recurring cycle underlay his cosmogony,[11] but cataclysms in it were restrained to occasional intensified natural

---

[7] *The Geography of Strabo*, loc. cit.
[8] A bibliography on cyclical time conceptions is given in Mircea Eliade, *The Myth of the Eternal Return*, tr., Willard R. Trask (Pantheon Books, New York, 1954), originally published in French, Paris, 1949. Detailed information on the Great Year is also given in Pierre Duhem, *Le système du monde* (Paris, 1913-17), I, 65-85, 275-96; II, 214-23, 447-53; v, 133-7, 223-6.
[9] See, e. g., Pierre Duhem, *Études sur Léonard de Vinci*, . . . (Paris, 1906-9), II, 286 ff. on the *Liber de Mundo* of Philo, or pseudo-Philo.
[10] *Meteorologica*, Bk. II, Ch. I, par. 1. Presumably Anaximander and Diogenes of Apollonia, see p. 124n, H. D. P. Lee's translation, Loeb Classical Library.
[11] E. g., ibid., I, XIV, p. 352a.

events, such as the heavy rainfall during the Deucalion Flood, which he considered to be a local phenomenon.[12] The even tenor, gradualism, and continuity of geological change in Aristotle's system suggest modern uniformitarianism. Springs shift their location and affect the moisture in an area, marshy land dries sufficiently for the support of agriculture, then it becomes too dry, and people move to more luxuriant regions, erosion carries off the dry land, sediments build up into offshore banks, behind which lakes are created, and these in turn become marshes, tillable land, and desert. Little by little all the lands and seas alternate their positions on the earth's surface, not once, but again and again. "But the whole vital process of the earth takes place so gradually and in periods of time which are so immense compared with the length of our life, that these changes are not observed, and before their course can be recorded from beginning to end whole nations perish and are destroyed."[13]

The *Meteorologica* was an almost ever-present source for ideas regarding the nature of change in cosmogony, whenever such problems came under consideration in Western thought, but Aristotle's eternalism made his cosmogony unacceptable to the generality of Christians. Aristotle wrote:

> It is therefore clear that as time is infinite and the universe eternal that neither Tanais nor Nile always flowed but the place whence they flow was once dry: for their action has an end whereas time has none. And the same may be said with truth about other rivers. But if rivers come into being and perish and if the same parts of the earth are not always moist, the sea also must necessarily change correspondingly. And if in places the sea recedes while in others it encroaches, then evidently the same parts of the earth as a whole are not always sea, nor always mainland, but in process of time all change.[14]

The catastrophists criticized Aristotle's cosmogony, holding that if the natural place of water was above earth, and the upraised parts of the land were eroded throughout eternity, at some point all the land must be transported to its natural place below the water. How then, could the proportion of exposed land area to sea area remain constant, as Aristotle maintained? But Theophrastus and other Aristotelians countered these criticisms by insisting that lands emerged constantly through the action of fiery gases enclosed

---
[12] Ibid.
[13] Ibid., p. 351b, E. W. Webster translation, Oxford edition (1923).
[14] Ibid., p. 353a. (H. D. P. Lee's translation.)

within the earth.¹⁵ Strabo even ingeniously deduced that the land of vast regions was slowly uplifted by the pressure of underlying hot gases which could not find a vent for escape.¹⁶ With the addition of the principle of the uplift of land masses to the effects of erosion, the Greeks were able to reach a general view of cosmogony which undoubtedly had an influence on the formulation of uniformitarian views in modern geology.

Important as the uniformitarianism in Aristotle's picture of changes on the earth's surface may have been to the development of geological science, it was offset by other aspects of his cosmogony. The agencies of change—rain, wind, floods, earthquakes, volcanoes, evaporation, condensation—were themselves caused by variations in the action of the sun as it moved closer or farther away from the earth along the path of the ecliptic. There was a measure of truth in such a conception, but when Aristotle assumed that the sun and other celestial bodies exerted a direct influence on the elements, a supernatural causation was resorted to which could not be verified. He thought that by means of celestial influences fossil bodies were generated in the earth. Whether or not it was his intent, this form of generation removed with one stroke the testimony of fossils in the argument of the creationists on the diminution of the seas. However, by putting the organic origin of fossils in doubt, he also lessened their importance as a witness of geological changes.

In antiquity the term fossil meant anything dug from the ground, and the distinction between organic fossils and minerals was not clearly made until the modern period. Aristotle must have been familiar with fossil remains, but the only probable reference to them in his writings occurs in *De Respiratione*, where he remarks that a great many fishes live in the earth motionless and are found when excavations are made.¹⁷ He apparently believed that most organic fossils were akin to minerals and were generated in the earth out of exhalations under celestial influences: "For there are, we maintain, two exhalations, one vaporous and one smoky; and there are two corresponding kinds of body produced within the earth, 'fossils' and metals." ¹⁸ Such at least was the development of thought on

---

[15] See discussion in Duhem, *Études*, II, 286 ff.
[16] *Geography*, Intro., Bk. I, Ch. III, par. 5.
[17] See Frank Dawson Adams, *The Birth and Development of the Geological Sciences* (Williams and Wilkins, Baltimore, 1938), p. 12.
[18] *Meteorologica*, Bk. III, Ch. VI, par. 2.

fossils among Aristotelians in the Middle Ages and the Renaissance. It was also expressed by his successor at the Lyceum, Theophrastus, who maintained that fossil ivory and large stone-like bones had been spontaneously generated in the earth. However, Theophrastus also gave the opinion that fish might have grown in the earth from spawn left by fish moving in underground passages connected with the sea. This theory could have stemmed from observing goby-like walking fish in the mud flats of the Black Sea,[19] and a similar observation may have been the basis of Aristotle's remark in *De Respiratione*.

Fossils continued to attract attention, even at the Lyceum. They were often casually collected as curiosities. Large bones were regarded as remnants of giants who lived in the heroic age, and marine fossils were thought to be witnesses of a diminution of the sea, or relics of the Deucalion Flood.[20] No attempt will be made here to assay the many reasons behind the failure of the ancients to develop a science of geology, but certainly one reason was that fossils, outside of presenting the enigma of changes on the earth's surface, were not systematically investigated. Such an investigation of the fossils would have soon disposed of the idea that they were spontaneously generated, but here again, the belief in archetypes in nature probably would have led the naturalists to dismiss extinct species as abberations from the norm. Such an attitude certainly lay behind the philosophy on fossils of a later period, immortalized in the designation of fossils as "sports of nature." Lacking a knowledge of the order of successive changes in the history of the earth contained in the fossil record, the geological process in ancient thought was left timeless.

III

With the advent of Christianity, celestial determination passed out of favor, at least in human destiny,[21] but the belief that fossils

[19] See Pliny, *Natural History*, Bk. IX, Ch. 83; Bk. XXXVI, Ch. 29.

[20] See Lasaulx, *Geologie*, pp. 522-30, for specific instances. Some of the ancients thought the giant bones were those of Orestes. Apuleius of Medaura thought fossil fish in the mountains were relics of the Deucalion Flood.

[21] Marshall Clagett, *Greek Science in Antiquity* (Abelard-Schuman, New York, 1955), p. 139, cites Augustine's sanction of a limited celestial influence in nature (*City of God*, v, 6).

12 FORERUNNERS OF DARWIN

were created in the earth by a plastic or mineralizing force under celestial influences survived and passed over into the medieval period. The fossil question had been raised by Arabic philosophers, and Avicenna revived the theory of an organic origin of fossils which was subsequently put in circulation by Albertus Magnus and Vincent of Beauvais. Avicenna thought that vegetable and animal bodies could be converted into stone, either by a *vis lapidificativa* to be found in stony places, or by the drying up of mud in which they were enclosed. His views on a mineralizing virtue were probably derived from Aristotle, and his cosmogony was unquestionably influenced by *Meteorologica*, since his account of geological changes was appended to his translation of this work of Aristotle.[22] Subsequently another theory of the organic origin of fossils was stated by Ristoro d'Arezzo in *La Composizione del Mondo* (1282),[23] but this time it was in terms of Mosaic history. D'Arezzo reasserted the belief, first expressed by Tertullian,[24] that marine animals had been left in the mountains by the Noachian Deluge.

The Flood theory of fossil origins became increasingly popular in Italy in the late Middle Ages and Renaissance, but along with it flourished the alternative theories that fossils were abortive attempts ("sports of nature" and "figured stones") of astral forces to generate in the earth imitations of the productions of nature on its surface, or that fossils were organic remains petrified by the mysterious "mineralizing virtue." These views also continued as the main explanations of fossil origins well into the eighteenth century; but before passing the Renaissance, a few exceptions are worthy of notice, although their ultimate influence is debatable.

The writings of Leonardo da Vinci present a remarkable illustration of the scientific potential of medieval Aristotelian cosmogony when infused with the modern spirit of observation. Leonardo asked himself "why the bones of great fishes and oysters and corals and various other shells and sea-snail are found on the high tops of mountains that border on the sea, in the same way in which they are found in the depths of the sea?"[25] Turning to nature for

[22] See Duhem, *Études*, II, 302-19, and Adams, op. cit., pp. 19, 82-3, 333-5 (includes excerpt of Avicenna's *de Congelatione*).
[23] See Duhem, *Études*, II, 319 ff., and Adams, op. cit., pp. 335-41 (includes excerpt from *La Composizione*).
[24] *De Pallio*. See Lasaulx, *Geologie*, pp. 529, 538.
[25] Edward MacCurdy, ed., *The Notebooks of Leonardo da Vinci* (Reynal and Hitchcock, New York, 1939), p. 342. The notebooks consist of fragmentary notes written at various periods of time, and it might be objected

an answer, Leonardo saw detail with the eye of an artist and looked at the operations of natural forces with the understanding of an engineer. From some lofty elevation he scanned the surrounding countryside, and saw that the topography near Florence bore the appearance of having been sculptured by running waters. He studied how the waters in flood-swollen rivers carried their burden of mud and debris from the land down to the sea. " When the floods of the rivers which were turbid with fine mud deposited this upon the creatures which dwelt beneath the waters near the ocean borders, these creatures became embedded in this mud, and finding themselves entirely covered under a great weight of mud they were forced to perish for lack of a supply of the creatures on which they were accustomed to feed." [26] The mud displaced the soft parts of these shell animals, and both the mud inside the shells and in the surrounding bed turned to stone in the process of time. Strata were formed in a succession in which " may be counted the winters of the years during which the sea multiplied the layers of sand and mud brought down by the neighbouring rivers, and spread them over its shores." [27] In such a manner, Leonardo thought, the fossil beds seen in the mountains had been formed.

By the study of the living counterparts of the fossils, Leonardo explained why the Deluge could not have deposited the fossils. The arrangement of the fossils in pairs and rows as in living colonies was proof that the deposition had taken place slowly and in quiet waters. The succession of layers itself indicated a long process of time, and ". . . between the various layers of the stone are still to be found the tracks of the worms which crawled about upon them when it was not yet dry." [28] Corals could be found in the fossil beds with worm holes in them, and mixed with the corals in the stone were stocks and families of oysters, as in the undisturbed waters of the seas today. Furthermore, the waters of the Deluge could not have carried the heavy live shell animals on their crest and thus

that it is unwise to press from such fragments an organized philosophy. This Duhem has done in his *Études sur Léonard de Vinci* and, on the whole, it seems to me, successfully. The bulk of the notes on fossils occur close together in the Leicester MS, some are extended and approach essays in their organization, there is little contradiction between them, and the whole MS appears to have been intended for the treatise on the nature of water, to which Rafaelle du Fresne referred (MacCurdy, p. 51). Some of the passages in MacCurdy's work have been reprinted in Kirtley F. Mather and Shirley L. Mason, eds., *A Source Book in Geology* (McGraw-Hill, New York, 1939), pp. 1-6.
[26] MacCurdy, *Notebooks*, p. 311.
[27] Ibid., p. 339.   [28] Ibid., p. 338.

to the tops of the mountains. Nor could the animals have transported themselves. The oysters were fastened to the bottom of the sea, while cockles could not have traveled from the Adriatic to the mountains of Lombardy in forty days, since their rate of travel was only three or four braccia a day.[29]

As to the alternative theory of the generation of fossils in the earth through celestial influences, Leonardo said it was maintained only by a set of ignoramuses,[30] and then he listed the anatomical and morphological evidences which rendered such a theory ridiculous. However, the keen analysis Leonardo gave to the processes of erosion, deposition of fossils, and petrifaction of strata still did not explain the elevation of the fossil beds into the mountains. Here his powers of observation were inadequate. Some theory was needed, and Leonardo fell back on the cosmogony of Albert of Saxony, a fourteenth-century Aristotelian.[31]

In trying to solve the old problem in the Aristotelian cosmogony of why the exposed land did not take its natural place beneath water as a result of erosion over the course of an eternity, Albert had postulated that the earth has two centers, one of weight and one of form, which do not coincide. The center of the earth's form, he thought, is at the fixed center of the universe, whereas the center of the earth's weight is set slightly off from the fixed center as a result of the irregularities of the earth's bulk. He also thought that every change of weight on the surface of the earth was transmitted all the way across the fixed center to the opposite side, so that the parts of the earth would be in equilibrium on each side of the center. Thus, as land moved across the surface of the earth by erosion, the adjustments of weight across the fixed center caused subsidence or elevation at the antipodes of the axis of movement, and the progressive advance of erosion on the surface of the earth would keep the center of weight slowly rotating around the fixed center. Since water is fluid, its circumference would remain equidistant from the fixed center, while the land surface was depressed beneath it in one area and elevated above it in another in the progressive rotation of the center of weight. This insured a constant proportion between the exposed land and the seas, at the same time allowing for a continuous alternation of their positions.

[29] Ibid., pp. 330-31.
[30] Ibid., p. 338.
[31] Duhem, *Études*, I, 1-50, presents an extensive discussion of Albert's theories and Leonardo da Vinci's knowledge of them.

Leonardo integrated this system of Albert of Saxony with his own observations of erosion to explain the position of fossils in the mountains, as can be seen in the following notes:

> Because the centre of the natural gravity of the earth ought to be in the centre of the world the earth is always growing lighter in some part, and the part that becomes lighter pushes upwards, and submerges as much of the opposite part as is necessary for it to join the centre of its aforesaid gravity to the centre of the world; and the sphere of the water keeps its surface steadily equidistant from the centre of the world.[32]
>
> [After rephrasing the above thought in another place:] And this may also be the reason why the marine shells and oysters that are seen in the high mountains, which have formerly been beneath the salt waters, are now found at so great a height, together with the stratified rocks, once formed of layers of mud carried by the rivers in the lakes swamps and seas; and in this process there is nothing that is contrary to reason.[33]
>
> And now these beds [of the sea] are of so great a height that they have become hills or lofty mountains, and the rivers which wear away the sides of these mountains lay bare the strata of the shells, and so the light surface of the earth is continually raised, and the antipodes draw nearer to the centre of the earth, and the ancient beds of the sea become chains of mountains.[34]
>
> If the earth of the antipodes which sustains the ocean rose up and stood uncovered far out of this sea but being almost flat, how in process of time could mountains valleys and rocks with their different strata be created? . . .
>
> The water which drained away from the land which the sea left, at the time when this earth raised itself up some distance above the sea, still remaining almost flat, commenced to make various channels through the lower parts of this plain, and beginning thus to hollow it out they would make a bed for the other waters round about; and in this way throughout the whole of their course they gained breadth and depth, their waters constantly increasing until all this water was drained away and these hollows became then the beds of torrents which take the floods of the rains. And so they will go on wearing away the sides of these rivers until the intervening banks become precipitous crags; and after the water has thus been drained away these hills commence to dry and to form stone in layers more or less thick

---

[32] MacCurdy, *Notebooks*, p. 688.   [34] Ibid., p. 321.
[33] Ibid., p. 356.

according to the depth of the mud which the rivers deposited in the sea in their floods.[35]

How the rivers have all sawn through and divided the members of the great Alps one from another; and this is revealed by the arrangement of the stratified rocks, in which from the summit of the mountain down to the river one sees the strata on the one side of the river corresponding with those on the other.[36]

In a single lifetime, busy with a host of other activities, and with few guides in geology, Leonardo da Vinci pursued the quandary of marine fossils until he had devised an amazingly good system of historical and dynamical geology. His uplift and subsidence of land masses working slowly over long periods of time was hardly more mysterious in its operation than the self-adjusting forces in James Hutton's *Theory of the Earth* (1788), which marks the beginning of modern theories of geological dynamics. Like Hutton, Leonardo was a uniformitarian who appealed directly to nature for an understanding of how geological processes had operated in the past. Leonardo could see that nature had a history, and its record was to be found, not in literary sources, but in the earth itself, or, as he expressed it:

> Since things are far more ancient than letters, it is not to be wondered at if in our days there exists no record of how the aforesaid seas extended over so many countries; and if moreover such record ever existed, the wars, the conflagrations, the changes in speech and habits, the deluges of the waters, have destroyed every vestige of the past. But sufficient for us is the testimony of things produced in the salt waters and now found again in the high mountains, sometimes at a distance from the seas.[37]

Although Leonardo failed to see the role of heat in the metamorphosis of strata, geology could have gone far on his principles. Unfortunately, his notes did not appear in print until after Hutton's theory was published. In the meantime, Leonardo's fundamental premises, drawn from Aristotelian cosmogony, were demonstrated

---

[35] Ibid., pp. 310-11.
[36] Ibid., p. 338. He had apparently recognized formations as well as strata: "... the shells of Lombardy are found at four levels. And so it is with all which are made at different periods of time; and these are found in all the valleys that open out into the seas" (p. 357). However, Leonardo does not seem to have noticed any alteration of species, nor to have developed any clear conception of how a chronology could be raised from the strata, outside of varve studies. [37] Ibid., p. 345.

as false by the Copernican astronomy, while the Reformation reaffirmed the authority of a literal interpretation of Genesis, so that the Noachian Deluge was accepted among naturalists as a real and universal catastrophe in natural history.

In the sixteenth century the Flood theory of fossil distribution prevailed among those who recognized the organic origin of the fossil remains, although there were exceptions. Girolamo Fracastoro, a contemporary of Leonardo, after studying the excavations made for building the citadel of San Felice in Verona, asserted that the marine animals whose remains were unearthed had lived and died where they were found, and that they could not have been left by the Flood. Jerome Cardan, who may have had access to the manuscripts of Leonardo, also rejected the idea that one universal flood could have distributed the fossils in the earth. The best reasoning displayed on the nature of fossils in the century, after Leonardo, was by the self-taught Huguenot potter, Bernard Palissy.[38] He was patronized by Catherine de Medici, but his death in the Bastille for refusing to give up his Calvinism emphasizes that in matters of doctrine there were points beyond which it was not safe to go. From the recognition of the succession of strata containing fossils and an appreciation that each period of deposit involved a considerable lapse of time, Palissy came to the conclusion that it was impossible for the Flood to have put the marine shells inside the stone of the mountains.[39]

There is no evidence that the views of Fracastoro, Cardan, or Palissy on fossils had any appreciable influence on their age. Palissy held lectures at which he displayed his collection of fossils, and the interest he helped to arouse in them as curiosities may have been ultimately more important than his theories about them. The sixteenth century was a period of increased activity in collecting fossils and assembling fossil collections. Alexandri, Sarayna, Agricola, Gesner, Moscardo, Fallopio of Padua, Kentmann, Cesalpino, Oliva of Cremona, Mercati, Mattioli, Camden, Imperato, Majoli, Aldrovandi, and Columna, among others, were collecting fossils and writing about them. Gesner's *De Rerum Fossilium* (1565) was the first

---

[38] Duhem believes that Palissy read a 1556 edition of Cardan's *De Subtilitate* translated into French by Richard le Blanc, and that Cardan had used the manuscripts of Leonardo da Vinci. See *Études*, I, 234 ff.

[39] *Recepte véritable* (La Rochelle, 1563) and *Discours admirables* (Paris, 1580) in *Oeuvres de Bernard Palissy*, ed., Benjamin Fillon (Niort, 1888). An English translation of the *Discours* by Aurèle La Rocque was published by the University of Illinois Press in 1957.

extensively illustrated work showing fossils; and others soon followed. This activity of collecting, establishing museums, printing illustrations, and focusing attention on fossils in general was a necessary first step in the identification and classification of fossil specimens. The theories that accompanied the activity make amusing reading today, and so too does the classification based on the theories. Nevertheless, collecting took the naturalist into the field, while the display and illustration of collections added to the observational experience of the individual naturalist and his ability to compare specimens. The variety of theories which fossils stimulated was to some extent the result of the conflict which deductions from field data presented to received ideas, classical and Christian, and was an indication of the vigor of the scientific movement. However, as witnesses of tremendous alterations in the earth's surface, the bearing of fossils on cosmogony was the first thing about them to be recognized, aside from their being curiosities, and speculative reasoning to reconcile their existence with philosophical and theologically based systems of cosmogony flew ahead of empirical knowledge about fossils.

IV

With the Reformation, Mosaic history, particularly its great catastrophe of the Deluge, had to be reckoned with in cosmogony, and fossils were only one of many difficulties presented to literalist theology by the expansion of natural knowledge. The interplay of the critical spirit of rationalism with theological authority in connection with the literal interpretation of Noah's Flood has been brilliantly narrated by Don C. Allen, which he has summarized as follows:

> . . . So we have noticed that the first truly important question about the Noah story was concerned with the adequacy of the Ark's size, and we shall observe that almost at the moment that the question was answered in what seemed to be a sure and scientific way, seafarers and zoologists arrived bearing large burdens of strange birds and unknown animals that had to be enrolled on the Ark's list of passengers. We have also been spectators while the rational exegetes attempted to iron out the chronological difficulties of the great watery event, and we have been as dismayed by their flounderings among a variety of calendars and conflicting

dates as they were themselves. Finally, we have read the fluttering attempts of the commentators to assemble enough gallons of water to flood the whole earth, and we have not been surprised to see them retract the size of their maps, first to the inhabited world and then to Palestine itself. Such a process of attempted rationalization and subsequent consternation occurred when the literal commentators attempted to expound other sections of the Bible. When to this intellectual defeat was added the uncertainties arising from canonical and textual studies, the reputation of the Bible as an inspired book was seriously threatened.[40]

Threatened, but the reputation of the Bible as an inspired book was not so readily overthrown. The rational exegetes often attacked each other's interpretations of the text of Genesis, rather than the literal accuracy of the text as a history of nature. The practice developed among the rationalists, all in the name of literal accuracy, of redefining the words of the text and of making interpolations where needed to bring Mosaic history into harmony with natural history. In this movement the role of Descartes was of paramount importance for cosmogony.

Descartes wrote the substance of his cosmogony in *Le monde, ou traité de la lumière*, which he withheld from publication upon learning of the condemnation of Galileo's work.[41] The essence of it was, however, included in his *Principes de la philosophie* (1644). Anxious to avoid antagonizing the Church, he unfolded his theory of the creation of the universe with the book of Genesis clearly in mind, although only mechanical principles were utilized. The dualism between matter and spirit which Cartesianism brought into philosophy helped free the new mechanical philosophy from theology, but it also tended to remove the time process from the operations of nature. An exception was in the unfolding of his cosmology, following the model of Genesis. He preserved the genetic view, showed the universe evolving out of chaotic matter, and provided concrete principles in cosmogony which were useful in geology, although their first application was apt to be fanciful, and was to furnish interpolations in Mosaic history.

The Cartesian cosmogony gave the religious rationalist, temporarily at least, a means of extricating the waters of the Flood out

---

[40] Don Cameron Allen, *The Legend of Noah, Renaissance Rationalism in Art, Science, and Letters* (Univ. of Ill. Press, Urbana, Ill., 1949), p. 90.
[41] See letter of Descartes to R. P. Mersenne, July 22, 1633, *Oeuvres de Descartes*, ed., Victor Cousin (Paris, 1824-6), VI, 238-9.

of the difficulties in which they were mired. Picking up the evolution of the Cartesian cosmos at the point where the earth had taken shape, we see it as a core of fiery material surrounded by concentric layers of other kinds of material. The outer surface was a hard, opaque crust and under it was a layer of watery substance. The heat of the summer sun dried out the crust until it was checkered with fissures, while the heat also expanded the liquid in the next layer until part of it escaped through the pores of the crust. In the ensuing winter, however, the fissures were contracted until they blocked the return of the watery element. The watery layer had also contracted, and the result was that a vault-like space was left under the crust. Weakened by fissures and with no support under it, the crust collapsed. Because of the smaller circumference at the base of the waters, large sections of the crust came to rest at angles and their upraised edges formed the mountains of the earth, while the water that was forced out of its place became the seas of the world.

This "broken-crust" cataclysm was pregnant with possibilities for explaining the Deluge, and Thomas Burnet made the most of it in his *Sacred Theory of the Earth*, first published in Latin in 1681. The work was tremendously popular, and the robust figurative prose of Burnet endeared him to generations of readers long after his science was known to be fiction. For his own age, Burnet's *Sacred Theory*, in its prolix and labored way, brought up heavy artillery in support of Milton's graceful and inspired exposition of Genesis in *Paradise Lost*.

In the *Sacred Theory*, Paradise existed when the outer crust of the earth had newly formed and was smooth and fertile. It was lost at the Deluge, when the crust fell into the abyss of the water layer below. The falling pieces hurled waves over everything, their upturned edges formed the mountains of the earth, and between them were left the seas. Air was trapped under the curved sections of the crust, and as water worked its way into these cavities, the seas were lowered, signs of which still remain on the land. By this theory of the disruption of the earth, Burnet thought he had also solved the riddle of the longevity of the patriarchs. When the cataclysm occurred, the unevenly distributed debris tilted the axis of the earth, introduced the seasons, which had a baleful effect on the health of man, and shortened the span of his life. And now this wreck of a world, ugly, scarred and wrinkled, was a grim reminder of the punishment inflicted on man for his sins.

The method employed by Burnet in harmonizing natural history with Sacred History was essentially scholastic. Nature (the Work of God) was treated like Scripture (the Word of God) as a text. To prove a point in either area Burnet could smother it with citations from classical and patristic authors until all contradictions were obscured. It was a dazzling method, but for those investigating fossils, it did not come to grips with the main challenge to sacred cosmogony. Another method, that of directly interrogating nature, was used by Nicolaus Steno in seeking the concordance of the two histories.

After gaining a reputation in Paris as a physician, Steno went to Florence in 1665 and received an appointment as physician to Grand Duke Ferdinand II. He was also admitted to the brilliant company of scientists at the Accademia del Cimento, and it is probably through this influence that his attention was drawn to the problem of fossils. In 1669, after studying the fossil-rich area around Arezzo in Tuscany, Steno published a *Prodromus* to a *Dissertation concerning a Solid Body enclosed by Process of Nature within a Solid*, a work which displayed keen powers of observation and reasoning. Steno presented a persuasive argument that those bodies dug from the earth which looked like the parts of plants and animals were in fact the parts of plants and animals and were therefore extraneous bodies in the strata enclosing them. His analysis of the processes of petrifaction and the formation of strata enunciated for the first time some of the basic principles of the science of stratigraphy, and by means of them he demonstrated that in the formation of the earth's crust, the strata which contained extraneous bodies could not have existed from the beginning of things but must have been laid down in succession, one on top of the other.

On the assumption that each stratum was formed in a fluid medium, Steno pointed out that the fluid had to be bounded below and at its sides (unless it surrounded the earth) by solid material. Although the under and side surfaces would conform to the shape of the enclosing material, because of its fluid origin, the stratum's top surface would be level. If the top surfaces of strata were no longer parallel to the horizon, an alteration of position must clearly have taken place since the solidification, and the mountains of the earth contained the proof that such alterations had occurred. Steno suggested that the strata were uplifted by the pressure of gases in the earth. Erosion or the further action of gases then ate away the under strata in places until there was a collapse of the upper ones, creating a valley, and leaving the edges of strata at the point of fracture exposed in the resulting mountains on either side of the valley. The rubble of the collapsed parts then became the building materials for secondary mountains with a heterogeneous composition.

In a year's time Steno had been able to discover the historical character of geological process, but when he turned to the elaboration of the history, the pattern of Genesis was invoked. Because of the succession of strata, the Flood was ruled out as a means of depositing the fossils, but the last division of his work was devoted to the Universal Flood, to which he gave a large role in disrupting the strata. And the entire investigation was justified by Steno as an attempt to "set forth the agreement of Nature with Scripture by reviewing the chief difficulties which can be urged regarding the different aspects of the earth." [42] To Leibniz he confided "that he congratulated himself with having come to the aid of piety in supporting the faith of the Holy Scriptures and the tradition of the universal deluge on natural proofs." [43]

Leibniz, too, was interested in the natural history of the earth and the enigma of fossils. In a preamble to a history of the Brunswick-Lüneberg family, he expressed his views on cosmogony and fossils. This *Protogaea* was written in 1691. It was not published until 1749, but a résumé appeared in the *Acta Eruditorum* for January, 1693. Leibniz was influenced by Steno's treatment of sedimentation and stratification,[44] but he also drew directly from Cartesian cosmogony the principle of igneous action. Descartes, in unfolding his cosmos, had pretended that the earth was originally a small luminous star. Gross materials collected as clouds on its

---

[42] John Garrett Winter, ed., *The Prodromus of Nicolaus Steno's Dissertation* . . . (Macmillan, New York, 1918), p. 263. An excerpt of the principles is given in *A Source Book in Geology* by Mather and Mason.

[43] Gottfried Wilhelm Leibniz, *Protogée* . . . , ed., Bertrand de Saint-Germain (Paris, 1859), p. 18. The Latin and a German translation are also given by W. U. Engelhardt in *Protogaea* (Kohlhammer Verlag: Stuttgart, 1949), *Leibniz Werke*, I, 28, 31.

[44] It is curious to read in the excellent work of Adams, ". . . *Prodromus* remained almost unnoticed at the time and it was not, Zittel remarks, until Élie de Beaumont and Alexander von Humboldt drew attention to it that the importance of his [Steno's] work received due recognition" (p. 364). The fact is that almost every work on fossils during the remainder of the 17th century mentioned Steno. Henry Oldenburg, Secretary of the Royal Society, even translated the *Prodromus* into English in 1671. It is doubtful if any other work in Steno's age on the subject of fossils and stratification was better known than *Prodromus*, and in addition Steno exerted a personal influence on naturalists.

surface, like the spots on the sun, until the earth's fiery matter was obscured and the earth lost its luminosity. It then ceased to be a star and descended into the orbit of the sun.

Leibniz was enabled by his intimate knowledge of German mining to envisage a fiery, incandescent mass of the earth becoming obscured by a crust of slag as scoria rose to its surface. Once the crust had formed, he conjectured, the surrounding water vapor would cool and fall upon the earth, dissolving chemicals, eroding the vitreous crust into sands, clays, and other inorganic substances needed for the support of life, and gathering itself into the seas of the earth. Plants and animals were created when the earth was capable of sustaining life, and as degradation of the crust proceeded, strata, enclosing the remains of plants and animals, formed on the floor of the seas. For the subsequent disruption of the strata, Leibniz resorted to the Deluge, after the manner of Burnet and Steno, but he also added the possible alternatives of a wrinkling of the crust through the cooling of the igneous core and the collapse of "bubbles" left in the crust at the first hardening.

The greater part of the *Protogaea* was devoted to the manner of deposition, distribution, and petrifaction of fossils, and it was a splendid amplification of Steno's work. Leibniz gave further proofs of the organic origin of fossils, the succession of strata, and the importance of nature as a source for filling the gaps of history. But despite his comprehensive views on the operations of nature, Leibniz displayed no conception of the amount of time needed for their fulfillment. His frame of reference was Scripture, "from which we must not deviate." The igneous principle of the *Protogaea* did contribute at a later date, however, to the expansion of the time scale through the use made of it by Buffon.

Although the principal interest in fossils during the latter half of the seventeenth century was in connection with their meaning for cosmogony, the investigation of them as subjects of scientific interest spread throughout the international community of scholars. The scientific societies helped to focus attention on fossils and to encourage a new spirit in their study. Fossils were not only collected ever more diligently, agents even being hired to collect them in the field,[45] but they were brought under the experimental

[45] Charles E. Raven, *English Naturalists from Neckham to Ray* (Cambridge, 1947), includes much information on the activity among naturalists in England in studying fossils. Thomas Willisell was appointed an official collector of the Royal Society in 1668, and Raven thinks he was the first paid field-naturalist in England. Among his commissions was the gathering of fossils.

methodology. In 1663, Robert Hooke was slicing and polishing cross-sections of petrified wood and examining them under the microscope.[46] Fossilized substances were subjected to comparative tests with their living counterparts or similar kinds of materials. One such comparison investigated was the constitution of various kinds of stones, including those produced in the human body, in connection with the question of whether or not stones could grow. Kidney stones, gall stones, heart stones, and every kind of unusual hard growth in the body were excised, measured, weighed, described, and often illustrated. This question was intimately allied to that of the generation of fossils and fossil strata in the earth.

The progress of the incipient field of paleontology in the latter part of the seventeenth century can best be seen by comparing the attitudes of Robert Hooke about stones in 1663 with his later views. The following is an excerpt from a letter written by him to Robert Boyle about a meeting of the Royal Society:

> There happened an excellent good discourse about petrefaction; upon which occasion several instances were given about the growing of stones: some, that were included in glass viols; others, that lay upon the pasture ground; others, that lay in gravel walks; which was known by putting a stone in at the mouth of a glass viol, through which, after a little time, it would by no means pass. Next, the story of a field's being filled with stones every third year, was confirmed by some instances. And that the stones in gravel walks grows greater, had been often proved by sifting those walks over again, which had formerly passed all through the sieve, and finding abundance of stones too big to pass through the second time. Upon this, mention was made of the production of stones or lapidious concretions in the bodies of animals, and abundance of very strange instances were alledged of the finding of stones in several parts of a man's body, as in the joints of his fingers and toes, and of other parts of his body; and it was generally agreed to by all, that those people, that drink petrifying waters, are extremely subject to the stone. A place was mentioned in *Oxfordshire*, where there is such a water, and the people round about are extremely plagued with that disease. Mr. PELL and some others mentioned to have read somewhere an observation, there were more such concretions taken from one man, than the weight of his whole body amounted to.
>
> Mr. PALMER related a story of a French physician (whose

---

[46] R. W. T. Gunther, *The Life and Work of Robert Hooke* in *Early Science in Oxford*, VI (1930), 131-2.

name I have forgot) who landing sick at *Dover*, and taking a glister, voided an incredible number of small and great cockle-shells. The matter of fact was confirmed by very many of the Society, who had either had very good relation of it, or seen some of the shells. Dr. CHARLTON added, that they had lain a good while upon sea, and fed upon nothing but cheese (made of the milk of goats, which fed upon the mountains of *Bononia*, which are very full of such shells) and brandy.[47]

Continued study of fossils convinced Hooke that they could not have grown in the earth, and that the growth of stones in the body was not relevant to the petrifactions. He had observed great quantities of marine animal shells hundreds of fathoms above the level of the sea, hundreds of miles inland, in the depths of the earth, in the midst of stone, and even constituting the stone, species among the shells of a size and character no longer to be found, and the figures of tropical plants in coal. Hooke was brought to the conclusion: "That a great part of the Surface of the Earth hath been since the Creation transformed and made of another Nature; namely, many Parts which have been Sea are now Land, and divers other Parts are now Sea which were once a firm Land; Mountains have been turned into Plains, and Plains into Mountains and the like."[48] The exotic and extinct species provoked him to think that "this very land of England and Portland, did, at a certain time for some ages past, lie within the torrid zone."[49]

Hooke does not seem to have doubted the actual occurrence of the Flood of Noah, but he argued at length against it as the means of placing the fossils in the depths of the earth or of imbedding them in mountains. He thought this was done by earthquakes, which could elevate or depress large areas of the earth's surface. The fiery gases accompanying earthquakes would also provide a source of heat for liquefaction, baking, calcining, petrifaction, sublimation, distillation, and other transformations of a chemical nature.

Extinct fossil species further suggested to Hooke the idea of evolution, or perhaps devolution would be more precise, for Hooke was fond of Ovid's *Metamorphoses* and its thesis of a world in decline. He thought there was a Golden Age when the earth was soft, flexible, and smooth-skinned like a child, that it had passed

[47] Ibid., pp. 132-3.
[48] Richard Waller, ed., *The Posthumous Works of Robert Hooke* (London, 1705), "Discourse of Earthquakes," p. 290.
[49] Ibid., p. 343.

through a Silver Age, or maturity, when it dried and hardened, and an Iron Age, when the shell of the earth became petrified, crossed with wrinkles, scars, and furrows, and fell heir to the ailments of earthquakes, floods, and other debilitating disasters of old age. Noting that particular species seemed to thrive better in one climate than another, he wondered if a change of environment would have an influence on species.

> We will, for the present, take this supposition to be real and true, that there have been in former times of the world, divers species of creatures, that are now quite lost, and no more of them surviving upon any part of the earth. Again, that there are now divers species of creatures which never exceed at present a certain magnitude, which yet, in former ages of the world, were usually of a much greater and gygantick standard; suppose ten times as big as at present; we will grant also a supposition that several species may really not have been created of the very shapes, they now are of, but that they have changed in great part their shape, as well as dwindled and degenerated into a dwarfish progeny; that this may have been so considerable, as that if we could have seen both together, we should not have judged them of the same species. We will further grant there may have been, by mixture of creatures, produced a sort differing in shape, both from the created forms of the one and other compounders, and from the true created shapes of both of them. . . .
> As we see that there are many changings both within and without the Body, and every state produces a new appearance, why then may there not be the same progression of the Species from its first Creation to its final termination? [50]

Hooke's study of fossils had thus brought him to an awareness that there had been great alterations in the earth's surface and "That there have been many other species of creatures in former ages, of which we can find none at present; and that 'tis not unlikely also but that there may be divers new kinds now, which have not been from the beginning." [51] Hooke also found in such "a trivial thing as a rotten shell" a memorial of nature's history:

> Now these shells and other bodies are the medals, urnes, or monuments of nature, whose relievos, impressions, characters, forms,

---

[50] Ibid., p. 435. Much of the capitalization and italicization and a few typographical errors in the original have been eliminated in these citations from Hooke's "Discourse."
[51] Ibid., p. 291.

substances, etc. are much more plain and discoverable to any unbiassed person, and therefore he has no reason to scruple his assent: nor to desist from making his observations to correct his natural chronology, and to conjecture how, and when, and upon what occasion they came to be placed in those repositories. These are the greatest and most lasting monuments of antiquity, which, in all probability, will far antedate all the most ancient monuments of the world, even the very pyramids, obelisks, mummys, hieroglyphicks, and coins, and will afford more information in natural history, than those other put altogether will in civil. Nor will there be wanting *Media* or *Criteria* of chronology, which may give us some account even of the time when, as I shall afterwards mention.[52]

There is no evidence that Hooke actually attempted a chronology of natural history based on fossil evidence, but he clearly saw the possibility of it. Again speaking of the shells as monuments of antiquity, he remarked,

And tho' it must be granted, that it is very difficult to read them, and to raise a *Chronology* out of them, and to state the intervalls of the times wherein such, or such catastrophes and mutations have happened; yet 'tis not impossible, but that, by the help of those joined to other means and assistances of information, much may be done even in that part of information also.[53]

The views of Hooke and Steno were available to the public, but neither the intrinsic merit of their views, nor the reputation of the authors, swept paleontology into precipitous progress. Indeed, the distinction of Father of British Paleontology has been assigned by R. W. T. Gunther to Hooke's contemporary Edward Lhwyd,[54] who came to the conclusion that fossils grew in the earth from seeds dispersed by vapors of the sea. Lhwyd was unhappy with this theory, it is true, and adopted it reluctantly in opposition to his good friend, John Ray, who had probably discussed the fossil problem with Steno, and who held out firmly for the organic origin classification, had rejected the principle of spontaneous generation of fossils. Ray, an eminent naturalist and forerunner of Linnaeus in in nature both on the basis of the researches of Redi, Malpighi, Swammerdam, Lister, and Leeuwenhoek, and also on the basis of

---

[52] Ibid., p. 335. [53] Ibid., p. 411.
[54] R. W. T. Gunther, *Life and Letters of Edward Lhwyd* in *Early Science in Oxford*, XIV (1945), iii.

the theological argument that all creation was completed on the sixth day, after which life was passed down from one individual to another.[55] Lhwyd agreed with Ray that spontaneous generation in the earth by plastic forces was impossible, but he kept pressing Ray on the extinction of species and how the masses of animal life represented by fossil remains could have existed and been orderly embalmed in the strata during the short time span of Sacred History. Robert Plot, Martin Lister, William Cole, and other naturalist associates who took up the fossil problem had, after extensive studies, also doubted the organic origin of fossils. The seed theory of Lhwyd was an attempt to avoid spontaneous generation and still explain the position of exotic species and the quantity of shells in a way which would avoid an extension of the time scale of Scripture. Ray, in contrast, toward the end of his life wavered between rejecting the natural organic origin of the fossils or the time scale of Scripture.

> Such a diversity as we find of figures in one leaf of Fern and so circumscribed in exact similitude to the plants themselves, I can hardly think to proceed from any shooting of salts or the like.... Yet on the other side there follows such a train of consequences as seem to shock the Scripture-history of the novity of the world; at least they overthrow the opinion generally received, and not without good reason, among Divines and Philosophers that since the first creation there have been no species of animals or vegetables lost, no new ones produced.[56]

The celebrated Dr. John Woodward, who went out of his way to antagonize Ray and Lhwyd, claimed he had resolved the entire mystery of fossils by a new principle. In his *Essay Towards a Natural History of the Earth* (1695) he simply dissolved all the upper crust of the earth in the Deluge and let the fossils settle out in strata according to their specific gravities. His naiveté surprised Lhwyd and Ray, but the *Essay* received much acclaim and was translated and republished on the Continent. Woodward was unequivocal about the organic origin of fossils, and based some of his proofs on Steno's *Prodromus*, but the inspiration of his system, and its plausibility, were derived from Newton's "Laws of Gravity,"

---

[55] Charles E. Raven, *John Ray Naturalist, His Life and Works* (Cambridge, 1942), p. 375.
[56] R. W. T. Gunther, ed., *Further Correspondence [of John Ray]* (Ray Society, London, 1928), p. 259. Cited in Raven, ibid., p. 437.

though the good Doctor wrote that there was a mighty collection of water in the bowels of the earth, in contradiction to Newton's conclusion that the center of the earth must be five or six times as heavy as water.

Another widely acclaimed attempt to explain Mosaic history on rational principles was that of William Whiston, who succeeded Newton as Lucasian professor of mathematics at Cambridge. He wrote an elaborate account, ostensibly incorporating Newton's principles, to explain how a passing comet had created the Flood and disrupted the surface of the earth. His *New Theory of the Earth, from its Original to the Consummation of all Things, Wherein the Creation of the World in Six Days, the Universal Deluge, and the General Conflagration, as Laid Down in the Holy Scriptures, are Shown to be Perfectly Agreeable to Reason and Philosophy* (1696) won the praise of John Locke, the philosopher of reason, and of Newton, who was by then absorbed in trying to unravel the prophecies of Daniel and the apocalypse of St. John.

By the end of the seventeenth century the fossil enigma had been brought into the mainstream of thought about cosmogony. In the humanist tradition, the writings of the ancients, the Church Fathers, and medieval scholars were searched for light on the problem, and the interpretation of observations of natural phenomena was mixed with previous ideas, but in snowball fashion, the history of ideas on the subject of fossils was continued from writer to writer throughout the seventeenth century. The prejudices were perpetuated, but so too were the accumulating knowledge and the welter of solutions advanced to solve the fossil enigma. The first step towards such a solution was the unequivocal recognition of the organic origin of fossils, and this was all but achieved through the continuous study of specimens, especially by comparing them with their living counterparts, until the second step became clear, the step of recognizing fossils and fossil strata as the product of time and natural processes. It thereupon became unavoidably apparent that fossil evidence contradicted the time pattern of Mosaic history. The naturalists hesitated, then retreated before theologically supported tradition, and turned to a reexamination of the conclusions of their first step. Thus the eighteenth century inherited a fully developed and pressing issue over the nature and formation of fossils. If they were the product of natural processes, how great an extent of time did their mute presence in the rocks tell forth?

TWO

THE GERMINATION OF THE IDEA OF BIOLOGICAL SPECIES

BENTLEY GLASS

Evolutionary concepts inevitably center on the idea of the biological species—on its hereditary nature, the limits to its variability, the very meaning of a biological *kind* of plant or animal. Yet strangely enough, the historical germination of this concept of biological species has been rather neglected. In most accounts of evolutionary thought, Linnaeus is heralded as the great protagonist of the species concept, and John Ray, a century earlier, as his prophetic forerunner. No account of the history of the idea of species could possibly afford to minimize the roles of either Ray or Linnaeus in clarifying and defining the concept; but the idea itself was a very old one, its genesis shrouded in the mists of time, its variations barely adumbrated here and there in ancient and medieval literature. To trace fully the early history of this idea is beyond the scope of the present essay, which will merely sketch the outline of what may some day be a more definitive treatment.

The period of the earliest scientific refinements in the conception of the species may be taken to conclude with the end of the seventeenth century. The time of Linnaeus and his contemporaries marks a new era and will be treated more fully in a subsequent essay, "Heredity and Variation in the Eighteenth Century Concept of the Species." In that later century the modern conception of the species approached definition through the gradual resolution of two great controversies that conditioned the concept of the nature of species and their transformability. One was the controversy over the spontaneous generation of living organisms: did it exist or not?

The other was the controversy over preformation: were the new individuals of successive generations preformed within the one or the other parent, or were they shaped anew out of formless matter? As we shall see, a belief in the transformation of species was clearly associated in the minds of eighteenth century naturalists with the positions they held in respect to these controversies. The controversies themselves were not new. They grew out of the observations made by naturalists of the seventeenth century, and out of the conclusions they reached. This introductory essay will consequently concern itself not only with the germination of the concept of biological species, but also with the earliest attacks on the problem of spontaneous generation and with the genesis of the doctrine of preformation. These form a conceptual whole.

The term *species* means *kind*, and either term is precise or vague in proportion to the effort made to define it. It has certainly been recognized from antiquity—and indeed from prehistoric times, as cave paintings bear witness—that there were different kinds of plants and animals, and that in the normal course of nature each kind reproduces its own sort. The book of *Genesis*, in relating the Creation Story, says that each creature brought forth " according to its kind." [1] Likewise Aristotle, when discussing hybrids in the *Generation of Animals*, says: " The partners in copulation are naturally and ordinarily animals of the same kind; but beside that, animals that are closely allied in their nature, and are not very different in species, copulate, if they are comparable in size and if their periods of gestation are equal in length." [2] Again: " In the normal course of nature the offspring which a male and a female of the same species produce is a male or female of that same species—for instance, the offspring of a male dog and a female dog is a male dog or a female dog." [3] These ideas of species, if not in the form of definitions, are nevertheless almost equivalent to the concept adopted by John Ray, and provide evidence that from the very beginning of human thought reproduction after its own kind was considered to be a basic criterion of a species.

Aristotle, however, like most men of his time and long afterwards, also believed in spontaneous generation. Thus we find, in the *History of Animals*, the following remarks: " Now there is one

---

[1] *Genesis*, 1: v. 21, 24, 25.
[2] Aristotle, *Generation of Animals* (tr. A. L. Peck), Loeb Classical Library, Harvard University Press, Cambridge, 1943, 46a29.
[3] Ibid., 47b31.

property that animals are found to have in common with plants. For some plants are generated from the seed of plants, whilst other plants are self-generated through the formation of some elemental principle similar to a seed. . . . So with animals, some spring from parent animals according to their kind, whilst others grow spontaneously and not from kindred stock; and of these instances of spontaneous generation some come from putrefying earth or vegetable matter, as is the case with a number of insects, while others are spontaneously generated in the inside of animals out of the secretions of their several organs." [4] Clearly, the conception that reproduction and heredity define the character of species is beclouded by this simultaneous belief in spontaneous generation. If animals and plants could be generated from mud or slime or filth, or as parasites from the inward parts of their hosts, then the transformation of one species into another was an even milder sort of deviation from normal heredity and reproduction.

Before men could fully accept the constancy of species, the ghost of spontaneous generation had to be laid. Yet that was not to be until the seventeenth century, and even then exclusive of the bacteria. Meanwhile the need to classify plants and animals, from practical as well as scientific reasons, was rapidly growing. During the Renaissance, first the unsystematic compilation of all the known kinds of plants and animals of the Old World, and later the voyages of discovery, particularly in North and South America, had already greatly increased the number of living forms needing classification. Many a book was written in the seventeenth century to explain in terms of the Biblical story who the American Indians were and whence they came. More and more new species of animals and plants from foreign parts were sent back by collectors to their compatriots in Europe; and even the elasticity of Noah's ark had its limit. Could these novel forms in the New World be familiar European, Asiatic, or African species carried thither and thereafter subjected to modification? But, as one of these authors said, "Who can imagine that in so long a voyage men woulde take the paines to carrie Foxes to Peru, especially that kind they call 'Acias,' which is the filthiest I have seene? Who woulde likewise say that they have carried Tygers and Lyons?" It was too far for most of the animals or birds to swim or fly, and besides—what would they have eaten

---

[4] Aristotle, *Historia Animalium* (tr. D'Arcy W. Thompson), Clarendon Press, Oxford, 1910, Bk. v, 539a16-26.

on the way? As for the fishes, this posed a particularly difficult problem, for how could freshwater fishes have reached the New World? Some therefore urged that there had been a Special Creation for America, or that species had originated there by spontaneous generation—yet such beliefs conflicted dangerously with the doctrine of the universality of the Great Deluge. Controversies raged interminably. Not a voice suggested evolution.

To these problems deriving from the voyages of discovery in the sixteenth century the seventeenth century added its own distinctive contribution, the great variety of microorganisms discovered through the observations of Antoni van Leeuwenhoek and others. Old systems of classification were splitting at the seams, and new ones had to be fashioned.

Because Aristotle had left a workable classification of animals, whereas plants had been classified almost wholly on the basis of their usefulness to man, the first efforts to achieve a " natural " classification, that is, one based on structure, were devoted to the plants. In the beginning of the seventeenth century Gaspard (Caspar) Bauhin, and later the German Joachim Jung, had developed the art of describing the parts of plants accurately.[5] By the end of the century, Joseph de Tournefort, in France, worked out and defined the concept of the *genus*, as being a group of species closely related in fruit and flower structure; and in addition he outlined a system of major categories more inclusive than the genus.[6] It was, however, the Englishman John Ray who in 1686 gave the concept of the species its first clear definition, by seeking out those characteristics which are the sharpest and the most constant.

Ray had embarked with his close friend, Francis Willughby, upon a complete classification of living organisms. Ray himself undertook to describe and classify all of the known species of plants—a considerable increase over the 6000 of Bauhin. Willughby was to treat the animals, but his early death left the project incomplete, and it was finished by Ray. Willughby was responsible for the ornithology and a part of the fishes; Ray completed the natural

---

[5] See E. Guyénot, *Les sciences de la vie aux XVIIe et XVIIIe siécles: l'idée d'évolution*, Ed. Albin Michel, Paris, 1941, pp. 13-14; E. Nordenskiøld, *The History of Biology* (tr. L. B. Eyre), Alfred A. Knopf, New York, 1928 (reprinted, Tudor, New York, 1942), pp. 194-195; and C. Singer, *A History of Biology*, 3rd ed., Henry Schuman, New York, 1950, pp. 174-181.

[6] J. de Tournefort, *Eléments de botanique, ou méthode pour reconnaitre les plantes*, 1694, 3 vol.

history by a *Synopsis of the Quadrupeds and Serpents* (1693) and a *History of Insects* (posthumous—1710). The classification of the animals was clearly modeled on Aristotle's, although with definite improvements, especially in the classification of the vertebrates. All animals provided with blood were divided into those with respiration by lungs or by gills. Having thus separated off the true fish from the cetaceans, Ray divided the vertebrates with pulmonary respiration into those with a heart possessing a single ventricle (reptiles) and those having a heart with two ventricles. The latter were again subdivided into those with hair and viviparous (quadrupeds) and those with feathers and oviparous (birds). The bats were thus for the first time correctly assigned to their true class instead of being confused with birds. The hippopotamus, seal, and beaver were no longer reckoned as being fishes because of their habitats, even though the whales and porpoises were left in an intermediate category. Mythical animals were rigorously excluded.

Ray's classification of plants was less farsighted, especially since he retained the ancient and confusing custom of separating herbs from trees. After setting aside the imperfect herbs, the algae (including corals), fungi, mosses, and ferns, he turned to the flowering plants and here had the merit of introducing the basic distinction between the monocotyledons and dicotyledons. The latter, in their turn, were classified according to the presence or absence of a corolla, the placement of the flowers, the number of petals, the substances and form of the fruit. In this respect Tournefort had done rather better.

But the real problem is that of the species. Classification at the higher levels may be artificial or natural, static or evolutionary, and still be useful and logical within its frame of reference. But at the level of the species the major questions of heredity and variation cannot be evaded. The species is a group of individuals or populations of similar individuals. Examination shows that while they are alike, they are not absolutely alike. Their likeness is owing to their common descent; in other words, it is hereditary. What, then, about the differences between them? Are these mere transient variations produced by differences of climate and nourishment? Are not some of them likewise hereditary? And if so, what keeps them from being regarded as meriting the rank of species?

To some naturalists of the seventeenth century, and indeed of the eighteenth century too, the answer was that of the *type*. Just as Plato had sought reality in the pure idea rather than the imperfect

John Ray

form of a thing, so the naturalists conceived of the real species as an ideal morphological type, from which individuals deviated in a variety of superficial ways. The species was thus quite definite and immutable; the variations, evanescent and insignificant.

Other naturalists had seized upon the existence, or at least the probability, of hereditary variations. Were the limits between species so impassable that within a species there might actually be several, or even numerous, hereditary varieties—like human races, for example? Or did these actually from the moment of their origin constitute new species? Or, if one believed wholeheartedly in the Great Chain of Being, with no real gaps to mar its perfect gradations, did the inherited variations constitute bridges between the species?

We must now see how Ray grappled with this problem and to what conclusions he came, for upon such answers depended future attitudes toward the transformation of species. We may begin by quoting a remarkable passage from the Preface to the *Methodus plantarum* of 1682.

> A Method seemed to me useful to botanists.... But I would not have my readers expect something perfect or complete; something which would divide all plants so exactly as to include every species without leaving any in positions anomalous or peculiar; something which would so define each genus by its own characteristics that no species be left, so to speak, homeless or be found common to many genera. Nature does not permit anything of the sort. Nature, as the saying goes, makes no jumps and passes from extreme to extreme only through a mean. She always produces species intermediate between higher and lower types, species of doubtful classification linking one type with another and having something in common with both—as for example the so-called zoophytes between plants and animals.[7]

Were species themselves then real and distinct? Upon what basis could one suppose that the links in the Great Chain of Being maintained their individuality, instead of constantly flowing into one another?

Ray pointed out that among animals the male and the female of the same species were often very different, and yet that fact does not prevent them from belonging to the same species, as their common parental origin makes clear. One must therefore search

---

[7] Quoted from C. E. Raven, *John Ray, Naturalist: His Life and Works*, Cambridge, at the University Press, 1942, p. 193.

for characteristics that are truly essential to the species, that persist among the differences that may mark individuals of common descent. "The specific identity of the bull and the cow, of the man and the woman," said Ray, "originate from the fact that they are born of the same parents, often of the same mother." Filiation, common descent, this it is which constitutes the surest criterion of a species ("Non aliud certius indicium convenientiae specificae est"). Among plants, Ray regarded as belonging to the same species any plants which produced similar new plants. "One species does not grow from the seed of another species," he said, and for this reason concluded that the species through time is unalterable ("speciem suam perpetua servant").[8]

Yet John Ray did not hold this general law to be perpetual or absolute. Observations showed that sometimes certain kinds of seeds would degenerate and produce plants different from the maternal species. In this way there might arise among plants "*transmutations of species.*" In so saying, Ray did not give undue credence to ancient beliefs that maize could become transformed into wheat, barley into oats, or *Sisymbrium* into *Mentha*. What he thought reliable was a case in which a London gardener had sold authentic seed from a cauliflower and these seeds had grown into ordinary cabbages. Such transmutation of species, Ray thought, could take place only between very similar species; and one might in fact question whether the two types of *Brassica* do actually represent two distinct species. Ray does not seem to have envisaged other types of hereditary variation.

An even earlier figure had, however, looked closely at this biological problem. Francis Bacon, in the *Novum Organum*, the title he proudly selected for his magnum opus, emphasized among other matters the great value of investigating nature's own deviations from the normal, common run of things. Living creatures are particularly suitable for such a study, he thought, "for it would be very difficult to generate new species, but less so to vary known species . . . for if nature be once seized in her variations, and the cause be manifest, it will be easy to lead her by art to such deviation as she was at first led to by chance. . . ."[9] Bacon clearly recognized, and was perhaps the first to do so, the importance of transitional forms, or "bordering instances," as he called them, "for they point out

---

[8] John Ray, *Historia plantarum generalis*, 1686.
[9] Francis Bacon, *Novum Organum*, Bk. II, sect. 29.

admirably the order and constitution of things, and suggest the causes of the number and quality of the more common species. ..."[10] His examples, it is true, were as ludicrous as those of Benoît de Maillet a century later. For instance, he regarded flying-fishes as intermediate between birds and fish, a mistake Aristotle would never have made. Yet he directed naturalists toward the study of variation among living organisms, and so opened the way to one of the great primary fields of exploration upon a knowledge of which the modern theory of evolution is based. Bacon appears without question to have believed in the mutability of species. In his *New Atlantis* he projected a great institution to be devoted to experimental investigation, among the objectives of which were to be the discovery of what causes species to vary, and of how to modify them by art or to create fertile hybrids between them.[11]

Descartes, too, seems to have glimpsed the importance of hereditary variations, although he expressed himself with great reserve and caution, no doubt in order to avoid the antagonism of the Church, already sufficiently aroused by his views of the cosmic evolution of a universe governed by mechanical principles and natural law. But though he spoke in veiled terms of the living inhabitants of his mechanistic universe, no one who reads carefully the fifth part of the *Discourse on Method* will be left with much doubt that inwardly he firmly believed that plants and animals were part of the natural order—and in consequence the product of a gradual evolution produced by the uninterrupted operation of purely physical causes.[12]

Later in the century Leibniz, fusing the Aristotelian concept of the Great Chain of Being with the recognition of the fullness and diversity of nature, arrived at his famous Principle of Continuity. "All the orders of natural beings," he wrote, "form but a single chain in which the various classes, like so many rings, are so closely linked one to another that it is impossible for the senses or the imagination to determine precisely the point at which one ends and the next begins."[13] And elsewhere: "All advances by degrees in

[10] Ibid., sect. 30.
[11] Bacon, *The New Atlantis*, 1627, World's Classics, Oxford University Press, London, p. 268. Cf. H. F. Osborn, *From the Greeks to Darwin*, 2nd ed., Chas. Scribner's Sons, New York, 1924, pp. 90-94.
[12] R. Descartes, *Discourse on the Method of Rightly Conducting the Reason and Seeking Truth in the Sciences* (tr. in P. Valéry's *The Living Thoughts of Descartes*, David McKay Company, Philadelphia, 1947), Pt. v.
[13] G. W. Leibniz, Letter, text in Buchenau and Cassirer, *Hauptschriften zur*

Nature, and nothing by leaps, and this law as applied to each, is part of my doctrine of Continuity."[14] He compared the fossil ammonites with the living nautilus to show how the mutability of species, brought about as he supposed by great changes of habitat, might afford unbroken continuity; and he even hazarded the thought of species intermediate between man and the apes, although he cautiously relegated such "missing links" to some other world than ours. He went on to say, "I strongly approve of the research for analogies; plants, insects, and Comparative Anatomy will increase these analogies, especially when we are able to take advantage of the microscope more than at present."[15]

Thus Leibniz, looking at species not as fixed types but as parts of a continuum which admitted no gaps, played tentatively with the idea of evolution. His thought was expressed most clearly in the following words:

> Perhaps, at some time or somewhere in the universe, the species of animals are, or were, or will be, more subject to change than they are at present in ours; and several animals which possess something of the cat, like the lion, tiger, and lynx, could have been of the same race and could be now like new subdivisions of the ancient species of cats. So I always return to what I have said more than once, that our determinations of physical species are provisional and proportional to our knowledge.[16]

Leibniz obviously possessed a scientific background for such thoughts that earlier natural philosophers had lacked: an appreciation of the true nature of fossils, animal and plant classification including a wide knowledge of variants and transitional forms, a science of comparative anatomy, and the newly revealed world of the microscope. Without these, as well as that appreciation of the continuity of life in time which Leibniz himself did so much to generate, there could have been no *scientific* theory of evolution.

This continuity of life embraced, as Guyénot says, two related notions:

> the continuity by generation with or without transformation

---

Grundlegung der Philosophie, II, 556-559. Quoted from A. O. Lovejoy, *The Great Chain of Being*, Harvard Univ. Press, Cambridge, 1953, p. 145. Cf. Osborn, op. cit., pp. 95-96.

[14] Leibniz, *Protogoea*, XXVI. Cf. Osborn, op. cit., p. 96.
[15] Ibid.
[16] Leibniz, *Nouveaux essais*, t. III, chap. VI. Cf. J. Rostand, *L'évolution des espèces*, Libr. Hachette, Paris, 1932, p. 21.

(*continuité dans le temps*) and the present continuity of living forms (*continuité dans l'espace*) [from the interplay of which] the idea of Evolution must naturally result. If our species, our genera, our orders impinge on each other, is it not precisely because Life, being able to transform itself in the course of the generations (*variation*), has little by little engendered, by insensible gradations, that infinitely gradated world of present living forms? Thus, the idea of Evolution was, from the beginning, indissolubly tied to the notion of continuity.[17]

In the long run, undeniably so! But one ought to be on guard against the elementary error of assuming that because the idea of evolution inevitably rests upon and grows out of the concept of continuity, the latter on the other hand necessarily implies a well-conceived idea of organic evolution. Nothing is clearer, as Lovejoy has amply demonstrated in his discussion of the subject in *The Great Chain of Being*, than that the concept of continuity scarcely ever, from Aristotle to Leibniz, signified any sort of transformation of species. Guyénot is thus in error in speaking of this fusion of two great ideas basic to the concept of organic evolution as having occurred in the seventeenth and eighteenth centuries. So fundamentalist a believer in the Continuity Principle as Charles Bonnet (see Chapter Six, pp. 164-170) drew from it no notion of evolutionary transformation of one species into another. In fact, to him such an idea seemed inescapably to violate the Principle of Continuity and consequently was abhorrent. Only with the nineteenth century did the concepts of the reproductive continuity of life and of intergradations between species become fused in the idea of evolution. Before that could occur, it had to be shown that species are in fact distinct, generally immiscible biological entities; and the search for intermediate forms was to shift from a belief in their present existence (" ce monde infiniment nuancé des formes vivantes actuelles ") to a search for missing links among the fossil relics of past life.

By the time of Leibniz' death, in 1716, his idea of continuity had already been strongly affected by the increased understanding of the continuity of life through the reproductive process. There was, in the first place, a notable weakening in the belief in spontaneous generation, although this was by no means universal. This was after all the time when so notable an opponent of Aristotelianism and so

---

[17] Guyénot, op. cit., p. 382.

true an experimentalist as van Helmont could in all seriousness suggest the following recipe for the spontaneous generation of mice.

> Compress a dirty shirt in the mouth of a pot containing some grain. After about twenty days a ferment coming from the dirty shirt combines with the effluvium from the wheat, the grains of which are turned into mice, of both sexes and able to reproduce among themselves or with other mice. These mice from the fermented wheat do not come out from the grain dwarfed or malformed, but are perfectly grown, without needing, like others, the mother's breast.[18]

The first devastating blow to the belief in spontaneous generation is universally acknowledged to have been dealt by Redi. The year of Francesco Redi's great experiment was 1668. In his own words:

> I placed in four flasks with large mouths a snake (coluber), some freshwater fish, four eels from the Arno, and a slice of veal. Then I closed the mouths very exactly with paper tied and hermetically sealed. In four other flasks, I disposed the same objects, but left these receptacles open. At the end of a short time, the fish and meat of the second series were filled with worms [maggots], the flies being able to enter and leave at will. On the contrary, in the stoppered flasks I have not seen a single worm born, even at the end of several months....[19]

Later, to answer the possible objection that the sealing of the flasks had prevented free access of air, Redi repeated the experiment while replacing the paper by fine Naples netting. The result was identical. Flies swarmed about the odorous vessels, which Redi gave double protection by placing them within a second shelter of net. Flies laid their eggs on the mesh and the eggs developed into maggots, but not a single worm appeared in the putrid meat.

Redi, although convinced by his experiments that maggots always originate from flies' eggs, nonetheless retained a belief in other aspects of spontaneous generation. Like William Harvey, he found the source of internal parasitic worms and the grubs inside of oak-galls too difficult to account for by the ordinary processes of generation through eggs.

---

[18] Guyénot, op. cit., p. 211.

[19] F. Redi, *Esperienze intorno alla generazione degl' Insetti*, 5th ed., Florence, 1688. Cited from Guyénot, op. cit., p. 214. A lengthier English translation by M. Bigelow (Chicago, 1909) is excerpted in T. S. Hall, *A Source Book in Animal Biology*, McGraw-Hill, New York, 1951, pp. 362-368.

Leeuwenhoek, however, the person of his time most qualified to speak concerning microscopic organisms, declared that even the tiniest animalcules possessed reproductive powers. No microscopic forms of life, he held, can arise either through putrefaction or fermentation. In a letter of October 17, 1687, he confirmed Redi's earlier experiment, though apparently without knowing of it. A surgeon brought to him a ganglion, removed from the leg of a woman, in which were some worms that he thought had originated spontaneously. Leeuwenhoek examined them, instantly recognized them as insect larvae, and affirmed that "the maggots came from eggs laid by flies" on the anatomical specimen. The surgeon remaining skeptical, Leeuwenhoek demonstrated his point. The larvae were removed to a piece of beef, grew and transformed into pupae, and finally hatched into flies. A pair of these having been mated, the female deposited 115 eggs from which larvae like the original ones emerged. The cycle of generation was complete.[20] Hence, in 1694, we find Leeuwenhoek stating that "no creature takes birth without generation." [21]

The more difficult problems of the generation of insects that hatch from galls were attacked by Jan Swammerdam, Marcello Malpighi, and Antonio Vallisneri. Swammerdam, in a letter of 1670, asserted that while it was hopeless to try to prove by experiment that insects can be engendered within the tissues of plants, at the same time there was good evidence that these tiny creatures were enclosed for the sake of nourishment. They come from insects which "insinuate their semen or their eggs previously into the plants." The opening once being closed, the larvae are thenceforth imprisoned in the fruit or the gall.[22] Malpighi, with his student Vallisneri, made further observations, and discovered the gall-flies which with their amazing ovipositors lay eggs in buds or inside leaves and subsequently produce the galls.[23]

As for intestinal worms, the problem, as Guyénot points out, was philosophically far more difficult.[24] If spontaneous generation was to be denied, must Adam then have harbored at least one of every sort of human parasite, and this even before the fall from grace?

---

[20] Guyénot, op. cit., pp. 214-215.
[21] A. v. Leeuwenhoek, *Epistola 83* (May 1694). Cf. Guyénot, op. cit., p. 215.
[22] J. Swammerdam, Supplément pour les lecteurs qui s'appliquent sérieusement à la recherche de la vérité (*Histoire générale des Insectes*, Utrecht, 1685, p. 158). Cf. Guyénot, op. cit., p. 216.
[23] See Guyénot, op. cit., pp. 216-218.
[24] Guyénot, op. cit., pp. 218-219.

Before so horrible a conclusion who would not draw back? Yet we find Vallisneri, true to his conviction that living creatures arise only from others of their own kind, writing in 1712 that these worms must have existed from the beginning and must reproduce in some mysterious, still unsuspected fashion. The mystery was not solved; not until 1832 would it be known that the cysticerci to be found in meat are the encysted larvae of tapeworms. But the thought was there, already at the close of the seventeenth century. The belief in spontanous generation, though still held by common folk and by some scientists, as indeed it would be for centuries, was disproved in the gross, was suspect in entirety. Redi, Leeuwenhoek, Swammerdam, Malpighi, and Vallisneri had achieved a revolution in biological thought hardly second to the nineteenth century theory of organic evolution; and they had unwittingly paved the way for the latter.

This was so in spite of their own views, for the first impact of *biogenesis* (not to be confused with Haeckel's so-called Biogenetic Law) was anti-evolutionary. It led to a concept of the fixity of species. The generation of living things from parents of their own kind, and only by that means, strengthened in Swammerdam and Malpighi the view that the young generation must reside within the old. From Aristotle to Harvey, observers of the origin of the embryo or foetus had been epigenesists, that is to say, had viewed the embryo as taking form from unorganized material, a mixture of the two semens of the parents. Harvey, in 1651, had described in the developing hen's egg a successive formation of entirely new structures as the embryo took shape; and he had looked in vain for embryos in the uteri of does killed at short intervals after insemination. But Swammerdam and Malpighi now conceived of a preformation of the embryo within the parent organism, and attacked the observations and interpretations of Harvey. It is true that actually Harvey's observations on the chick embryo were not very good, having been made without the benefit of magnification; in fact, they were not as good as those of Aristotle in some respects. Thus Harvey had seen the young heart beating for the first time on the fourth day of incubation of the egg, but Aristotle had seen this occurring earlier, on the third day. Malpighi himself, in 1669, was able to see the heart beating and the blood circulating already on the second day, together with the inflexion of the head and the first appearance of the eyes. After only six hours incubation of the egg he could discern the head region and the spine, while the somites, which he called "vertebrae," were visible at the twelfth hour. Clearly, he was not impressed by Harvey's inferior observations and preferred to form his own views. Similarly, in 1669, we find Swammerdam, on the basis of his meticulous observations of insect anatomy and development, generalizing that "there is never generation in nature, but only a lengthening or increase of parts."[25] Development was like the enlargement of the little leaves or flower parts in a bud was supposed to be, simply a process of growth and an unfolding of preexisting parts. The black spot in the center of a frog's egg was a complete little frog, if we could but see it. Adam and Eve contained all humanity in their loins. (Swammerdam perceived the basis for the doctrine of original sin in this principle!)

Thus there germinated the famous version of preformation known as the encasement (emboîtement) theory. It offered a mechanistic explanation for the fixity of species and the impassibility of the discontinuities between them. It was utterly foreign in spirit to the philosophy of the Great Chain of Being, as Leibniz had conceived it; and it made inconceivable any evolutionary transformation of one species into another, although of course one might still expect some sucessive modification of species to occur, if the Creator had so desired it in the beginning and had packed it into the successively enclosed embryos of the future. (It is truly extraordinary that, in spite of this logical contradiction, Leibniz himself was a notable preformationist.) In modern terms, microevolution (within the limits of the species) could occur within the framework of preformation, but macroevolution (transcending the limits of the species) could not. An adherence to evolutionary views from this time onward consequently implied both resistance to the theory of preformation and reluctance to abandon the belief in spontaneous generation, for preformation and biogenesis mutually supported the concept of the fixity of species.

Guyénot has pointed out how unscientific epigenesis must have seemed in the seventeenth century.[26] In spite of Descartes' effort to explain the formation of the foetus by virtue of purely mechanical principles, through the agitation of particles by heat and kinetic force, cartesian mechanics seemed hopelessly inadequate for the

[25] J. Swammerdam, *Miraculum naturae*, 1672, p. 21. Cf. Guyénot, op. cit., p. 297.
[26] Guyénot, op. cit., pp. 306-309.

purpose. As Maupertuis said in the next century, "An attraction uniform and blind, diffused in all matter, could not serve to explain how those particles become arranged to form even the simplest body. If all have the same tendency, the same force to unite with one another, why are these going to form an eye and those an ear? Why this marvelous action? And why don't they unite pell-mell? One will never explain the formation of any organized body solely through the physical properties of matter: and from the time of Epicurus to Descartes one has only to read the writings of all the philosophers who have attempted it to be persuaded of the fact." [27]

Each epigenesist tried in his own way to resolve the difficulty. Harvey, for example, postulated a *formative faculty*. A century later Buffon assumed that his hypothetical "organic molecules," whether present in the semens of parents or in the nourishment, were already heterogeneous, having had impressed upon them the character of that organ from which they had come; but he evaded the question of how the organic particles find the right places and produce the right species of organism. Caspar Friedrich Wolff actually demonstrated epigenetic development, but to account for it he called upon a *vis essentialis*. Even Hans Driesch, in the twentieth century, was unable to face the mystery of organic development without postulating an autonomous life factor, to name which he turned once again to Aristotle for the term "*entelechy*." In short, the problem of the factors controlling development has remained to this day perhaps the most difficult in biology; and until the recent synthesis of genetics, biochemistry, and embryology it has remained the final stronghold of vitalism. To the seventeenth century biologist, imbued with the scientific desire to seek a physical, a material cause for all phenomena, and strengthened by the vanquishment of spontaneous generation, the encasement theory must have seemed far more scientific than the *entelecheia* of Aristotle or the *formative faculty* of Harvey. It removed the need to explain anything more than the unfolding and enlargement of parts already there; and if it did not undertake to explain the postulated beginning, that was wrapt in the mystery of Creation. Science does not investigate Final Causes.

It is really of interest, therefore, that the basic observations of both Malpighi and Swammerdam, on which the preformation theory

[27] P. L. M. de Maupertuis, *Système de la nature*, xiv, in *Œuvres*, Lyon, 1756, T. i, pp. 146-147.

rested, were actually erroneous. Swammerdam had investigated insect metamorphosis, especially the transformation of a caterpillar into a chrysalis and the latter into a butterfly. He observed that when chrysalids are immersed in warm water or alcohol it becomes possible to see the future butterfly already formed within the chrysalis. Upon dissecting the caterpillar, he thought he could discern within it, already formed, such structures as the wings and the six legs, and he therefore concluded that the butterfly was formed already within the caterpillar and needed only to be released, unveiled, disenveloped. Carried away by enthusiasm, he supposed that since the caterpillar comes from an egg, the butterfly must be already inside the egg, and that must be so even before the egg is laid by the mother butterfly. From this to the endless emboîtement required but a stretch of the imagination, although Swammerdam did not assume that infinite smallness was possible. On the contrary, the number of encasements within encasements must surely be limited, and in the fullness of time the supply would run out and the species would abruptly terminate. Mankind, too, must expect to disappear with the ultimate exhaustion of the supply of future generations encased within the eggs of our Mother Eve. Could there be a more perfect example of the risk of extrapolating indefinitely from initial observations which are sound?

Malpighi, say most writers, fell into the same pit. Utilizing good microscopic magnification, he saw much more adequately than Harvey the very early stages in the development of the chick embryo, located in the center of the blastoderm; and he prepared excellent plates showing the early state of the vertebrae, liver, heart, and other structures. He took pains to point out that the heart was clearly present in advance of the time when it commenced to beat, and so previously had been thought to arise. Looking at the unfertilized egg, Malpighi here, too, observed the cicatricula, or blastoderm, which was in his estimation smaller than in an egg that had been fertilized. He was, however, unable to find any trace of the embryo in the blastoderm of the unfertilized egg. Further observations were made upon " unincubated " but fertilized eggs, which he appears to have thought critical for the question of the initial presence of the embryo. Looking at such eggs, he discerned the embryo even before incubation had started, and hence concluded that the lineaments of the chick preexist in the egg prior to birth, just as the plant embryo is to be found within the seed. His own words were as follows: " Quare pulli stamina in ovo praeexistere, altioremque originem nacta esse fatere convenit, haut dispari ritu, ac in Plantarum ovis." [28] (Actually, it seems, these eggs had been incubated for two days by the considerable heat of the Italian August sun!) What did Malpighi think, then, in regard to the events occurring at fertilization? Was it, as Cole suggests,[29] conceived to be a sort of precipitation, a rendering visible, of what was already there? Or was there at that time a formation of structures in a real epigenetic sense? We cannot be sure. Adelmann (cited by Oppenheimer [30]) thinks that we do not know just what Malpighi meant by the preexistence of the animal in the egg. Yet whatever Malpighi himself held to be the case, his followers—his student Vallisneri, and such others as Malebranche and Maître-Jan—had no doubts. Preformation seemed demonstrated; and Malpighi became regarded as one of the prime founders of the doctrine. But the conclusions drawn from these observations were nevertheless faulty and the extrapolation to the time of fertilization (and before) totally unsound, as C. F. Wolff was to show a century later. By that time, however, the theory of preformation seemed, at least to its supporters, to have been unshakably established.

One thing remained. There was always, in sexually reproducing forms, both a father and a mother. The doctrine of the two semens, derived from the ancients, was thoroughly epigenetic in conception. Preformation, however, could reside only in a single lineage, male or female, not both. Preformation at first seized upon the egg. But this was to suppose that the male parent provided only a stimulus to impel the egg, with its encased generations, to release to development the outermost one. Then in November, 1677, Leeuwenhoek wrote to the Royal Society that he had confirmed the discovery of a young compatriot, Louis Dominicus Hamm, that human semen contained "spermatic animalcules." Leeuwenhoek almost at once came to the conclusion that these sperms were the authors of reproduction, the sources of embryos, the very larvae of men. Semen of various male animals was examined and the regularity of the presence of spermatozoa was established. "Conceptio non fit per

---

[28] M. Malpighi, *De formatione pulli in ovo*, London, 1673. [Quoted by F. J. Cole, *Early Theories of Sexual Generation*, Oxford, at the Clarendon Press, 1930, p. 49.]

[29] Cole, ibid., p. 50.

[30] J. Oppenheimer, "Problems, Concepts, and their History," p. 6, in *Analysis of Development*, edited by B. H. Willier, P. A. Weiss, and V. Hamburger, W. B. Saunders Company, Philadelphia and London, 1955.

ovum . . . ex ovis imaginariis, sed ex animalculis vivis, seu vermiculis in semine virili contentis . . ." [81]

The reference to "imaginary eggs" was perfectly understood at the time. In 1667, Nicolaus Steno (Nils Steenson), the first great interpreter of the true nature of fossils, had in his anatomical studies at Florence investigated the reproduction of the dogfish, which from the time of Aristotle had been of interest because, unlike most other fish, it brings forth its young alive. The embryos develop in a duct quite like a uterine horn in a female mammal. Steno had found that the female dogfish has large eggs in her ovaries, a proof that the embryo does not form in the duct de novo but comes from an egg released by one of the ovaries. Forming the audacious hypothesis that the same situation must obtain in the female mammal, Steno conjectured that the ovarian follicles might be the eggs, and in 1675 he published his observations on the "eggs" of various mammals. In the meantime Regnier de Graaf had reached the same conclusion and reported furthermore that he had actually seen the eggs arrive in the uterus. What he had really observed, in female rabbits opened by dissection at various intervals after copulation, was the filling of the empty, erupted follicles with corpora lutea. He saw, too, in one rabbit opened three days after copulation, three altered follicles in the right ovary and three tiny eggs, the size of a mustard seed, in the right oviduct, besides four empty follicles in the left ovary and one tiny egg in the left oviduct. In rabbits opened at later stages after copulation, the implanted little blastocysts could be seen, and their growth to embryonic form observed. De Graaf concluded that all viviparous females produce eggs which they lay internally. Consequently, all beings reproduce alike.

The battle raged, animalculists on one side, ovists on the other—but this is not the place to recount that often recited tale. By the end of the century the ovists had largely won the ascendancy, though in either case it would have been a victory for preformation. The fixity of species seemed assured.

Yet at last, as the sequel will show, all these ideas were prerequisite for a scientific theory of evolution. Epigenesis and preformation, apparently irreconcilable, were to reach a higher synthesis embracing them both; and the mutation of the elements of

---

[81] Observationes D. Anthonii Lewenhoeck de natis è semine genitali animalculis. *Philos. Transactions*, vol. XII, pp. 1040-43, 1679. Letter dated Nov. 1677. Included in Letter 113. Cf. Cole, op. cit., pp. 9-12.

preformed pattern would become the basis of the hereditary variations governing the epigenetic pattern of development. Neither sperm nor ovum would be found to contain an embryo, but each would have within it the hereditary determinants of the species, and jointly they would provide the continuity of life between generations without which species could have neither that heritage of stable morphological and physiological characters upon which the entity of the species depends, nor that supply of hereditary variations which endows the species with adaptability and makes possible its evolution.

THE EIGHTEENTH CENTURY

THREE

MAUPERTUIS, PIONEER OF GENETICS AND EVOLUTION *

BENTLEY GLASS

The mid-eighteenth century was a period almost unexampled in the vigor and advancement of science. Newton's physics had finished remaking the Heaven and the Earth. In biology, the new classification of plants and animals, introduced in 1735 by Linnaeus and developed into the binomial system in 1753, gave a vast stimulus to the discovery and description of new species. Louis Leclerc, Comte de Buffon, was beginning his tremendous *Natural History*, which ran to 36 volumes, yet was a best-seller to be found in the library of every European with any pretension to culture. Réaumur was engaged in detailed and fascinating studies of the life of insects. The Abbé Spallanzani had performed the first artificial fertilization of eggs by collecting semen from a male frog and applying it to freshly laid frog eggs. In Switzerland, Charles Bonnet had discovered parthenogenesis in plant-lice and thus had lent to the Virgin Birth an aura of scientific credibility; and while serving as a tutor in Holland, Bonnet's young relative, Abraham Trembley, had described the little Hydra and its powers of budding and regeneration, which awakened the greatest amazement. Was this supposed plant-animal not a living witness to the fact that, as Leibniz had asserted, "all advances by degrees in Nature, and nothing by leaps"? Here indeed was substantiation for the rightness of the concept of " The Great Chain of Being," which according to Arthur

* A revision of the two articles, "Maupertuis and the Beginning of Genetics" (*Quart. Rev. Biol.*, 22:196-210, 1947) and "Maupertuis, a Forgotten Genius" (*Scientific American*, 193:100-110, 1955).

O. Lovejoy was as much of a key to the thinking of the 18th century as the word "evolution" was to that of the 19th.[1]

Preeminent among natural philosophers of the time was Pierre Louis Moreau de Maupertuis, President of Frederick the Great's Academy of Sciences in Berlin. As a young man, Maupertuis was the first person on the Continent to understand and appreciate Newton's laws of gravitation; and indeed, it was through Maupertuis that Voltaire first became convinced of their truth. When thirty years of age, in 1728, Maupertuis had visited London. It was the year of Newton's death. Here Maupertuis became a member of the Royal Society and a disciple of Newton. Upon his return to France, at a time when Newton's theory of gravitation was still violently opposed by the generality of believers in Cartesian vortices, Maupertuis became the open defender and expounder of the new scientific doctrines, just as Huxley over a century later sprang to the defense of Darwinism. D'Alembert, in his introduction to the Encyclopedia (1778),[2] praised Maupertuis' courage in being the first to declare himself openly (in the *Discours sur la figure des astres*, 1732) an adherent of Newton.[3] In that same year Voltaire, more and more interested in scientific pursuits by the fascinating Mme. du Châtelet, who was undertaking to translate Newton's work into French, wrote to Maupertuis, asking him in flattering terms for his judgment upon Newton's theory. Voltaire's conversion followed, and later, his own work on the subject, which was by far his most serious scientific production.

Four years later, in 1736, Maupertuis headed one of two expeditions sent out to test the flattening of the earth toward the poles, by accurately measuring a degree along a meridian of longitude in two places, in the one instance at the equator, and in the other, just as far to the north as feasible. Maupertuis directed the expedition into Lapland, La Condamine the expedition to Peru; and between them they provided the first convincing proof to the world that Newton was correct, just as the expeditions to measure the gravitational deflection of light rays around the sun during an

---

[1] A. O. Lovejoy, *The Great Chain of Being*, p. 184. Harvard University Press, Cambridge, 1936.

[2] J. LeR. d'Alembert, Discours preliminaire des editeurs. [In the *Encyclopédie*, Diderot and d'Alembert, editors, v. 1, i-lxxviii] Soc. Typographiques, Berne et Lausanne, 1778.

[3] P. L. M. de Maupertuis, *Oeuvres*, nouvelle edition, corrigée et augmentée. 4 vol. J.-M. Bruyset, Lyon, 1756.

eclipse were in a later day and age to confirm the relativity theory of Einstein. There was no Nobel prize in those days, or surely Maupertuis would have earned one, at the age of 38.

Upon his return from Lapland, Maupertuis was addressed by Voltaire in these flattering words (letter of 19th Jan., 1741): "M. Algarotti is count; but you—you are marquis of the arctic circle, and you have won for yourself one degree of the meridian in France and one in Lapland. Your name covers a good part of the globe. I find you really a very great seigneur. Remember me in your glory." It was, in fact, on Voltaire's recommendation that, in 1740, Maupertuis was invited by Frederick the Great to come to Berlin as head of the reorganized Academy of Sciences.[4] [Fig. 1].

Maupertuis, however, came to consider that his work as a champion of Newton on the Continent afforded him no great personal glory, and therefore in later years he laid great weight on his discovery in 1746 of the Principle of Least Action, which is all too commonly credited to one of the three great mathematicians, Euler, Lagrange, and Hamilton, who further developed it. This Principle is indeed one of the greatest generalizations in all physical science, although not fully appreciated until the advent of quantum mechanics in the present century.[5] Maupertuis arrived at this principle from a feeling that the very perfection of the universe demands a certain economy in nature and is opposed to any needless expenditure of energy. Natural motions must be such as to make some quantity a minimum. It was only necessary to find that quantity, and this he proceeded to do. It was the product of the duration (time) of movement within a system by the "vis viva," or twice what we now call the kinetic energy of the system. Having found the quantity that tends to a minimum, Maupertuis regarded the principle as all-inclusive: "The laws of movement and of rest deduced from this principle being found to be precisely the same as those observed in nature, we can admire the application of it to all phenomena. The movement of animals, the vegetative growth of plants . . . are only its consequences; and the spectacle of the uni-

[4] P. Brunet, *Maupertuis*. Vol. I. *Étude biographique*. Vol. II. *L'œuvre et sa place dans la pensée scientifique et philosophique du XVIIIe siècle*. A. Blanchard, Paris, 1929. Cf. also *Les physiciens hollandais et la méthode expérimentale en France au XVIIIe siècle*, A. Blanchard, Paris, 1926; and *L'introduction des théories de Newton en France au XVIIIe siècle*, 2 vols., A. Blanchard, Paris, 1931.

[5] J. Fee, Maupertuis, and the principle of least action, *Sci. Monthly*, 52: 496-503, 1941.

verse becomes so much the grander, so much the more beautiful, the worthier of its Author, when one knows that a small number of laws, most wisely established, suffice for all movements." This sort of talk aroused vigorous opposition. Translated into recent biological terminology, however, what is it other than Claude Bernard's principle of the maintenance of the internal environment, W. B. Cannon's principle of homeostasis, or the principle of Le Chatelier: "In a system in equilibrium, when one of the factors which determine the equilibrium is made to vary, the system reacts in such a way as to oppose the variation of the factor, and partially to annul it"? In these days when homeostasis has spread from physiology into embryology and genetics, and when social scientists, emulating ecologists who talk about the balance of nature, themselves talk about the balance of economic forces or social equilibria, the confidence of Maupertuis in the universality of his principle seems justified.

In any event, it seems unfair to say, as Sir James Jeans does, that Maupertuis arrived at his principle from reasons "theological and metaphysical rather than scientific."[6] That is to decry the theoretical approach which has proved so fruitful in modern science, and which, to cite only one example, led Einstein first to the General Theory of Relativity and in after years to a search for its union with electromagnetic phenomena, in a single, magnificently simple, Unified Field Theory. Of the validity of the latter we are still uncertain, although Heisenberg now claims to have reached the goal; but that this is one valid type of scientific approach to nature who can doubt, when the evidence that $E = mc^2$ is so perpetually before us?

The magnificent irony is that it was through this very discovery of the Principle of Least Action that Maupertuis, so eminent and so highly respected in his own day, became a forgotten scientist, for whose name one must hunt assiduously to find more than a mention in the histories of scientific thought and achievement. The villain in this history was none other than Voltaire. It happened in the following way. A young scientist named Samuel Koenig, a friend and former fellow student of Maupertuis, moved by unknown motives, charged that Maupertuis' Principle of Least Action was in the first place erroneous, and in the second had been proposed by

[6] Sir J. Jeans, *The Growth of Physical Science*, p. 233. Cambridge, at the University Press, 1948.

Fig. 1. Pierre Louis Moreau de Maupertuis

Leibniz before his death in a letter, of which Maupertuis was supposed to have had knowledge. Koenig sent the Berlin Academy a fragment of the letter, and later transmitted to Maupertuis, upon his insistence, a complete copy of it. The two did not agree, and the original could not be found, for its reputed owner had literally lost his head some time previously, having been executed by the ministers of justice. After considering the charges, the Berlin Academy by unanimous vote, with Maupertuis abstaining, declared Koenig's letter to be a fabrication, and he was expelled from the Academy.

Voltaire reacted in characteristic fashion, whether because he was always eager to defend the injured against the tyranny of those in high places, or because, as some said, Maupertuis had refused him a requested favor. The world has heard largely Voltaire's side of this famous dispute, and a sober reappraisal, such as Fee has made, had long been overdue. Koenig seems to have been a sincere, well-meaning man; but one cannot credit him with very good judgment in attacking Maupertuis upon such flimsy evidence; and nothing but a smile seems appropriate for the logic that led him in one breath to claim first that the discovery was based on error and was without worth and then that it had already been made by another person. It seems quite probable that Leibniz did arrive at the Principle of Least Action, and he may have written such a letter as Koenig claimed. But at any rate he did not publish it, and Maupertuis was well within his rights in demanding that the original letter be produced. No unpublished discovery can take from Maupertuis that recognition due his priority on publication. Nor was his clear formulation of the law at all, as some have claimed, "a generalization vaguely conceived and ill formulated "[7] or " proposed in metaphysical shape."[8] It was far in advance of the understanding it was to receive during the next century and a half, and Maupertuis had worked out specifically its applications to both inelastic and elastic impact and to the principle of the lever (cf. Fee, 1941). To Voltaire, however, the Principle of Least Action seemed so much nonsense, and Maupertuis a vain man using his prestige and position to suppress criticism and to salve his injured pride. So he took up his pen to write against Maupertuis some of his finest invective, of which

[7] A. O. Lovejoy, Some eighteenth century evolutionists, *Pop. Sci. Monthly*, 65: 238-251; 323-340, 1904. Reprinted, *Sci. Monthly*, 71: 162-178, 1950.

[8] H. T. Pledge, *Science Since 1500*, p. 81. Philosophical Library, New York, 1947.

56   FORERUNNERS OF DARWIN

Macaulay was to say: "Of all the intellectual weapons which have ever been wielded by man, the most terrible was the mockery of Voltaire."

In the *Diatribe du Docteur Akakia*,[9] Voltaire showered his ridicule on the proposals for the advancement of the sciences which Maupertuis had published in 1752 in his *Lettres*[10] and in the *Lettre sur le progrès des sciences*.[11] Those ideas, looked at from the standpoint of today, strike us as mostly extraordinarily well-conceived suggestions for systematic experimentation and investigation, only too far ahead of the times. Maupertuis' comments on the nature of the chemical elements, for example, and the possibility yet impracticability of their transmutation, might have been written today. But to Voltaire the *lettres* were further evidence of Maupertuis' vanity and incompetence. Maupertuis had suggested, among other things, that it might be a good idea not to pay doctors when they failed to heal, so Voltaire chose a physician as the mouthpiece of his sarcasm, to ask·

> What you think a man would say, I pray you, who had, for example, 1200 ducats pension for having talked of mathematics and metaphysics, for having dissected a couple of toads, and for having had himself painted in a fur bonnet [see Fig. 1], if the treasurer should come to him with this language: "Monsieur, there is a deduction of 100 ducats for your having written that stars are made like windmills; 100 ducats more for having written that a comet will come to steal away our moon and carry its attacks to the sun itself; 100 ducats more for having imagined that comets all of gold and diamond will fall on the earth; you are taxed 300 ducats for having affirmed that babes are formed by attraction in the abdomen of the mother, that the left eye attracts the right leg, etc.? One cannot deduct less than 400 ducats for having imagined understanding the nature of the soul by means of opium and by dissecting the heads of giants, etc., etc."[12]

What of this is not gross misinterpretation merely shows that Voltaire was not gifted with scientific prevision; but it had its effect. Frederick, supporting Maupertuis at least in public, although snick-

---

[9] F. M. A. de Voltaire, *Diatribe du Docteur Akakia*, 1752. [In *Oeuvres complètes* (Mélanges II, v. 23).] Garnier Fr., Paris, 1879.
[10] Maupertuis, *Lettres*, 1752. [In *Oeuvres*, v. 2, pp. 185-340.]
[11] Maupertuis, *Lettre sur le progrès des sciences*, 1752. [In *Oeuvres*, v. 2, pp. 341-400.]
[12] Voltaire, op. cit.

ering in private, made Voltaire leave Potsdam and retire temporarily to France; but the reputation of Maupertuis never recovered, particularly since from that time until Maupertuis' death in 1759 Voltaire did not relent in his flood of unmerciful, unscrupulous, and myopic ridicule. Particularly interesting is the relation of this quarrel to the allegories of *Micromégas*, which Voltaire also published in 1752.[13] Literary critics seem not to have observed that it was unquestionably Maupertuis' expedition to Lapland that the giant Saturnian and the super-giant from Sirius were supposed to have discovered and conversed with—Voltaire's effective way of putting Maupertuis' philosophy and science in a Lilliputian scale. Maupertuis had fallen in love while in Lapland, and brought the young girl back with him to the South. On the way the expedition met with a shipwreck, which Voltaire introduces into the story, while making of Maupertuis and his little "Lapp" maiden the butt of some of his nastiest witticisms. These are hardly relieved by more favorable comments on the "follower of Locke," who might possibly be identified as Maupertuis also, since in a letter of 1734, Voltaire had addressed Maupertuis as a "disciple of Locke and Newton." But it is more probable that Voltaire was simply thinking of himself! Clearly, the significance of *Micromégas* and its attack upon pseudo-scientists and false philosophers is entirely missed if its connection with Voltaire's quarrel with Maupertuis is overlooked.

Actually, in philosophy as in physical science, Maupertuis proved himself a powerful and original thinker, reproached by Voltaire for views the latter was himself later to embrace. As Lovejoy has pointed out, Maupertuis anticipated Beccaria and Bentham, and along with Helvetius represents "the head-waters of the important stream of utilitarian influence which became so broad and sweeping a current through the work of the Benthamites."[14] Maupertuis also considered anew the favorite argument of the eighteenth century for the existence of God—the argument from the apparent Design of Nature—and formulated in terms very like those of Romanes well over a century later a beautifully clear statement of the survival of the fittest:

> May we not say that, in the fortuitous combination of the productions of Nature, since only those creatures *could* survive in

---

[13] Voltaire, *Micromégas*, 1752. [In *Oeuvres complètes* (Romans), v. 21, Garnier Fr., Paris, 1879.]
[14] Lovejoy, op. cit.

whose organization a certain degree of adaptation was present, there is nothing extraordinary in the fact that such adaptation is actually found in all those species which now exist? Chance, one might say, turned out a vast number of individuals; a small proportion of these were organized in such a manner that the animals' organs could satisfy their needs. A much greater number showed neither adaptation nor order; these last have all perished. . . . Thus the species which we see today are but a small part of all those that a blind destiny has produced.[15]

While he did not deny the possibility of ultimate purpose in the universe, Maupertuis thus pointed out that biology has no need of teleological explanations. In this he reminds us of David Hume, his Scotch contemporary, who argued, in similar fashion: " It is in vain . . . to insist upon the uses of the parts of animals or vegetables and their curious adjustment to each other. I would fain know how an animal could subsist, unless its parts were so adjusted." [16]

Maupertuis soon came to realize that Newton's principle of gravitation affords an insufficient basis for explaining the phenomena of chemistry, and especially the behavior of living organization. Here he turned to the ideas of Leibniz, who saw at the basis of life properties of consciousness as well as mass. Hence arose the fact that, as Cassirer says, " It was especially Maupertuis who brought Leibniz to France. His personal relation to Leibniz is not indeed free from contradictions, but the objective kinship of his metaphysics, his philosophy of nature, and his theory of knowledge with Leibniz's basic ideals is undeniable." [17] In fact, Maupertuis' great work, the *Système de la Nature*, regarding which more will be said later, may properly be considered an attempt to " reconcile the two great opponents in the philosophy of nature of the seventeenth century," Newton and Leibniz. It would be a mistake, however, not to regard this work as one of the highest order, which goes far beyond either of the great originals; and Cassirer's evaluation of Maupertuis' ideas suffers fatally from a failure to understand the genesis of those ideas not solely in the philosophers named, but equally in Maupertuis' investigations of heredity.

Another significant aspect of Maupertuis' philosophy was his

---

[15] Maupertuis, *Essaie de cosmologie*, 1750. [In *Oeuvres*, v. 1, pp. i-xxviii, 1-78.]

[16] D. Hume, Dialogues concerning natural religion (posthumous). [In *Essays and Treatises on Various Subjects*, v. 2.] Edinburgh, 1793.

[17] E. Cassirer, *The Philosophy of the Enlightenment*, trans. by F. C. A. Koelln and J. P. Pettegrove, p. 86. Princeton University Press, Princeton, 1951.

attempt to introduce a calculus of pleasure and pain in order to evaluate the "good life" and to measure happiness. Here again his effort to quantify and to apply mathematical methods was characteristic of his whole approach to nature, for he proposed that the amount of pleasure or pain was a product of intensity and duration, a formulation strongly reminiscent of his Principle of Least Action. Maupertuis concluded pessimistically that in general the sum of the evils in life always exceeds the sum of the goods. In this philosophy his thinking leads straight to Kant, who spent much effort in fruitless argument against Maupertuis' ideas, ultimately to set aside the whole problem of objective measurement when developing his own ethical theory in which "there is evidently nothing left but the value which we ourselves place upon our lives, not only through what we ourselves do, but through what we perform independently of nature in a purposive manner...."[18] In so doing, Kant places the aims of human life within a subjective area forever barred to science. But his solution ignores the later development of the theory of evolution, including the origin of man from lower forms of life, a theory which has inescapable implications for the sources of mental, emotional, and behavioral phenomena. To Maupertuis, these implications could not be set aside. His theory of human evolution was consistent with his philosophy, and only too far ahead of his time.

Maupertuis as Epigenesist

Eminent as were these contributions to physical science and to philosophy, it is in his biological ideas that Maupertuis was most clearly gifted with prevision. Here he must be reckoned as fully a century or a century and a half before his time. His biological ideas may be considered under the three heads of the formation of the individual, the nature of heredity, and the evolution of species, although obviously these are so closely interrelated that the division is largely artificial.

He began with an interest in the formation of the embryo, and quickly put his finger on the soundest argument against the preformationists, who believed the embryo to be fully preformed be-

[18] Kant, *Werke*, ed. by Cassirer, v. 5, p. 514n, *Critique of Judgment*, sect. 83. [Quoted from Cassirer, op. cit., p. 151.]

fore conception, and present either in the sperm or the egg. This argument led him to a study of heredity, and he may be justly claimed as the first person to record and interpret the inheritance of a human trait through several generations. He was also the first to apply the laws of probability to the study of heredity. He was led by the facts he had uncovered to develop a theory of heredity that astonishingly forecast the theory of the genes. He believed that heredity must be due to particles derived both from the mother and from the father, that similar particles have an affinity for each other that makes them pair, and that for each such pair either the particle from the mother or the one from the father may dominate over the other, so that a trait may seemingly be inherited from distant ancestors by passing through parents who are unaffected. From an accidental deficiency of certain particles there might arise embryos with certain parts missing, and from an excess of certain particles could come embryos with extra parts, like the six-fingered persons or the giant with an extra lumbar vertebra whom Maupertuis studied. There might even be complete alterations of particles—what today we would call "mutations"—and these fortuitous changes might be the beginning of new species, if acted upon by a survival of the fittest and if geographically isolated so as to prevent their intermingling with the original forms. In short, virtually every idea of the Mendelian mechanism of heredity and the classical Darwinian reasoning from natural selection and geographic isolation is here combined, together with De Vries' theory of mutations as the origin of species, in a synthesis of such genius that it is not surprising that no contemporary of its author had a true appreciation of it.[19]

Even in his earliest biological essays, "Observations et expériences sur une des espèces de salamandre" (1727),[20] and "Expériences sur les scorpions" (1731),[21] Maupertuis evinced an absorbing interest in reproduction and those problems that it inevitably generates. Having observed both eggs in the ovaries and live young to the number of 54 in the oviducts of single female salamanders, Maupertuis remarked that "these animals appear very suitable for enlightening the mystery of generation." In the study of scorpions

---

[19] Buffon, it is true, seems to have derived his own ideas of the seminal particles at least in part from Maupertuis (see pp. 77, 81).

[20] Maupertuis, *Mémoires de l'Académie des Sciences*, 1727, 27-32. [Published, 1729.]

[21] Maupertuis, *Mémoires de l'Académie des Sciences*, 1731, 223-229. [Published, 1733.]

it was again the presence of living young (up to 65 in number) within the female, and the fantastic courtship of the sexes which excited his interest, as it did that of Fabre a century and a half later.

The prevailing ideas regarding the origin of the individual in Maupertuis' time were those of preformation. William Harvey, in 1651, had plainly declared his conviction, based on the study of the development of the embryo chick, in favor of epigenesis, that is, of the view that the parts of the embryo are formed in succession out of unorganized material; but the application of the microscope to the reinvestigation of the matter, by Malpighi, led to the common opinion that Harvey had been wrong. Malpighi was perhaps led astray by his inability to find any but a fairly well-developed embryo in freshly laid eggs, not recognizing that the embryo might have begun its development while the egg was descending the oviduct. Swammerdam's investigations of insect metamorphosis—in particular, his demonstration that a perfectly formed butterfly is to be found within the chrysalis—seemed convincing demonstrations of a like sort, although the idea was pushed far beyond demonstrable limits. The caterpillar must mask a perfect butterfly, within which there would be eggs containing, *multum in parvo*, caterpillars, chrysalids, and butterflies of a generation still to come; and so on (see chap. II, pp. 43-46).

Another notable scientific doctrine was involved in the common acceptance of the preformation theory in the seventeenth and eighteenth centuries. This was the belief in spontaneous generation. As long as it was believed that living creatures could arise spontaneously from mud, filth, and putrescent matter, either epigenesis or some preformation of germs disseminated throughout nature seemed inescapable, at least for such organisms. The demonstration by Redi, in 1668, that maggots hatch only from flies' eggs, and do not appear in putrefying meat except when the flies have access to it, cleared the way for the encasement ("*emboîtement*") theory of preformation.

The predominance of the preformationist doctrine led to heated claims in behalf of the ovum by the followers of Malpighi and Swammerdam, on the one hand, and of the spermatozoa by the disciples of Leeuwenhoek, on the other; since, if the embryo is preformed, it could not logically be preformed in both parents alike. The discovery of the mammalian egg (or, in reality, of the ovarian follicle) by Regnier de Graaf, in 1672, was a great victory for the ovists, who claimed that the spermatic animalcules possessed only the function of stirring and mixing the two seminal fluids. The animalculists countered with purported observations of homunculi within the heads of the spermatozoa. But this is not the place to trace the exaggerated and frequently ridiculous claims and counter-claims of these two schools of thought. Suffice it to say that by 1740 there remained very few epigenesists indeed, and the encasement theory of preformation prevailed almost universally.

In 1745 a small anonymous volume appeared from the press entitled *Vénus Physique, contenant deux dissertations, l'une sur l'origine des hommes et des animaux: et l'autre sur l'origine des noirs*.[23] (The second part had already been printed in the preceding year, with the interesting title *Dissertation Physique à l'Occasion du Negre Blanc*.)[24] This book was written in a popular style for gentlemen and ladies of the court. The author, who later turned out to be Maupertuis, dared to adopt the discredited epigenetic theory of development. He was led to this by a consideration of the plain facts of biparental heredity, which had been pointed out by Aristotle and again by Harvey, but which the ovists and animalculists attempted to explain away. According to them, heredity might be regarded as a spiritual essence, transferrable in the seminal fluids, and impressing itself upon the preformed embryo provided by one or the other of the parents. This essentially vitalistic idea was abhorrent to Maupertuis. A mathematician trained in Cartesian views, an early convert to Newton's teachings, Maupertuis belies the remark made by Driesch that all epigenesists were vitalists, for Maupertuis was a consistent mechanist. In criticizing the views of the preformationists, Maupertuis would have nothing to do with essences and spiritual "virtues." He argued against the possibility that maternal impressions can affect the foetus, and to him biparental heredity necessarily implied *corporeal* contributions from each of the two parents.

Maupertuis' Investigation of Polydactyly

One may well wonder what led Maupertuis to these views. The

---

[23] Maupertuis, *Vénus Physique, contenant deux dissertations, l'une sur l'origine des hommes et des animaux; et l'autre sur l'origine des noirs*. La Haye, 1745. [Also in *Oeuvres*, v. 2, pp. 1-134.] Hereafter abbreviated as *V. P.*

[24] Maupertuis, *Dissertation Physique à l'Occasion du Negre Blanc*. Leyde, 1744.

answer is sufficiently clear. He had already, prior to 1745, conducted a careful study in human genetics that is matched by only one other of its time. It seems that a young albino negro had been brought to Paris and naturally enough created a great stir. This started Maupertuis to thinking about heredity. Surely the albino condition must be hereditary, for although it seemed to appear sporadically among Negroes, it was reportedly not rare in Senegal, where whole famiiles were said to be "white." Of the famous albino Indians of Panama he had also heard, and discoursed about them with no little sentimentality. "The black color," he said, "is just as hereditary in crows and blackbirds, as it is in Negroes: I have nevertheless seen white blackbirds and white crows a number of times." [25]

Shortly after coming to Berlin, Maupertuis began to hunt for a case he could study at first-hand. He was fortunate in his search: he ran across the polydactylous Ruhe family. The investigation had been completed some time before the *Vénus Physique* was written, and from the internal evidence seems to have been the main directive to the thoughts and arguments presented in that book. When it is remembered that this was Maupertuis' first extensive venture into biological investigation, that the collection of human pedigrees was a novel enterprise, and that even in the present century the necessity of making a complete record of all normal as well as all abnormal members of a family has frequently been neglected, one can only marvel at his perspicacity. Maupertuis described his study of this hereditary trait in the *Système de la Nature*, and in the *Lettres*. It is in the latter (Letter XIV) that we find the fullest account, which follows:

> A great physician proposes in a useful and inquiring work (L'art de faire éclorre des oiseaux domestique, par M. de Réaumur, t. II, mém. 4) to perform some experiments on this question [of biparental inheritance]. In the race (*genre*) of fowls it is not rare to see types (*races*) which bear five toes on each foot; it is hardly more to see those which are born without rumps. M. de Réaumur proposes to mate a hen with five toes with a four-toed cock, a four-toed hen with a five-toed cock; to do the same experiment with the rumpless cocks and hens: and [he] regards these experiments as able to decide whether the foetus is the

---

[25] *V.P.*, Part II, Chap. IV.

product solely of the father, solely of the mother, or of the one and the other together.

I am surprised that that skilful naturalist, who has without doubt carried out these experiments, does not inform us of the result. But an experiment surer and more decisive has already been entirely completed. That peculiarity of the supernumerary digits is found in the human species, extends to entire breeds (*races*); and there one sees that it is equally transmitted by the fathers and by the mothers.

Jacob Ruhe, surgeon of Berlin, is one of these types. Born with six digits on each hand and each foot, he inherited this peculiarity from his mother *Elisabeth Ruhen*, who inherited it from her mother *Elisabeth Horstmann*, of Rostock. Elisabeth Ruhen transmitted it to four children of eight she had by Jean Christian Ruhe, who had nothing extraordinary about his feet or hands. Jacob Ruhe, one of these six-digited children, espoused, at Dantzig in 1733, Sophie Louise de Thüngen, who had no extraordinary trait: he had by her six children; two boys were six-digited. One of them, *Jacob Ernest*, had six digits on the left foot and five on the right: he had on the right hand a sixth finger, which was amputated; on the left he had in the place of the sixth digit only a stump.

One sees from this genealogy, which I have followed with exactitude, that polydactyly (*sex-digitisme*) is transmitted equally by the father and by the mother: one sees that it is altered through the mating with five-digited persons. Through these repeated matings it must probably disappear (*s'éteindre*); and must be perpetuated through matings in which it is carried in common by both sexes.

The pedigree is diagrammed conventionally in Fig. 2.

Maupertuis was quite struck by this apparent weakening of the trait with time, and it led him to the conclusion that through repeated matings with normal individuals the trait might in time disappear. In other words, the deviations of nature tend to fade out, and " her *works* always tend to resume the upper hand." [26] Maupertuis had evidently a clear idea that most if not all abnormalities are disadvantageous, and as we shall see later, this became an important part of his evolutionary thought.

It is truly astonishing to discover that in the second edition of that very book by Réaumur which served as inspiration to Mau-

---

[26] *V.P.*, Part II, Chap. v, italics added.

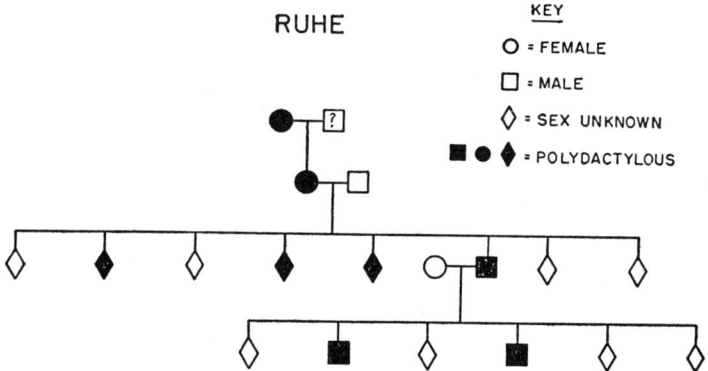

Fig. 2. Polydactyly in the Ruhe family, according to Maupertuis

pertuis (*L'art de faire éclorre les oiseaux domestiques*),[27] there is a celebrated account of the inheritance of human polydactyly that is as circumstantial and complete as that of the Ruhe family. Charles Bonnet, in 1762, struggled vainly to deal with its implications, so embarrassing to the theory of preformation and to ovists in particular.[28] A Maltese couple, named Kellaia, whose hands and feet were constructed upon the ordinary human model, had a son, Gratio, who possessed six perfectly moveable fingers on each hand and six toes, deformed and crownlike, on each foot. This man married a normal woman and had by her four children. The eldest, Salvator, a boy, had six fingers and six toes; the second, George, had five fingers and toes, but his hands and feet were slightly deformed; the third, André, was normal; the fourth, a girl, Marie, had five fingers and toes, but her thumbs were slightly deformed. All of these children grew up and married normal persons. The eldest had four children, two boys and a girl polydactylous, and one normal boy. The second, George, had three girls, all polydactylous, and a normal son. Two of these girls had six fingers and six toes on each side, but one had only five toes on his left foot. The fourth of Gratio Kellaia's children, his daughter Marie, had one boy with six toes and three normal children (one boy; two girls). The normal son of Gratio had only normal children. This pedigree is represented in Fig. 3.

How did Maupertuis come to overlook this account by Réaumur,

[27] R. A. F. de Réaumur, *L'art de faire éclorre et d'enlever en toute saison des oiseaux domestiques des toutes esspeces*, 2 vol. Impr. Royale, Paris, 1749. Second edition, 1751.

[28] C. Bonnet, *Considérations sur les corps organisées*, Part II, Chap. VIII, CCCLV. In *Collection complète des Oeuvres*, T. III, pp. 518 ff. Neuchatel, 1779.

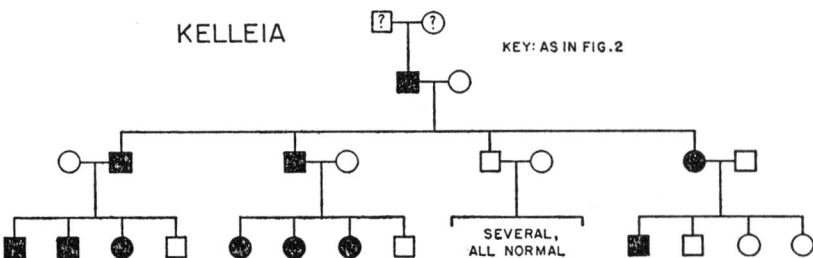

Fig. 3. Polydactyly in the Kellaia family, according to Réaumur

which so closely resembles his own investigation? An examination of Réaumur's rather rare book shows that the edition of 1749 does not contain the account. It was first added in the second edition, of 1751, in an appended section in square brackets at the very end of the book. It must be presumed that Maupertuis never saw it.

This well-known pedigree was cited by T. H. Huxley in his essay of April, 1860, entitled " [Darwin on] the Origin of Species," [29] but with one egregious addition to the original. Huxley stated: " A Maltese couple, named Kelleia [sic], whose hands and feet were constructed upon the ordinary human model, had born to them a son, Gratio. . . ." Nowhere in Réaumur's account, however, is it said that both the parents of Gratio Kellaia definitely had normal hands and feet. I was consequently misled in my original essay on Maupertuis when I relied on Huxley's account for this point, which would instantly be recognized by a geneticist as highly significant. For if both parents had been carefully examined and had been found to have normal digits, one might suppose that a mutation had occurred in one of the germ cells from which Gratio issued, although the very mild expression of the deformity in his daughter Marie is sufficient warning of a highly variable manifestation of the condition and a possibly reduced penetrance, such that occasionally a seemingly normal person might transmit the deformity. It would be necessary, therefore, to establish that Gratio Kellaia's ancestry for more than one generation back had been free of the trait before the assumption of a mutation would be warranted. However, Maupertuis might well have seized upon the sudden

---

[29] T. H. Huxley, [Darwin on] the Origin of Species, April, 1860, *Westminster Review*. In *Darwiniana* (Collected Essays, II), pp. 22-79, Appleton, New York and London, 1912. Also in *Man's Place in Nature and Other Essays*, Everyman's Library, Dent, London, 1906.

appearance in the son of normal parents of a hereditary condition that thereafter descended in unbroken lineage as a striking example of one of those fortuitous and tenacious alterations of the hereditary particles which he thought might constitute the beginning of a new species (see below).

Maupertuis' Theory of Heredity and the Origin of the Foetus

This investigation of polydactyly was the work upon which Maupertuis founded his theory of the formation of the foetus and the nature of heredity, a theory that was more than a century before its time and that brilliantly anticipated the discoveries of Mendel and de Vries. Maupertuis started out with the idea of chemical attraction, that attraction between the particles of different elements which results in the formation of compounds between them. This new chemical concept of Geoffroy and other French chemists of the time was much discussed, and Maupertuis was predisposed to it on account of his early devotion to the concept of gravitational attraction, which seemed to him to be an analogous phenomenon. As Maupertuis conceived it, two substances possess a tendency to unite by virtue of their chemical affinity (*rapports*); but if a third appears on the scene with a greater affinity for one of the two, it " unites with it while making it take leave of the other." [30] Maupertuis boldly applies these laws of chemistry to living beings. " Why," he asks, " if this force exists in Nature, would it not operate in the formation of the body of animals? Let us suppose that there are in each of the semens particles destined to form the heart, the head, the intestines, the arms, the legs; and these particles may each have a greater uniting power (*rapport d'union*) with that one which, in order to form the animal, has to be its neighbor, than with any other; the foetus will form, and were it yet a thousand times more organized than it is, it would form." [31] Several years later, when Maupertuis, under the pseudonym of Doctor Baumann, wrote the *Système de la Nature* (1751),[32] his ideas regarding these hereditary particles had apparently been influenced both by Leibniz's famous monads and by Buffon's theory of organic particles disseminated throughout nature, the latter theory having appeared

---

[30] *V. P.*, Part I, Chap. XVII.
[31] Ibid.
[32] Maupertuis, *Système de la Nature*, 1751. In *Œuvres*, v. 2, pp. 135-184. Hereafter abbreviated as *S. N.*

in the meantime; and Maupertuis attributed to these postulated particles a property " akin to that which in us we term desire, aversion, memory." [33] Moreover, although in the *Vénus Physique* he proposed pangenesis with cautious reservation and only as a hypothesis worthy of investigation, in the *Système de la Nature*, no doubt encouraged in this direction by Buffon's embracement of the idea, he spoke without reserve. In the *Vénus Physique* the 1746 edition reads simply: " As to the manner whereby there form in the semen of each animal particles analogous to those of that animal, I do not at all examine it here." The same passage in the later version included in the *Oeuvres* (1756) is amplified as follows: " As to the matter of which, in the semen of each animal, particles like that animal are formed, it would be a very bold conjecture, but one perhaps not destitute of all truth, to think that each part furnishes its own germs. Experiment could perhaps clear up this point, if one tried over a long period to mutilate certain animals generation after generation; perhaps one would see the parts cut off diminish little by little; perhaps in the end one would see them annihilated." [34] In the *Système de la Nature* we find: " The elements suitable for forming the foetus swim in the semens of the father and mother animals; but each, extracted from the part like that which it is to form, retains a sort of recollection of its old situation; and will resume it whenever it can, to form in the foetus the same part." [35]

This particulate theory of heredity and the formation of the foetus is logically analogous to Mendelian theory, and far in advance of those ideas that heredity is determined by indivisible entities, fluids or vapors, subject to irrevocable blending through intermating, that were current until late in the nineteenth century. If in one direction Maupertuis' hereditary particles therefore look back to the monads of Leibniz, in the other they look forward to the concept of the genes; and they have really much more in common with the latter than the former. Of course, Maupertuis confused the hereditary particles with the effects they produce and with the parts whose development they control, but that was hardly avoidable at the time. After all, this was nearly a century before the formulation of the Cell Theory. To see that at bottom heredity must depend on a sort of organic, chemical memory, and to attribute this to separable

[33] *S. N.*, XXXI.
[34] *V. P.*, Part II, Chap. V.
[35] *S. N.*, XXXIII.

particles that maintain their nature in combination is extraordinary enough. And what can one say of the perspicacity that proposed, as a test of pangenesis, the very experiment that Weismann was to perform nearly 150 years later?

Next to the particulate nature of the hereditary material, the most important of Mendel's principles is that of segregation. The hereditary units, or genes, as we now call them, are present in pairs as a consequence of fertilization, and the members of each pair segregate from one another in the reproductive cells prior to the production of another generation of offspring. This principle, too, was foreshadowed by Maupertuis. However the particles might previously be combined in each one of the parents, the particles from the two semens, he supposed, would unite *separately* in accordance with their affinities and, since corresponding particles from the mother and the father would be most alike, they would unite two by two and exclude other combinations. " One ought not to believe," he says," that in the two semens there are only precisely the particles which are needful to form one foetus, or the number of foetuses that the female is to bear: each of the two sexes without doubt furnishes a great many more than are necessary. But the two particles which are to be adjacent once being united, any third, which could have made the same union, would no longer find its place, and would remain useless. It is this, it is by these repeated operations, that the child is formed from the particles of the father and the mother, and often bears visible marks that it partakes of the one and of the other." [36] " In the seminal fluid of each individual the particles suitable for forming traits like those of that individual are the ones which are ordinarily most numerous, and which have the greatest combining power (*affinité*); although there are a great many others for different traits. . . . The particles analogous to those of the father and the mother being the most numerous, and having the most combining power, will be those which most commonly unite; and they will ordinarily form animals like those from which they are come." [37]

This is further clarified by Maupertuis' consideration of hybrids: ". . . For as soon as there is a mixing of species, experience teaches us that the child resembles both the one and the other." [38] " If the elements come from animals of different species, but in which there still remains sufficient affinity between the elements; these

---

[36] *V.P.*, Part I, Chap. XVII.
[37] *V.P.*, Part II, Chap. V.
[38] Ibid.

more attached to the form of the father, those to the form of the mother, [they] will produce hybrid animals." [39] "Finally if the elements come from animals who no longer have between them sufficient analogy, the elements not being able to assume, or not being able to retain a suitable arrangement, generation becomes impossible." [40] Thus, Maupertuis goes on, one can explain the sterility of hybrids such as the mule. "One of the most singular phenomena, and one of the most difficult to explain, is the sterility of hybrids. Experiment has shown that any animal born of the coupling of different species cannot reproduce. Could one not say that in the parts of the hinny and of the mule, the elements having taken a particular arrangement which was neither that which they had had in the ass, nor that which they had had in the mare; when these elements pass into the semens of the hinny and of the mule, the habitute of this last arrangement being most recent, and the habitute of the arrangement which they had had in the ancestors being stronger, because contracted over a greater number of generations, the elements remain in a certain equilibrium, and unite neither in one manner nor in the other? " [41] To anyone familiar with the causes of hybrid sterility this last paragraph has a startlingly modern ring. For Maupertuis was essentially correct. A set of chromosomes is indeed an arrangement of hereditary elements, and very often in hybrids the chromosomes derived from the two pure species are incapable of normal segregation, because of differences in the arrangements of their genes. The sterile hybrid between the radish and the cabbage furnishes an example. Radish chromosomes cannot pair with cabbage chromosomes. Segregation produces no effective germ cells, for neither an array wholly of radish elements nor one wholly of cabbage elements is likely to recur, and random mixtures of radish and cabbage chromosomes are physiologically ill-assorted. In spite of the crudity of Maupertuis' idea, it is impossible not to admire his insight.

Mendel is often credited with having discovered genetic dominance, although actually it was known long before his day, having been described by Knight in 1823. Maupertuis also arrived, though vaguely, at the idea of dominance. This came about from two considerations. In the first place, as we know today, polydactyly is a dominant trait. In the second place, the other genetic character in

[39] *S.N.*, xxxviii.
[40] *S.N.*, xxxix.
[41] *S. N.*, xliii.

which Maupertuis became interested, and to which he devoted the second part of the *Vénus Physique*, was albinism in negroes, and this is a recessive. Maupertuis was therefore aware that whereas polydactyly descends regularly from affected persons, married to normals, to some but not all of their offspring, albinism, on the other hand, seemed to appear sporadically among negroes, albino negroes being born of parents both of whom were black. Yet Maupertuis was convinced that albinism was hereditary. Maupertuis was thus faced with the problem of accounting in his theory for the different modes of inheritance, and concluded: " There could be, on the other hand, arrangements so tenacious that from the first generation they dominate (*l'emporte*) over all the previous arrangements, and efface the habitude of these." [42]

Had Maupertuis evolved a metaphysical system of heredity and embryogenesis, and done nothing more, he could hardly be ranked above his contemporary biologists. Metaphysical systems, in the eighteenth century, were " a dime a dozen." Georges Hervé, the first person to have considered the work of Maupertuis in the light of Mendelian genetics, has said: " His right to figure in a gallery of pre-Mendelian genetics would in that case have been disputable; if he holds a place there, it is because he knew enough to carry his researches into experimental realms." [43] He diligently pursued the collection of further evidence on polydactyly, and with some success, since he could write (later): " I have found in Berlin two six-digited persons, and I have given the genealogy of one. I have not been able to follow with sufficient exactitude the genealogy of the other, who is a foreigner, and who has concealed himself from me: but he had six-digited children, and I have been assured that this polydactyly has been hereditary in his family for a long time. A scientist illustrious in Germany and Minister to the Duke of Württemberg, M. de Bulfinger, was of such a family, and born with a sixth finger that his parents had cut off as a monstrosity." [44]

Like Mendel, Maupertuis applied mathematics to genetic investigation, by calculating the probability that the polydactyly observed might be only a sporadic accident and not really inherited. He wrote:

---

[42] *S. N.*, XLIV.
[43] G. Hervé, Maupertuis génétiste, *Rev. anthropol.*, 22: 217-230, 1912.
[44] *Lettres*, XIV.

> But if one wished to regard the continuation of polydactyly as an effect of pure chance, it would be necessary to see what the probability is that this accidental variation in a first parent would be repeated in his descendants. After a search which I have made in a city which has one hundred thousand inhabitants, I have found two men who had this singularity. Let us suppose, which is difficult, that three others have escaped me; and that in 20,000 men one can reckon on one six-digited: the probability that his son or daughter will not be born with polydactyly at all is 20,000 to 1; and that his son and his grandson will not be six-digited at all is 20,000 $\times$ 20,000 or 400,000,000 to 1: finally the probability that this singularity will not continue during three generations would be 8,000,000,000,000 to 1; a number so great that the certainty of the best demonstrated things of physics does not approach these probabilities.[45]

This is not only an excellent example of scientific caution, but also represents what is without doubt the first application to genetics of one of the most important of the principles of the mathematics of probability, that of the probability of coincidence of independent items. It was this very principle that Mendel applied so effectively in his analysis of segregation, random recombination, and independent assortment.

Breeding Experiments

But Maupertuis was not content to make only this analysis. He undertook actual breeding experiments with animals to test out his theories, although of the results of these he has unfortunately left us only the account of a single one. It is related that he "adored animals and lived surrounded by them." "You are more pleased with Mme. d'Aiguillon than with me," wrote Mme. du Deffand to him one day, "she sends you cats." And Frederick wrote, too: "I know that at Paris just as at Berlin you are enjoying the delights of good company. . . . I am only afraid that Mme. la duchesse d'Aiguillon is spoiling you. She loves parrots and cats, which is a prodigious merit in your eyes . . ."[46] Maupertuis had established himself in the outskirts of Berlin, in a spacious house adjacent to the royal park, near the present Tiergarten; and this house he had con-

---

[45] Ibid.
[46] Letter of Frederick, Potsdam, 15 Nov., 1748. Quoted from Hervé, op. cit.

verted into a virtual Noah's ark. Samuel Formey, permanent secretary of the Berlin Academy, has left us the following description: " The house of M. de Maupertuis was a veritable menagerie, filled with animals of every species, who failed to maintain the proprieties. In the living-rooms troops of dogs and cats, parrots and parakeets, etc. In the fore-court all sorts of strange birds. He once had sent from Hamburg a shipment of rare hens with a cock. It was sometimes dangerous to pass by the run of these animals, by whom some had been attacked. I was especially afraid of the Iceland dogs. M. de Maupertuis amused himself above all by creating new species by mating different races together; and he showed with complaisance the products of these matings, who partook of the qualities of the males and of the females who had engendered them. I loved better to see the birds, and especially the parakeets, which were charming. . . ." [47]

It was of the Iceland dogs that Maupertuis has left us the account of his breeding experiment: " Chance led me to meet with a very singular bitch, of that breed (*espèce*) that is called in Berlin the Iceland Dogs: she had her whole body the color of slate, and her head entirely yellow; a singularity which those who observe the manner in which the colors are distributed in this sort (*genre*) of animals will find perhaps rarer than that of supernumerary digits. I wished to perpetuate it; and after three litters of dogs by different fathers which did not yield anything of the sort, at the fourth litter she gave birth to one who possessed it. The mother died; and from that dog, after several matings with different bitches, there was born another who was exactly like him. I actually have them both." [48] His breeding of dogs led him to wonder particularly about the supernumerary fifth digit which is not uncommon on the hind foot: " There are no animals at all upon whom supernumerary digits appear more frequently than upon dogs. It is a remarkable thing that they ordinarily have one digit less on the hind feet than on those in front, where they have five. However, it is not at all rare to find dogs who have a fifth digit on the hind feet, although most often detached from the bone and without articulation. Is this fifth digit of the hind feet then a supernumerary? or is it, in the regular course, only a digit lost from breed (*race*) to breed throughout the entire species, and which tends from time to time

[47] Quoted from Hervé, op. cit.
[48] *Lettres*, XIV.

to reappear? For mutilations can become hereditary just as much as superfluities." [49] Were all dogs, in other words, once five-toed on both front and back feet? Have we here a remnant, a vestige of a once functional structure? These observations might well have been made by Charles Darwin.

His Theory of Evolution

Maupertuis' studies thus led him to evolution. Here with certainty he must be ranked above all the precursors of Darwin. To begin with, he was faced with the problem of accounting for supernumerary digits, albinism, and other hereditary anomalies on the basis of his theory of generation. This he solved ingeniously. "If each particle is united to those that are to be its neighbors, and only to those, the child is born perfect. If some particles are too distant, or of a form too little suitable, or too weak in affinity to unite with those with which they should be united, there is born a monster with deficiency (*monstre par défaut*). But if it happens that superfluous particles nevertheless find their place, and unite with the particles whose union was already sufficient, there is a monster with extra parts (*monstre par exces*)." [50] Even Mendel did not foresee that deficiencies and duplications of the hereditary material might constitute a basis of abnormal development, a sort of mutation! Maupertuis comments on the remarkable fact that in monsters with extra parts, these are always to be found in the same locations as the corresponding normal parts: two heads are always on the neck, extra fingers are always on the hand, extra toes on the foot. This is very difficult to explain on the basis of the theory that monsters come from the union of two foetuses or eggs, which was the explanation forced upon the preformationists by the nature of their views; but it was not at all difficult to explain on the basis of Maupertuis' concepts. He described the skeleton of a giant man, preserved in the Hall of Anatomy of the Academy in Berlin, with an extra vertebra in the lumbar region, inserted in a regular fashion between the ordinary vertebrae. How could this be the remains of a second foetus fused with the first? he asked.

But on Maupertuis' particulate theory, "chance, or the scarcity of family traits, will sometimes make rarer assemblages; and one

[49] Ibid.   [50] *V.P.*, Part I, Chap. XVII.

will see born of black parents a white child, or perhaps even a black of white parents...."[51] "... there are elements so susceptible of arrangement, or in which recollection is so confused, that they become arranged with the greatest facility...;"[52] elements which represent the condition in an ancestor rather than that in the immediate parent may enter into union in forming the embryo, producing resemblance to the ancestor rather than to the parent,[53] but also "a total forgetfulness of the previous situation" may occur.[54]

Maupertuis thus came to the conclusion that hereditary variants are sudden, accidental products—mutations, to use the modern term. Moreover, since negroes could by mutation produce "whites" (i.e., albinos), it was clear that racial, or species, differences—the distinction was not too clear in the eighteenth century—are produced by mutations. To Maupertuis, exactly as to Hugo de Vries a century and a half later, a species was merely a mutant form that had become established in nature. The evidence for this was clear from the artificial breeds of domestic animals. As in the case of Charles Darwin a century later, it was in particular the pigeons that clinched the argument. "Nature contains the basis of all these variations: but chance or art brings them out. It is thus that those whose industry is applied to satisfying the taste of the curious are, so to say, creators of new species. We see appearing races of dogs, pigeons, canaries, which did not at all exist in Nature before. These were to begin with only fortuitous individuals; art and the repeated generations have made species of them. The famous Lyonnés [Lyonnet] every year created some new species, and destroyed that which was no longer in fashion. He corrects the forms and varies the colors: he has invented the species of the harlequin, the mopse, etc."[55] And then Maupertuis wonders seditiously why this art should be restricted to animals. Might sultans in their seraglios practise a similar art? Had not Frederick William of Prussia built an armed force of giant soldiers, and thereby, thought Maupertuis, singularly increased the stature of the Prussian people?

If the ingenuity of man can produce species, why not nature, either by "fortuitous combinations of the particles of the seminal fluids, or effects of combining powers too potent or too weak among the particles"[56] or by the action of the environment, such

[51] *V.P.*, Part II, Chap. v.
[52] *S.N.*, XL.
[53] *S.N.*, XLI.
[54] *S.N.*, XLII.
[55] *V.P.*, Part II, Chap. III.
[56] *V.P.*, Part II, Chap. VII.

as the effect of climate or nutrition, on the hereditary particles. It is worth emphasizing, for it has been misunderstood, that Maupertuis raises the latter possibility only as one worthy of investigation; but clearly at this point he anticipated both Erasmus Darwin and Lamarck in suggesting the possibility of evolution through an inheritance of environmentally modified characters. Even so, it is the direct mutational action of heat or other factors on the hereditary material itself that Maupertuis seems most to have had in mind. "For the rest," he says, "although I suppose here that the basis of all these variations is to be found in the seminal fluids themselves, I do not exclude the influence that climate and foods might have. It seems that the heat of the torrid zone is more likely to foment the particles which render the skin black, than those which render it white: and I do not know to what point this influence of climate or of foods might extend, after long centuries of time." [57]

It is likewise clear that Maupertuis understood that most mutant forms are deleterious and at a disadvantage in comparison with the normal or wild types. "What is certain is that all the varieties which can characterize new species of animals and plants, tend to become extinguished: they are the deviations of Nature, in which she perseveres only through art or system. Her *works* always tend to resume the upper hand." [58]

How, then, account for the distribution of different races and species? The "thousands" of human varieties are insuperable difficulties for the preformationist; but by mutation, migration, and isolation they are readily accounted for by Maupertuis. Perhaps, he suggested, in the tropics all the peoples are dark of skin in spite of the interruptions caused by the sea, because of the heat of the torrid zone over a long period of time. The geographical isolation of Nature's deviations must play a part here, for "in travelling away from the equator, the color of the people grows lighter by shades. It is still very brown just outside the tropics; and one does not find complete whiteness until one has reached the temperate zone. It is at the limits of this zone that one finds the whitest peoples." [59] Well, "men of excessive stature, and others of excessive littleness, are species of monsters; but monsters which can become peoples, were one to apply himself to multiplying them." [60] Are there not

[57] *V.P.*, Part II, Chap. V.
[58] Ibid. Italics added.
[59] *V.P.*, Part II, Chap. I.
[60] *V.P.*, Part II, Chap. VII.

races of giants and dwarfs? These "have become established, either by the suitability of climates, or rather because, in the time when they commenced to appear, they would have been chased into these regions by other men, who would have been afraid of the Colossi, or disdain the Pygmies.

"However many giants, however many dwarfs, however many blacks, may have been born among other men; pride or fear would have armed against them the greater part of mankind; and the more numerous species would have relegated these deformed races to the least habitable climates of the Earth. The Dwarfs will have retired toward the arctic pole: the Giants will have inhabited the Magellanic lands: the Blacks will have peopled the torrid zone."[61] However naïve these anthropological conceptions may be—and they were an easy target for the sharp gibes of Voltaire—there is nevertheless a groping here for a truth that was only to be captured fully by Charles Darwin and Alfred Russel Wallace in a later day.

There is no naïvety, only pure genius, in these final words: "Could one not explain by that means [mutation] how from two individuals alone the multiplication of the most dissimilar species could have followed? They could have owed their first origination only to certain fortuitous productions, in which the elementary particles failed to retain the order they possessed in the father and mother animals; each degree of error would have produced a new species; and by reason of repeated deviations would have arrived at the infinite diversity of animals that we see today; which will perhaps still increase with time, but to which perhaps the passage of centuries will bring only imperceptible increases."[62]

Maupertuis and His Contemporaries

Among his contemporaries in biology, the work of Maupertuis was most highly regarded. With Buffon, in particular, Maupertuis had much in common. Maupertuis was the author who seemed to Buffon "to have reasoned better than all those who have written before him on this subject" (of the generation of the individual); and Buffon spoke of the *Vénus Physique* as a treatise which, "although very short, assembles more philosophical ideas than there are alltold in several great volumes on generation.... This author,"

[61] Ibid.   [62] *S. N.*, XLV.

he continued, "is the first to have commenced to draw near to the truth, from which we were farther than ever since eggs had been supposed to exist and the spermatic animals were discovered." [63]

It is equally significant that Charles Bonnet, the great exponent of preformation, directed his main arguments against the "internal molds" (*moules intérieurs*) of Buffon and the "attractive forces" and the "memory of the seminal molecules" of Maupertuis fully as much as against the epigenetic *vis essentialis* of Caspar Friedrich Wolff. What is often overlooked today is the fact that Bonnet was profoundly right in holding as senseless the view that "that unity, that organic whole one calls an animal" can arise from an *amorphous* semen, and in affirming that a gel (*glu*) which appears to become organized in development must possess some organization to start with. Years later, writing in the *Palingénésie* (1769), Bonnet arrived at a view of preformation that is as close to our modern understanding of heredity and development as was the epigenetic view of Maupertuis. It is not necessary, he said, to limit the significance of preformation "to express an organic corpuscle that actually encloses in extreme miniature all the parts which characterize the species." It may be extended to include "every organic preformation from which an animal can result as its immediate principle." [64] And again, in the *Contemplation de la Nature*, he wrote: "I do not affirm that the buds which produce separate young polyps (*les rejetons d'un Polype à bras*) were themselves miniature polyps, hidden under the skin of the mother, but I affirm that there are, under the skin certain particles which have been preorganized in such a manner that a little polyp results from their development." [65] Every genesis, in other words, must issue from a "predetermination," a "secret preorganization," a "primordial design." Seen in this light, there is a close rapprochement between the ideas of Maupertuis and Bonnet, such that the modern theory of the genetic organization of the fertilized egg covers them both. As so often in the history of scientific ideas, the ultimate truth includes those antithetical ideas over which earlier men were embroiled in controversy.

---

[63] Buffon, *Histoire naturelle générale et particulière*, nouvelle édition, III, p. 244, Paris, 1769. [Quoted in Brunet, op. cit., Vol. II, p. 329.]
[64] Bonnet, *Palingénésie* [Quoted from J. Rostand, *Esquisse d'une histoire de la biologie*, 12th edition, p. 77. Gallimard, Paris, 1945.]
[65] Bonnet, *Contemplation de la Nature*.

The Reputation of Maupertuis in the 19th and 20th Centuries

It remains to account for the reputation of Maupertuis to the present day. There is little indeed to say. Darwin, it would seem, never heard of him. Thomas Henry Huxley, who was the first to make a somewhat careful study of Darwin's predecessors, makes only a cursory reference to Maupertuis' "curious hypothesis as to the causes of variation, which he thinks may be sufficient to account for the origin of all animals from a single pair." [66] Quatrefages, Clodd, Packard, Nordenskiøld, Singer and other historians of the theory of evolution have ignored him completely. Henry Fairfield Osborn, in *From the Greeks to Darwin*, almost completely misunderstood Maupertuis, classifying him with those who made speculations without the support of observation or the least deference to inductive canons.[67] Osborn apparently read only the *Système de la Nature*, which is admittedly difficult to interpret without prior knowledge of the *Vénus Physique*. In 1904, Lovejoy published a scholarly study of Maupertuis and other early evolutionists.[67a] Although Osborn knew of this study, he did not make the slightest use of it in preparing the revised edition of his book (1928), and he may consequently be regarded, after Voltaire, as the person chiefly responsible, at least in the English-speaking world, for the misapprehension of Maupertuis' significance as a scientist. Yves Delage, the great French historian of biological theories, was just to Maupertuis in discussing the latter's views in *L'hérédité et les grands problèmes de la biologie générale*. He considered Maupertuis' theories regarding ontogenesis, heredity, variation, and the formation of species in a spirit of admiration, and summed up by saying, "Geoffroy Saint-Hilaire could subscribe to his law of teratogenesis; Lamarck, to his ideas on variation; Darwin, to his system of representative germs. And all that had been said by him a century before them!" [68] And Delage compared Maupertuis with Buffon not en-

---

[66] Huxley, Evolution in Biology, 1876. In *Darwiniana* (Collected Essays, II), pp. 187-226. Appleton, New York and London, 1912.

[67] H. F. Osborn, *From the Greeks to Darwin*, Scribner's, New York, 1894. Second edition, 1928.

[67a] See footnote 7, this essay. Lovejoy's essay was the first, at least in English, to do justice to Maupertuis' theory of organic evolution, and to show that it was based on an empirical study of the inheritance of polydactyly and upon a considered epigenetic view of the nature of development, in contradistinction to currently prevailing preformationist views.

[68] Y. Delage, *L'hérédité et les grands problèmes de la biologie générale*, Schleicher Fr., Paris, 1896. Second edition, 1903.

tirely to the latter's advantage. Nevertheless, it must be remembered that Delage's book was written before the rediscovery of Mendel's laws and before the mutation theory of de Vries was published. In many respects the criticisms of Delage therefore sound antiquated—in matter of fact, some of the imperfections with which he charged Maupertuis have turned out to be the imperfections of Delage. Nor was Delage always correct, as when, for instance, he charged Maupertuis with having been a purely theoretical critic of preformation, or with having believed in the inheritance of acquired characters in a Lamarckian sense. Delage also seems to have regarded Maupertuis' theories as succeeding those of Buffon instead of preceding them, an error due to his listing the *Vénus Physique* as first published in 1748, instead of 1744-45. As Lovejoy pointed out, it was more likely Buffon who was influenced by the theories of Maupertuis than the reverse; although the differences between the views expressed in the *Vénus Physique* and those in the *Système de la Nature* may very well represent a counter-influence.

In this period the common judgment of Maupertuis was that derived from Voltaire. It is enlightening, for example, to read, in the light of D'Alembert's praise of Maupertuis' clarity and literary style (1778), the dictum of such a critic as Leon Sagnet, writing in *La Grande Encyclopédie* (1896): "Certainly, Voltaire was unjust. Before covering Maupertuis with insults, moreover, he had praised him above measure. One is none the less obliged to state that the favorite of the king of Prussia was a very mediocre writer, with a style stiff and pretentious, and who, inflated with pride and with his spirit warped by vanity, produced as a scientist, outside of his measurement of the arc of Tornea, neither any work nor discovery of the first rank, that, consequently, the great celebrity which he enjoyed among his contemporaries and the honors with which he was overwhelmed were as little justified as the mockeries of Voltaire were themselves exaggerated." [69]

In 1912 Georges Hervé wrote a brief paper entitled *Maupertuis génétiste*.[70] This is a very fine critique of Maupertuis' theories in the light of Mendelian genetics, paying tribute to his experimental approach. But it was by no means exhaustive, and, since it appeared in a journal little consulted by experimental biologists, it seems to

---

[69] L. Sagnet, Maupertuis. In *La Grande Encyclopédie*, v. 23, p. 241, 1896.
[70] Hervé, op. cit.

have been almost completely overlooked. The comments of Curtis [71] on Maupertuis relied wholly upon the previous analysis by Lovejoy (1904).

Coming to a more recent period, we find a number of biologists referring to Maupertuis, but always in a partial light. Joseph Needham, in his *History of Embryology*,[72] pays fitting tribute to Maupertuis as an epigenesist, but was not interested in his genetics or his evolutionary theories, F. J. Cole, in *Early Theories of Sexual Generation*,[73] besides an interest in the same subject, seems to have been struck chiefly by the fact that Maupertuis failed to appreciate the importance of the spermatozoa, and by other inadequacies in his views. But he does state that there can be no doubt that Buffon's famous system of organic particles, "his elaborate system of pangenesis . . . was inspired by that of Maupertuis, of which it is a development and extension." Conway Zirkle, in several papers investigating the history of the ideas of pangenesis and the inheritance of acquired characteristics,[74] has discussed the ideas of Maupertuis on these subjects; but his work is restricted in point of view, and somewhat confuses Maupertuis' arguments in favor of epigenesis and his position as to pangenesis. Emile Guyénot, alone among modern students of genetics and evolution, has been aware of the significance of Maupertuis. His masterly treatise, *Les sciences de la vie au XVIIe et XVIIIe siècles*,[75] estimates Maupertuis' evolutionary and genetic contributions at full worth. It is thus disappointing to find, after reading Guyénot's comment that "after these two precursors indulging in fantasy [Robinet and de Maillet], the work of Maupertuis can only appear to us more serious and more profound," that in a very recent and generally excellent account of the origins of Darwinism, the fantasies of Benoit de

[71] W. C. Curtis, *Science and Human Affairs*, Harcourt, Brace, New York, 1922.
[72] Needham, *A History of Embryology* (Part II of *Chemical Embryology*), Cambridge, at the University Press, 1931.
[73] F. J. Cole, *Early Theories of Sexual Generation*, Clarendon Press, Oxford, 1930.
[74] C. Zirkle, The inheritance of acquired characters and the provisional hypothesis of pangenesis, *Amer. Naturalist*, 69: 417-445, 1935; Further notes on pangenesis and the inheritance of acquired characters, *Amer. Naturalist*, 70:529-546, 1936; The early history of the idea of the inheritance of acquired characters and of pangenesis, *Trans. Amer. Philos. Soc.*, n. s., 35(2):91-152, 1946.
[75] E. Guyénot, *Les sciences de la vie aux XVIIe et XVIIIe siècles: l'idée d'évolution*, Albin Michel, Paris, 1941.

Maillet receive seven pages and Buffon a like amount, while Maupertuis, mentioned as rescued from undeserved obscurity, is dismissed in one brief paragraph.[76] Guyénot's summation is more just. He writes: "Geneticist from the very start, Maupertuis sketched, in all, an evolutionary conception that was based at once on heredity, on the fortuitous production of variations, on a prophetic vision of the phenomena of selection and of preadaptation. We feel him infinitely closer to us than the great theoreticians of Transformism. Already he was reasoning more on the basis of facts than of conceptions *a priori*. He was one of the founders of Evolution and appears to us, in addition, as the most remarkable precursor of contemporary Mutationism."[77] A paper, on the uncelebrated bicentennial of the publication of the *Vénus Physique*, by J. and M.-L. Dufrenoy hails Maupertuis not only as geneticist and evolutionist, but as also a brilliant precursor of comparative pathology.[78]

Conclusion

Here was a man who argued on genetic grounds against preformation fifteen years before the work of Caspar Friedrich Wolff which was eventually to dispose of the encasement theory of preformation; who investigated human heredity in a manner calculated to draw the admiration of any geneticist of the present day, and who applied the mathematical theory of probability to genetics over a century before Mendel; who undertook experiments in animal breeding to throw light on his theories; who formulated a theory of heredity that was particulate and involved the mutual attraction of analogous particles provided by each parent, and that implied segregation, dominance, and independent assortment; and finally, who formed a theory of organic evolution based upon mutation, natural selection, and geographic isolation. Surely he ranks above his contemporaries in biology. Instead, we must compare him with the mighty figures of a later time.

Thus, in the company of Lamarck, of Karl Ernst von Baer, who

---

[76] L. Eiseley, *Darwin's Century: Evolution and the Men Who Discovered It*. Doubleday and Co., Garden City, N. Y., 1958.

[77] Guyénot, op. cit., p. 393.

[78] J. and M.-L. Dufrenoy, Un bicentenaire oublié: La Vénus Physique, 1746; ou Maupertuis, précurseur de la Pathologie comparée, *Rev. Pathol. comp.*, 48: 107-115, 1948.

eventually disposed of the preformation theory, of Blumenbach, the father of anthropology, and of Darwin and Wallace, Mendel and Hugo de Vries, Maupertuis at last takes his rightful place. A man too far before his time, he was the most many-sided genius of them all. Only now, when the separate strands of genetics, embryology, anthropology, and evolution are woven together, and the Principle of Least Action is seen to be basic to them all, are we in a position to recognize in Maupertuis one of the greatest luminaries of eighteenth century science.

FOUR

UFFON AND THE PROBLEM OF SPECIES *

ARTHUR O. LOVEJOY

I

There is no chapter of the history of the theory of organic evolution more confused or more controverted than that which relates to the position of Buffon. Upon one point, indeed, nearly all expositors of the *Histoire Naturelle* are agreed—namely, that Buffon's own expressions on the subject, if taken at their face value, contradict one another. But upon the questions whether his utterances were meant to be taken at their face value; whether, by a due consideration of dates, the contradictions cannot be regarded as consecutive steps in a logical progress of doctrine; whether he was in the main a partisan or an opponent of transformism: upon these questions both the biographers of Buffon and the historians of evolutionism are greatly at variance.

The rival interpretations fall into six groups. (1) Older writers of the anti-evolutionary school, such as Cuvier and Flourens, while admitting that (in the words of Flourens) "the ideas of Buffon were constantly subject to profound mutations," were wont to maintain that in the last analysis and in the long run he must be counted among the defenders of the doctrine of immutability of species. Among recent writers Packard gives a similar account; while he recognizes "tentative" evolutionistic utterances in the

* Revised from the essay of the same title originally printed in the *Popular Science Monthly*, 79: 464-473, 554-567 (1911).

Histoire Naturelle, he opines that Buffon himself " did not always take them seriously, but rather jotted them down as passing thoughts." (2) One of the earlier French evolutionists, Isidore Geoffroy St. Hilaire,[1] contended that there was no mere fluctuation in Buffon's teaching, but simply an orderly movement of thought from one position to another.

> Buffon does but correct himself; he does not fluctuate. He goes forward once for all from one opinion to another, from what at the outset he accepted on the authority of another to what he recognized as true after twenty years of research.

The successive phases of opinion through which, according to Isidore Geoffroy, Buffon passed were three. At the beginning of his work (1749) and down to 1756 or later, he "still shared the views of Linnaeus" and affirmed consistently the theory of immutability. From 1761 to 1766 he asserted the hypothesis of variability in an extreme form. Later he became convinced that " in setting himself free from the prevailing notions," he had, " like all other innovators, gone somewhat to the opposite extreme "; and in all his writings subsequent to 1766 he held to a doctrine of "limited mutability," to the "permanence of the essential features" of species and "the variability of details." This division of Buffon's opinions into three periods, of which the middle one was characterized by an extreme evolutionism, has been accepted by a number of later writers. It is apparently adopted by Osborn, though not to the exclusion of other interpretations inconsistent with it.[2] (3) By several later writers—such as Samuel Butler, de Lanessan, Giard, and Clodd—Geoffroy's scheme of three periods is rejected, and Buffon is declared to have been an evolutionist throughout virtually his whole career as a writer. Those who take this view explain away his apparent self-contradictions by various suppositions. Giard, for example, holds that Buffon began as a transformist, but was led by his difficulties with the ecclesiastical authorities (in 1751) to conceal his real position for a number of years, becoming bolder and more outspoken after 1761, when his fame was securely established. In other words, Giard proposes an alternative division into three periods, in which the middle phase is the *least* evolutionistic. Samuel Butler,

[1] In his *Histoire Naturelle Générale,* Vol. II, 1859. His account is translated in Butler's *Evolution Old and New,* 1879.
[2] *From the Greeks to Darwin,* 1894, pp. 130-135.

who has taken the most extreme ground of all in favor of the view that Buffon was a whole-hearted evolutionist, endeavors at great length and with much ingenuity to show [3] that all the anti-evolutionary passages in the *Histoire Naturelle* are ironical. According to this interpretation, Buffon must almost be said to have woven a sort of cryptogram into his work. " His irony is not the ill-natured irony of one who is merely amusing himself at other people's expense, but the serious and legitimate irony of one who must either limit the circle of those to whom he appeals, or must know how to make the same language appeal differently to the different capacities of his readers, and who trusts to the good sense of the discerning to understand the difficulty of his position, and make due allowance for it." In other words, Buffon threw in sufficiently frequent affirmations of the immutability of species to deceive, or at least to quiet, the doctors of the Sorbonne, and in the very act of doing so he made it evident to the judicious reader that the opposite conclusion was the one to be accepted.

The three remaining interpretations of Buffon's position are less subtle and ingenious. (4) The author of one of the most comprehensive recent histories of biological theories [4] tells us that, though Buffon " speculated about the origination of one species from another," he did not " especially interest himself in the question of the mutability of species; his too little developed sense for the historical [i. e., the genetic] aspect of nature did not permit him to put clearly before himself such a question as that concerning the origin of species. How should he have done so, since he did not even believe in the existence of species, but recognized only individuals? " (5) Dacqué, in what is at many points the least inaccurate of the histories of evolutionism,[5] declares that Buffon brought forward no more profound ideas than his contemporaries " upon the interconnection of the phenomena of organic nature," though he did something to clarify the conception of geological evolution, and " regarded species as variable within certain limits." (6) Another writer, Landrieu,[6] seems finally to give up as hopeless the attempt

[3] In *Evolution Old and New*, 1879.
[4] Rádl, *Geschichte der biologischen Theorien*, Vol. I, 1905, pp. 117-118.
[5] *Der Descendenzgedanke und seine Geschichte*, 1903—a little book less known than it deserves to be.
[6] In his *Lamarck, fondateur de l'évolution*, 1909, pp. 275-283. May I improve this occasion to express the hope that both French and English writers may some day be broken of the habit of talking of " evolution " when they mean

BUFFON.

to reduce Buffon's utterances to harmony and coherency. He adds, however, that in spite of these inconsistencies, "Buffon retains the indisputable honor of having been the first zoologist to admit the possibility of specific variations due to environmental influences and extending beyond the limits of species."

All of these accounts of the matter seem to me to be either inadequate or erroneous, though all may be said in some measure to be founded on fact. Most of them—especially the more recent ones—wholly ignore two essential considerations in relation to Buffon's biological conceptions, in the light of which all that he wrote must be interpreted. In attempting to present a more adequate and more correct analysis of Buffon's opinions, I shall be obliged to tax the reader's patience with many and lengthy citations. Where there has been so much disagreement, it is necessary to present the proofs for nearly every statement propounded. And where so much error has arisen through the citation of brief passages in disregard of their contexts, it is important that pains be taken to quote or summarize so much of each text as appears to be in any way relevant to the question under consideration.

1. The first volume of the great treatise (1749) opened with a preliminary disquisition on the methodology of the science, a " Discours de la manière d'étudier et de traiter l'histoire naturelle." In this Buffon gave a salutary emphasis to the demand for a more " philosophical " way of studying botany and zoology than had been exemplified by Linnaeus and Tournefort and the other great systematists. Description and classification, Buffon insisted, were the least part, though a necessary part, of " natural history."

> We ought to try to rise to something greater and still more worthy of occupying us—that is to say, to combine observations, to generalize the facts, to link them together by the force of analogy, and to endeavor to attain that high degree of knowledge in which particular effects are recognized as dependent upon more general effects, Nature is compared with herself in her larger processes, and thus ways are opened before us by which the different parts of physical science may be perfected. For success

---

" evolutionism? " Both languages chance to be provided with a suffix for distinguishing a theory which affirms, or relates to, a given fact from the fact itself; it seems a pity to throw away this instrument of linguistic precision. It is surely absurd (not to say profane) to speak of Lamarck or any other mortal as the " founder of evolution "; or of the eighteenth century as " the beginning of evolution."

in the former sort of study there are needful only a good memory, assiduity and careful attention; but for the sort of which we are here speaking other qualities are requisite: breadth of view, steadiness of vision, a power of reasoning formed by the practice of reflection even more than by learning. For such study, in short, a man must have that quality of mind which enables him to grasp remote relations between things, to bring them together, and thereby to form a body of reasoned conclusions, after having duly estimated similarities and weighed probabilities.

But these judicious and stimulating, if slightly vague, appeals for the conversion of natural history into a science of causal relations and generalized laws were not the principal purpose of the preliminary discourse. The thought of Buffon at the time when he wrote that essay seems to have been dominated above all by a single idea, which was also one of the two or three ruling ideas of the whole of the first half of the eighteenth century—namely, the Leibnitian "principle of continuity" (*lex continui*). In the intellectual fashions of this period, next to the blessed word "Nature" the most sacred phrase was "the Great Chain of Beings"; indeed, one of the truths that man was supposed to know most surely about Nature was that she "makes no leaps." In the form, especially, of the neo-Platonic and Spinozistic metaphysical assumption that all possible forms must exist, the principle was much older than the philosophy of Leibniz;[7] but it owed to him and his disciples a more definite formulation and a greatly increased popular currency. It declared that all entities are arranged in a graded scale of similarity, so that for every being that exists there also exists some other (in the strict version of the principle, one and only one other) from which its difference is infinitesimal, i. e., less than any assignable difference. A typical statement of the doctrine is Bonnet's:

> Between the lowest and the highest degree of corporeal or spiritual perfection there is an almost infinite number of intermediate degrees. The series of these degrees constitutes the *Universal Chain*. It unites all beings, binds together all worlds, embraces all spheres. One Being alone is outside of this chain, and that is He who made it . . . There are no breaks (*sauts*) in nature; all is graduated, everything shades off into the next thing. If,

---

[7] This implied that there must be one, and can be only one, sample of every possible kind or degree of entity. To consider Leibniz's attitude toward this form of the principle would involve too much technical metaphysics.

between any two beings whatever, there existed a gap, what would be the reason of the transition from the one to the other? There is, therefore, no being above or below which there is not some other that approximates it with respect to some characters and diverges from it with respect to others.[8]

All this (as Bonnet's language intimates) was held by the Leibnitian philosophy to be logically implied by the still more fundamental " principle of sufficient reason." For if the gradations found in nature were discontinuous, if between any two beings an intermediate type were logically capable of existing, but were actually non-existent, the universe would stand convicted of irrationality. A thing for the existence of which there was just as much " reason " as there was for the existence of certain other things would have failed of realization, while the others arbitrarily enjoyed the privilege of actuality. The principle of continuity owed its vogue in part to the influence of the Leibnitian calculus, which had brought infinitesimals and the notion of the continuum peculiarly into fashion.

It was, then, the application of this principle to natural history that was Buffon's main object in his preliminary discourse. The consequences of it, when it was applied in this field, were simple and evident and drastic: there can be no such thing as a " natural," or even a consistent " system " of classification, since there are no sharp-cut differences in nature, and since, therefore, species and genera are not real entities but only figments of the imagination. It is easy, Buffon wrote, to see the essential fault in the work of the systematists, the inventors of " methods " as a class.

> It consists in an error in metaphysics, in the very principle underlying these methods. This error is due to a failure to apprehend Nature's processes, which take place always by gradations (*nuances*), and to the desire to judge of a whole by one of its parts.[9]

Man, placing himself at the head of all created things and then observing one after another all the objects composing the universe,

> will see with astonishment that it is possible to descend by almost insensible degrees from the most perfect creature to the most formless matter; . . . he will recognize that these imperceptible

---

[8] *Contemplation de la Nature* (1764), 2nd ed., 1769, Vol. I, pp. 26-27.
[9] *Hist. Nat.*, Vol. I, 1749, p. 20.

shadings are the great work of Nature; he will find them—these gradations—not only in the magnitudes and the forms, but also in the movements, in the generations and the successions, of every species.[10] If the meaning of this idea be fully apprehended, it will be clearly seen that it is impossible to draw up a general system, a perfect method, for natural history ... For in order to make a system or arrangement, everything must be included, and the whole must be divided into different classes, these classes into genera, and the genera into species—and all this according to an order in which there must necessarily be something arbitrary. But Nature proceeds by unknown gradations, and consequently can not wholly lend herself to these divisions—passing, as she does, from one species to another species, and often from one genus to another genus, by imperceptible shadings; so that there will be found a great number of intermediate species and of objects belonging half in one class and half in another. Objects of this sort, to which it is impossible to assign a place, necessarily render vain the attempt at a universal system.[11]

In short, the whole notion of species is inconsistent with the conception of nature as a graded continuum of forms in which there are no breaks.

In general, the more one increases the number of one's divisions, in the case of natural products, the nearer one comes to the truth; since in reality individuals alone exist in nature, while genera, orders, classes, exist only in our imagination.[12]

The vogue of the principle of continuity in the eighteenth century was, unquestionably, an important influence tending to prepare men's minds for the acceptance of the conception of evolution; but the two doctrines were by no means synonymous, nor did the adoption of the former necessarily imply adherence to the latter. The *lex continui* is historically important because it led to one of the early notable departures in modern thought from what may be called a Platonistic habit of mind, that had, in a hundred subtle ways, dominated most European philosophy and science for many cen-

---

[10] These words are Buffon's nearest approach in the introductory discourse to a suggestion of the mutability of species. De Lanessan has interpreted them as an affirmation of transformism; but they are too vague to justify such a construction.

[11] *Hist. Nat.*, Vol. I, 1749, p. 13. Much the same thing had, however, been said by Ray over sixty years before; cf. *Works*, 1686, Vol. I, p. 50.

[12] *Ibid.*, p. 38.

turies; it meant, in some degree, the abandonment of the fashion of thinking of the universe as tied up in neat and orderly parcels, the rejection of rigid categories and absolute antitheses, as inadequate instruments for the description of the complexity and fluidity and individuatedness of things. In other words, the principle of continuity, though itself the product of the extreme of philosophical rationalism, tended in a mild way towards a sort of anti-rationalism, towards a distrust of over-sharp distinctions and over-simple conceptions, towards a sense of a certain incommensurability between the richness of reality and the methods of conceptual thought. And in the nineteenth century this same tendency, in vastly more extreme forms, has been far more conspicuously furthered by the influence of the doctrine of evolution. But the idea of continuity as generally held in the time of Buffon had no reference to temporal sequences and by no means involved, in the minds of those who accepted it, any definite belief in the descent of what are commonly called species from other species.[13] If the presupposition of continuous gradations and imperceptible transitions had been explicitly brought to bear upon genetic problems in biology, it would naturally though not necessarily have suggested some sort of theory of descent. It did, in fact, in combination with what I have called the principle of plenitude, lead several important eighteenth-century writers— Leibniz, Bonnet, Akenside, Robinet—to adopt a theory of the progressive advance of organic types in the course of cosmic history, which foreshadowed, though it did not amount to, the hypothesis of organic evolution.[14]

But it seems to have been taken in an essentially static sense by Buffon in the introductory discourse in his first volume. A single obscure phrase, which I have already quoted, might be regarded as hinting at the conception of organic evolution, if the general tenor of the essay lent any confirmation to such an interpretation. But

---

[13] This fact has often been overlooked by interpreters of eighteenth century writers. When we find such a writer saying that "nature passes from one species to another by gradual and almost imperceptible transitions," it is by no means safe to assume that the phrase contains any reference to genealogical transitions, or that the writer meant by his words to affirm the transformation of species through the summation of some slight individual variations. Misapprehension upon this point has caused some eighteenth century authors to be quite undeservedly set down as evolutionists.

[14] See Lovejoy, *The Great Chain of Being*, 1948, Lecture IX: "The Temporalizing of the Chain of Being."

nowhere else in this writing is it even remotely suggested that the conception of the continuity of forms involves the conception of the descent of so-called species from one another. It is scarcely conceivable that if Buffon had had before his mind so momentous a new idea as that of evolution, he should not have contrived to give a far plainer intimation of it than a single vague remark that imperceptible gradations are found not only in the forms but also in the generations and the successions of every species. At this time, at all events—whatever he may have been later—Buffon was fairly outspoken in the expression of even heterodox hypotheses; it was only subsequently that he was condemned by the Sorbonne, on account of opinions propounded in his *Théorie de la Terre*, contained in the same volume as the preliminary discourse. It is significant, moreover, that at this date he saw no hint of any evolutionary significance in the homologies of the vertebrate skeleton; he had as yet learned nothing from comparative anatomy. This is shown in the argument by which he defends his own method of arranging species—a method which wholly ignored anatomical considerations and merely proceeded from the more familiar to the less familiar animals.

> Is it not better to make the dog, which is fissiped, follow (as he does in fact) the horse, which is soliped, rather than have the horse followed by the zebra, which perhaps has nothing in common with the horse except that it is soliped? . . . Does a lion, because it is fissiped, resemble a rat, which is also fissiped, more closely than a horse resembles a dog? [15]

It is probable, then, that in writing the opening discourse of his great work Buffon was innocent of any idea of organic evolution; it is certain that he did not convey that idea in any such way that a reader of his time might be expected to recognize it. Nor did he make any use of the conception of the descent of species in his *Théories de la Terre*, of the same date—where he might naturally have been expected to introduce the doctrine, if he held it; on the contrary he implies (p. 197) the equal antiquity of all species—though he does so in a way which, I confess, might plausibly be taken as ironical. The truth is that when under the influence of the principle of continuity Buffon's mind overshot the problem of the origin of species altogether. There were no such things as

---

[15] *Hist. Nat.*, Vol. I, 1749, p. 36.

species: upon this point he was clear. There was therefore no need of explaining their genesis. As for the further question, how successive generations of offspring are related in form to their forebears, that was a question upon which the principle of continuity had, strictly speaking, nothing to say. That offspring varied somewhat, and usually slightly, from their parents every one knew; to this extent the conformity of the laws of heredity to the law of continuity was a commonplace of everyday observation. Beyond this, no definite genetic or embryological consequences seemed necessarily to follow from the maxim *natura non facit saltus*.

The most important thing, however, to remark concerning Buffon's position in his first volume is that it is a position which he speedily abandoned, and to which he never returned.[16] Its most characteristic point was the contention that nature knows only individuals and that species are *entia rationis* merely. The most characteristic point of nearly all his subsequent references to the subject is the contention that species are real entities, definable in exact and strictly objective terms, and necessary to take account of in any study of natural history.

This change already was manifest in the second volume, published in the same year as the preliminary discourse (1749). In this volume Buffon propounded his celebrated definition of species, which was destined to have so great an influence upon the biological ideas of the later eighteenth century.

> We should regard two animals as belonging to the same species if, by means of copulation, they can perpetuate themselves and preserve the likeness of the species; and we should regard them as belonging to different species if they are incapable of producing progeny by the same means. Thus the fox will be known to be a different species from the dog, if it proves to be a fact that from the mating of a male and a female of these two kinds of animals no offspring is born; and even if there should result a hybrid offspring, a sort of mule, this would suffice to prove that fox and dog are not of the same species—inasmuch as this mule would be sterile (*ne produirait rien*). For we have assumed that, in order that a species might be constituted, there was necessary a continuous, perpetual and unvarying reproduction (*une pro-*

---

[16] Rádl's account, already quoted, of Buffon's attitude towards transformism and towards the conception of species, is apparently based chiefly upon the first volume. For virtually all of Buffon's views, except his early and quickly repudiated one, Rádl's statement is almost the exact reverse of the truth.

duction continue, perpétuelle, invariable)—similar, in a word, to that of the other animals.[17]

This language, it will be observed, implies not only that species are real entities, but also that they are constant and invariable entities. The same implication may be found again later in the volume. Buffon thus concludes the exposition of his embryological hypotheses—which embraced a theory of pangenesis:

> There exists, therefore, a living matter, universally distributed through all animal and vegetal substances, which serves alike for their nutrition, their growth and their reproduction. . . . Reproduction takes place only through the same matter's becoming superabundant in the body of the animal or plant. Each part of the body then sends off (*renvoie*) the organic molecules which it can not admit. Each of these particles is absolutely analogous to the part by which it is thrown off, since it was destined for the nourishment of that part. Then, when all the molecules sent off by all the parts of the body unite, they necessarily form a small body similar to the first, since each molecule is similar to the part from which it comes. It is in this way that reproduction takes place in all species. . . . There are, therefore, no preexisting germs, no germs contained within one another *ad infinitum*; but there is an organic matter, always active, always ready to be shaped and assimilated and to produce beings similar to those which receive it. Animal or vegetable species, therefore, can never, of themselves, disappear (*s'épuiser*). So long as any individuals belonging to it subsist, the species will always remain wholly new. It is as much so today as it was three thousand years ago.[18]

The reference here is primarily to the continuance rather than the invariability of species. But the latter seems also to be implied; and certainly Buffon does not improve the opportunity to introduce a hint of the doctrine of mutability—as he could hardly have failed to do if he had at this time held that doctrine and had been desirous of propagating it. It must be remembered that these passages also were written before Buffon's opinions had been censured by the Sorbonne.

No account of Buffon's position in the history of biology can be other than misleading which fails to note the decisive significance,

---

[17] *Hist. Nat.*, Vol. II, 1749, p. 10.   [18] *Ibid.*, p. 425.

for nearly all of his positions from the second volume onward, of the peculiarly Buffonian criterion of identity and diversity of species. Unless this criterion (and the implied distinction between species and varieties, which latter term covers many Linnaean species) be borne in mind, most of the pages in the *Histoire Naturelle* which have an evolutionistic sound are likely to be misinterpreted. This is what has happened in a number of the studies of Buffon's relation to evolutionism. The error is especially conspicuous in Samuel Butler's *Evolution Old and New*. Butler has devoted nearly one hundred pages to a review of Buffon's utterances on the subject; yet he nowhere lets his reader know that Buffon was the propounder of a new definition of species, which set up a radical distinction between species and varieties, and implied that a species was a definite, objective, "natural" entity. The oversight is not due to any neglect of Buffon's to emphasize and reiterate his definition. He recurs to it frequently in later volumes. His sense of its importance was such that the question of hybridism and of the limits of fertility in cross-breeding was one of the few subjects which he can be said to have studied experimentally on his own account. He writes, for example, in 1755:

> We do not know whether or not the zebra can breed with the horse or ass; whether the large-tailed Barbary sheep would be fertile if crossed with our own; whether the chamois is not a wild goat; . . . whether the differences between apes are really specific or whether the apes are not like dogs, one species with many different breeds. . . . Our ignorance concerning these questions is almost inevitable, as the experiments which would settle them require more time, care and money than can be spared from the fortune of an ordinary man. I have spent many years in experiments of this kind, and will give my results when I come to speak of mules. But I may as well say at once that I have thrown but little light on the subject and have been for the most part unsuccessful.[19]

II

We have thus far noted three generally disregarded but fundamental facts concerning Buffon's opinions about the nature of species. The

[19] *Ibid.*, Vol. v, 1755, p. 63. The passage is given by Butler, but he shows no sense of its general significance.

first fact is that in his preliminary discourse in the first volume of the *Histoire Naturelle*, in which he sought to apply the Leibnitian principle of continuity to natural history, Buffon's emphasis upon the continuity of the gradations between species probably had no evolutionary implications. The second fact is that the principal doctrine of this discourse is to the effect that only individuals exist in nature, while species exist only by grace of the human imagination, which, aided by human ignorance, sees sharp lines of cleavage among organisms where no such lines are. The third fact is that this doctrine was already tacitly but decisively abandoned in Buffon's second volume, where he represents species as real and well-marked natural entities, their limits being determined by the test of the sterility of the products of cross-breeding. There are, indeed, many later passages where the old phraseology incongruously recurs; but it recurs in contexts in which the reality of species is expressly insisted upon.

2. When the fourth volume of the *Histoire Naturelle*—the first dealing specifically with the lower animals—appeared in 1753, four years after the first three, Buffon's departure from the notions set forth in the preliminary discourse became still more evident. He had by this time, in the first place, been impressed by the homologies in the structure of the vertebrates; he had come to see some significance in those facts of comparative anatomy which his own treatise—though more through the contributions of Daubenton than through his own—was for the first time setting in a clear light. The existence throughout at least all the immensely diverse vertebrate forms of an underlying unity of type, Buffon was, I suppose, the first to bring forcibly to the attention of naturalists and philosophers, as a fact calling for serious consideration and explanation.

> If we choose the body of some animal or even that of man himself to serve as a model with which to compare the bodies of other organized beings, we shall find that . . . there exists a certain primitive and general design, which we can trace for a long way. . . . Even in the parts which contribute most to give variety to the external form of animals, there is a prodigious degree of resemblance, which irresistibly brings to our mind the idea of an original pattern after which all animals seem to have been conceived. What, for example, can at first seem more unlike man than the horse? Yet when we compare man and horse point by point and detail by detail, is not our wonder aroused rather by

the resemblances than by the differences to be found between them? ... It is but in the number of those bones which may be regarded as accessory, and in the lengthening or shortening or mode of attachment of the others, that the skeleton of the horse differs from that of the human body. ... The foot of the horse (as M. Daubenton has shown), in appearance so different from the hand of man, is nevertheless composed of the same bones, and we have at the extremities of our fingers the same small hoof-shaped bone which terminates the foot of that animal. Judge, then, whether this hidden resemblance is not more marvelous than any outward differences, whether this constancy to a single plan of structure—which we can follow from man to the quadrupeds, from the quadrupeds to the cetacea, from the cetacea to birds, from birds to fishes, from fishes to reptiles—whether this does not seem to show that the Creator in making all these used but a single main idea, though varying it in every conceivable manner—so that man might admire equally the magnificence of the execution and the simplicity of the design.

But consideration of the anatomical homologies did not lead Buffon merely to pious reflections. He saw clearly and unequivocally declared that this unity of type forcibly suggests the hypothesis of community of descent. To one who considers only this class of facts, he wrote:

> Not only the ass and the horse, but also man, the apes, the quadrupeds, and all the animals, might be regarded as constituting but a single family. ... If it were admitted that the ass is of the family of the horse, and differs from the horse only because it has varied from the original form, one could equally well say that the ape is of the family of man, that he is a degenerate (*dégénéré*) man, that man and ape have a common origin; that, in fact, all the families, among plants as well as animals, have come from a single stock, and that all animals are descended from a single animal, from which have sprung in the course of time, as a result of progress or of degeneration, all the other races of animals. For if it were once shown that we are justified in establishing these families; if it were granted that among animals and plants there has been (I do not say several species) but even a single one, which has been produced in the course of direct descent from another species; if, for example, it were true that the ass is but a degeneration from the horse—then there would no longer be any limit to the power of Nature, and we should not be wrong

in supposing that, with sufficient time, she has been able from a single being to derive all the other organized beings.

Buffon thus presented the hypothesis of evolution with entire definiteness, and indicated the homological evidence in its favor. But did he himself regard that evidence as conclusive, and therefore accept the hypothesis? The passage cited is immediately followed by a repudiation, ostensibly on theological grounds, of the ideas which he has been so temptingly presenting.

> But no! It is certain from revelation that all animals have participated equally in the grace of direct creation, and that the first pair of every species issued fully formed from the hands of the Creator.[20]

This repudiation has been regarded as ironical, or as inserted merely *pro forma*, by those interpreters of Buffon who have made him out a thorough-going evolutionist. Unfortunately, nearly all these writers—dealing somewhat less than fairly with their readers—have failed to mention that his rejection of the evolutionary hypothesis was not put forth by him as resting *exclusively* upon these religious considerations. If the words just quoted stood alone, it would, indeed, be scarcely possible to take them seriously. But they do not stand alone; they are directly followed by arguments of quite another order against the possibility of the decent of one *real* species from another; and the essence of the most emphasized of these arguments lies in the Buffonian conception of the nature of species, already expounded in the second volume. In other words, the fact of the sterility of hybrids, and certain other purely factual considerations, were urged by Buffon as conclusive objections against the theory of descent.

Specifically, his arguments against evolution are three. (1) Within recorded history no new true species (in his own sense of the term) have been known to appear. (2) There is one entirely definite and constant line of demarcation between species: it is that indicated by the infertility of hybrids.

> This is the most fixed point that we possess in natural history. No other resemblances or differences among living beings are so constant or so real or so certain. These, therefore, will constitute the only lines of division to be found in this work.

---

[20] *Ibid.*, Vol. IV, 1753, p. 383.

But why, it may be asked, should the sterility of hybrids be a proof of the wholly separate descent of the two species engendering such hybrids? This question Buffon does not neglect to answer. An "immense and perhaps an infinite number of combinations" would need to be assumed before one could conceive that "two animals, male and female, had not only so far departed from their original type as to belong no longer to the same species—that is to say, to be no longer able to reproduce by mating with those animals which they formerly resembled—but had also both diverged to exactly the same degree, and to just that degree necessary to make it possible for them to produce only by mating with one another." The logic of this is to me, I confess, a trifle obscure; but it is evident that Buffon conceived that the evolution from a given species of a new species infertile with the first could come about only through a highly improbable conjunction of circumstances. (3) Buffon's third reason for maintaining the fixity of species is the argument from the "missing links."

If one species had been produced by another, if, for example, the ass species came from the horse, the result could have been brought about only slowly and by gradations. There would therefore be between the horse and the ass a large number of intermediate animals. Why, then, do we not today see the representatives, the descendants, of these intermediate species? Why is it that only the two extremes remain?

Taking these three arguments into account, then, Buffon arrives at this conclusion:

Though it can not be demonstrated that the production of a species by degeneration from another species is an impossibility for nature, the number of probabilities against it is so enormous that even on philosophical grounds one can scarcely have any doubt upon the point.[21]

However plausibly Buffon's incidental expressions of deference

[21] These, the most definite and decisive words on the subject to be found anywhere in Buffon's writings, have been strangely disregarded by most of those who have discussed his attitude towards evolutionism. Samuel Butler can scarcely be acquitted of suppressing the passage, fatal to his theory. For he quotes in full the opening part of the passage, leaving off abruptly at the point where Buffon begins to introduce his serious objections to the theory of descent. Cf. *Evolution Old and New*, p. 91.

to the testimony of revelation may be regarded as perfunctory and insincere, it would be absurd to suppose that he was also ironical in these legitimate and ostensibly scientific (however poor) arguments for the fixity of species—arguments which are closely connected with that conception of the nature of species which was perhaps his most influential personal contribution to the biological ideas of his time. We must conclude, then, that, while he clearly envisaged the hypothesis of evolution as early as 1753, and recognized that there was some probable evidence in its favor, he then seriously believed that the preponderance of probability was enormously against it. It is certain that contemporary readers must have understood this to be his position.

The same doctrine—that true species, as determined by the sterility of hybrids, are real natural entities and constant units amid the otherwise infinitely variable phenomena of organic nature—is repeated and emphasized many times in subsequent volumes of the *Histoire Naturelle*. Thus in volume five (1775) Buffon—trying to retain as much of the principle of continuity as could be made consistent with his present view—writes as follows:

> Although *animal species are all separated from one another by an interval which Nature cannot overstep*, some of them seem to approximate one another by so great a number of relations, that there remains between them only so much of a gap as is necessary to establish the line of separation.[22]

In the same volume he insists upon the equal antiquity of all real species, in the very passage in which he emphasizes the possibility of a wide range of variation within the species:

> Though species were formed at the same time, yet the number of generations since the creation has been much greater in the short-lived than in the long-lived species; hence variations, alterations, and departures from the original type, may be expected to have become far more perceptible in the case of animals which are so much farther removed from their original stock.[23]

This is advanced as a partial explanation of the extreme diversity of breeds in the canine species: the dog is a short-lived animal and has therefore been capable of a relatively great degree of diversification.

[22] *Hist. Nat.*, Vol. v, 1755, p. 59 (italics mine).
[23] *Ibid.*, p. 194.

A little later,[24] Buffon declares that "nature imprints upon every species its inalterable characters." In 1765—that is, at precisely the period during which we are told that Buffon "was expressing very radical views on the mutability of species"—we find him (in his "Second View of Nature," Vol. XIII) giving his most extreme expression to the doctrine of the reality and constancy of genuine species. Here the language of the preliminary discourse concerning the relative significance in nature of the species and the individual has come to be completely reversed.

> An individual, of whatever species it be, is nothing in the universe; a hundred, a thousand individuals are nothing. Species are the only entities of nature (*les seuls êtres de la nature*)— perduring entities, as ancient, as permanent, as Nature herself. In order to understand them better, we shall no longer consider species as merely collections or series of similar individuals, but as a whole independent of number, independent of time; a whole always living, always the same; a whole which was counted as a single unit among the works of the creation, and which consequently makes only a single unit in nature. . . . Time itself relates only to individuals, to beings whose existence is fugitive; but since the existence of species is constant, it is their permanence that constitutes duration, the differences between them that constitute number. . . . Let us then give to each species an equal right at Nature's table; they are all equally dear to her, since she has given to each the means of existing, and of enduring as long as she herself endures.[25]

This sort of rhetoric is not the dialect of an evolutionist; it is almost that of a Platonist. And there is more in plainer language to the same effect:

> Each species of both animals and plants having been created, the first individuals of each served as models for all of their descendants. . . . The type of each species is cast in a mold of which the principal features are ineffaceable and forever permanent, while all the accessory touches vary.[26]

Many years later still, in 1778, there appeared the subdivision of the *Histoire Naturelle* which Buffon's contemporaries regarded as

[24] *Ibid.*, Vol. VI, 1756, p. 55.
[25] *Ibid.*, Vol. XIII, "Second View of Nature," p. i.
[26] *Ibid.*, pp. vii, ix.

his most brilliant and most significant work—the *Époques de la Nature*. This was a resumption on a grander scale, and upon new principles, of the task attempted in the "Theory of the Earth" in the first volume, thirty years before—an outline of planetary evolution. To the diffusion of evolutionary ways of thinking in the larger and vaguer sense, this treatise was a contribution of capital importance. Into the details of Buffon's geology I do not wish to enter here. But it is worth-while for our purpose to recall one or two striking facts about the *Époques*. In it the writer whom a recent German historian of biology has declared to have had a too little developed sense for the historical or genetic aspect of nature, attempted, in a far more comprehensive, more definite, and more impressive way than any of his predecessors, to write the history of the gradual development of our planet from the time when, an incandescent ball, it was separated from the sun. The task was, of course, undertaken prematurely; but Buffon not only made the need of its eventual achievement evident, but also indicated two of the essential means by which it was to be accomplished: the study of present phenomena which can throw light upon the past processes through which existing conditions have been brought about; and the study of those natural "monuments which we ought to regard as witnesses testifying to us concerning the earlier ages." He insisted, moreover, with the utmost plainness upon (as it was then regarded) the extreme antiquity not only of the earth, but also of organic life. And in doing so he showed himself not at all disposed any longer to permit "revelation" to settle scientific questions. "How," he writes, "some one will ask me, do you reconcile this vast antiquity which you ascribe to matter with the sacred traditions, which give to the world only some six to eight thousand years? However strong be your proofs, however evident your facts, are not those reported in the holy book more certain still?" Buffon replies that he has all possible respect for scripture, but that it always pains him to see it used in this way. Doubtless there is no real conflict between its testimony and that of science; and he thereupon introduces one of the long series of reconciliations of Genesis and geology. The six days were not really days, but long periods of time, and so forth. But in any case, he concludes, the Bible was originally addressed to ignorant men at an early stage of civilization, and was adapted to their needs and their intelligence. Its science was the science of the time, and ought not to be taken too literally. Finally, it is to be noted that in the *Époques* Buffon ceased to talk of the simultaneous creation of all species, and advanced the doctrine of the gradual appearance of different sorts of animals in conformity with geological conditions.

If, then, Buffon was desirous of inculcating the theory of the mutability of species, here was the place in which, above all others, he might be expected to do so fully and unequivocally. But here once more we find him reiterating the substance of his old doctrine:

> A comparison of these ancient monuments of the earliest age of living nature with her present products shows clearly that the constitutive form of each animal has remained the same and has undergone no alteration of its principal parts. The type of each species has not changed; the internal mold has kept its shape without variation. However long the succession of time may be conceived to have been, however numerous the generations that have come and gone, the individuals of each kind (*genre*) represent to-day the forms of those of the earliest ages—especially in the case of the larger species, whose characters are more invariable and whose nature is more fixed.[27]

By the "larger species" here, Buffon means those of greater size, such as the elephant and hippopotamus; and when he says that these are "especially" invariable, he means, as the whole context shows, not that any other species ever departs from its specific type, but that in these larger creatures even the "accessory touches" have been comparatively little altered.

Thus, in a long series of passages, from 1753 on, we find Buffon reiterating with explicitness and emphasis the same teaching, which has, for him, its principal bases in two of his most cherished conceptions: namely, in his conviction that the sterility of hybrids shows that species are real "entities of nature"; and in his embryological theory of "organic molecules" and of the "internal mold" which "casts into its own shape those substances upon which it feeds" and "can operate in the individual only in accordance with the form of each species." One of the first of modern naturalists to make the idea of organic evolution familiar to his contemporaries and to discuss it seriously, Buffon repeatedly rejected that theory, at all periods of his career; and he did so, not from timidity merely nor from an affectation of deference to scriptural authority, but

---

[27] *Hist. Nat.*, Supp., Vol. v, 1778, p. 27.

upon reasoned grounds which he plainly stated and had every appearance of presenting as conclusive. Yet it is also undeniable, as will presently be seen, that he did not maintain this negative position without occasional waverings and doubts and at least one clear, though possibly inadvertent, self-contradiction.

3. In spite of his habitual emphasis upon the constancy of true species, Buffon insisted more than any of his predecessors, and more, perhaps, than any of his contemporaries, except Maupertuis and Diderot, upon the variability of organisms and the potency of the forces making for their modification.

> Though Nature appears always the same, she passes nevertheless through a constant movement of successive variations, of sensible alterations; she lends herself to new combinations, to mutations of matter and form, so that today she is quite different from what she was at the beginning or even at later periods.[28]

The passage is from one of Buffon's later writings; but its close counterpart is to be found as early as 1756:

> If we consider each species in the different climates which it inhabits, we shall find perceptible varieties as regards size and form; they all derive an impress to a greater or less extent from the climate in which they live. These changes are made slowly and imperceptibly. Nature's great workman is Time. He marches ever with an even pace, and does nothing by leaps and bounds, but by degrees, gradations and successions he does all things; and the changes which he works—at first imperceptible—become little by little perceptible, and show themselves eventually in results about which there can be no mistake.[29]

For the most part these changes were clearly represented by Buffon as taking place only within the limits of species; they amounted merely to the formation of new "races" or varieties. Since his criterion of identity of species (the possibility of interbreeding) did not essentially depend upon morphological similarity, he could with consistency suppose the descendants of a given pair to have departed to a very great (though not to an indefinite) degree, in the course of ages, from the form and external characters of their

---

[28] *Ibid.*, p. 3.
[29] *Ibid.*, Vol. VI, 1756, pp. 59-60. I have borrowed Butler's excellent rendering of this passage.

ancestors. It was, in other words, characteristic of his biological system that he set up an absolute distinction between species and varieties, gave an extreme extension to the notion of a variety, and sought to reduce the number of separate species as much as possible, by assuming—until the establishment of the sterility of the hybrids should prove the contrary—that most of the Linnaean species were merely varieties descended from a relatively small number of original specific types. Near the close of his essay " De la dégénération des animaux " (1766), Buffon writes:

> To account for the origin of these animals [certain of those found in America] we must go back to the time when the two continents were not yet separated and call to mind the earliest changes which took place in the surface of the globe; and we must think of the two hundred existing species of quadrupeds as reduced to thirty-eight families. And though this is not at all the state of nature as we now find it, but a state much more ancient, at which we can arrive only by induction and by analogies ... difficult to lay hold of, we shall attempt nevertheless to ascend to these first ages of nature by the aid of the facts and monuments which yet remain to us.[30]

Here, clearly, is an evolutionary program, strictly limited by the assumption that there are irreducible ultimate species, yet tolerably ambitious: to regard all known kinds of quadrupeds as derived from thirty-eight original types, by modification in the course of natural descent; and to determine the general causes and conditions of the production of species in the ordinary sense, i. e., of relatively stable varieties. These ideas occurred to Buffon too late to be made use of in his general plan for the classification of the quadrupeds; that plan, it will be remembered, was formed while he was unluckily under the influence of the principle of continuity. But in the volumes on birds, of which the first appeared in 1770, he had the opportunity for a fresh start; and he took advantage of it to introduce a method of distinguishing and classifying species which— within the limits already indicated—is expressly evolutionary in its principles.

> For the natural history of the birds I have thought that I ought to form a plan different from that which I followed in the case of the quadrupeds. Instead of treating of the birds ... by distinct

---

[30] *Ibid.*, Vol. xiv, 1766, p. 358.

and separate species, I shall bring several of them together under a single *genus*. Except for the domesticated birds, all the others will be reunited with the species nearest to them and presented together as being approximately of the same nature and the same family. . . . When I speak of the number of lines of parentage, I mean the number of species so closely resembling one another that they may be regarded as collateral branches of a single stock, or of stocks so close to one another that they may be supposed to have a common ancestry and to have issued from that same original stock with which they are connected by so many points of resemblance common to them all. And these related species have probably been separated from one another only through the influences of climate and food, and by the lapse of time, which brings about all possible combinations and gives play to all the agencies that make for variation, for improvement, for alteration and for degeneration.[31]

Even the groupings which he gives, Buffon adds, cannot be regarded, in the existing state of knowledge, as correctly and exclusively enumerating all the apparent species which are really akin to one another. The number of separate species which he lists, he intimates, is probably much too great. But at all events, he concludes with pride, his work is the first real attempt at an *ornithologie historique*.

The purpose of the present inquiry does not call for any extended exposition of Buffon's views about the causes of modification in animals and the ways in which quasi-species are formed. In the essay "De la dégénération des animaux" the subject is discussed at the length of over sixty quarto pages; the theories there advanced have been sufficiently accurately summarized by many previous writers. In brief, the factors in modification which he mentions as the most important are changes of climate (in which the most potent influence is temperature), changes of food, and the effects of domestication. But it is evident that he also believed in a general tendency to variation in the germ, and in the influence of acquired habits, of the use and disuse of parts, and of acquired lesions and mutilations. Thus he explains the humps, and the callosities on the knees and chest, of the camel and the llama as due to the habits of those animals under domestication. Similarly, the callosities on the haunches of the baboons arise from the fact that "the ordinary position of

---

[31] "Histoire des Oiseaux," Vol. I, 1770, preface.

these animals is a sitting one—so that the hard skin under the haunches has even become inseparable from the bone against which it is continually pressed by the weight of the body." These theories, of course, take for granted the inheritance of acquired characters, which Buffon also (less cautious here than Maupertuis [32]) explicitly asserts. It is, I suppose, also well known that Buffon called attention (as Linneaus did independently) to the struggle for existence between species, due to the excessive multiplication of individuals, and pointed out how an equilibrium comes to be established (so long as external conditions remain constant) by means of this opposition.

> It may be said that the movement of nature turns upon two immovable pivots—one, the boundless fecundity which she has given to all species; the other, the innumerable difficulties which reduce the results of that fecundity, and leave throughout time nearly the same number of individuals in every species.[33]

Buffon, in fact, rather over-worked this notion of a stable equilibrium, which rested upon the assumption of an approximate equality among species in their endowment for the struggle for survival. This is perhaps one reason why it did not occur to him to think of that struggle as causing a process of natural selection, or to see in it a factor in the formation of so-called species.

4. It must be evident to the reader from all that precedes that Buffon's mind, throughout nearly the whole of his life, was played upon by two opposing forces. Quite apart from any illegitimate and external influences, such as fear of the ecclesiastics—of which too much has been made—his thought was affected by two conflicting sets of considerations of a factual and logical sort. He saw certain definite reasons for regarding species as the fundamental constants of organic nature; what those reasons were has been sufficiently indicated. But he also saw that there was some force in the argument from the homologies; and—what in his case was still more important—he was committed to the program of explaining the diversities of organisms, so far as might be, by the hypothesis of modification in the course of descent; he was deeply impressed by the fact of variability; and he held to a theory of heredity (namely, of the heritability of acquired characters) which acted

---

[32] Cf. Lovejoy, " Some Eighteenth Century Evolutionists," *Pop. Sci. Monthly*, July, 1904, p. 248 n.
[33] *Hist. Nat.*, Vol. v, 1755, p. 252.

as a sort of powerful undertow towards a generalized evolutionism. Add to this that he was little careful of consistency—and extremely careful of rhetorical effect—and it is not surprising that he occasionally forgot one side of his doctrine in emphasizing the other. There is one and, so far as I can discover, only one passage in which he seems categorically to contradict his ordinary teaching of the impossibility of the descent of really distinct species, sterile *inter se*, from common ancestors. This occurs at the end of the chapter on "Animals Common to Both Continents" (Vol. IX, 1761).

> It is not impossible that, without any deviation from the ordinary course of nature, all the animals of the New World may be at bottom the same as those of the Old—having originated from the latter in some former age. One might say that, having subsequently become separated by vast oceans and impassable lands, they have gradually been affected by a climate which has itself been so modified as to become a new one through the operation of the same causes which dissociated the individuals of the two continents from one another. Thus in the course of time the animals of America have grown smaller and departed from their original characters. This, however, should not prevent our regarding them today as different species. Whether the difference be caused by time, climate and soil, or be as old as the creation, it is none the less real. Nature, I maintain, is in a state of continual flux and movement. It is enough for man if he can grasp her as she now is, and cast but a glance or two upon the past and future, to endeavor to perceive what she may once have been and what she may yet become.

Here Buffon seems either to have forgotten or to have deliberately discarded his own usual criterion of diversity of species. He does not propose to inquire whether American species are capable of having fertile progeny when mated with their respective congeners in the Old World, but predicates difference of species solely on the ground of dissimilarity of form; and to the distinct species so determined he attributes an identical origin. But it is possible that he has here merely lapsed (as he apparently does occasionally elsewhere) into the terminology in which he was brought up, and is using the word species in the Linnaean sense rather than in his own.

More significant, perhaps, than this possibly inadvertent inconsistency is the fact that, in his fourteenth volume [34] (1766), Buffon seems to raise explicitly the question—though only *as* a question—whether, after all, descent with modification may not extend to species as well as varieties.

> After surveying the varities which indicate to us the alterations that each species has undergone, there arises a *larger and more important question, namely, how far species themselves can change* —how far there has been a more ancient modification, immemorial from all antiquity, which has taken place in every family, or, if the term be preferred, *in all the genera in which species that closely resemble one another are to be found.* There are only a few isolated species which are like man in forming at once a species and a whole genus. Such are the elephant, rhinoceros, hippopotamus and giraffe, which constitute genera, or simple species, and descend in a single line, with no collateral branches. But all other races have the appearance of forming families, in which we may perceive a common source or stock from which the different branches seem to have sprung.[35]

Even here one cannot be wholly sure that Buffon is not referring to Linnaean species, and using the word genera to indicate what he usually means by species in the strict sense. Assuming, however, that he is speaking of " true " species, it must be observed that while he raises the question of their mutability, he does not answer it finally in the affirmative. For the passage is shortly followed by that cited earlier in this paper, in which Buffon though he derives many species traditionally regarded as distinct from a common stock, yet finds even in " the first ages of nature " thirty-eight irreducible diversities of specific type among quadrupeds.

There is, however, one peculiarly interesting essay in which Buffon shows himself a little dubious even about that " most fixed point in nature " upon which his usual doctrine of the reality and constancy of species was based—namely, the fact of the sterility of hybrids. As I have already mentioned, this seemed to him so central a point in natural history that he for many years assiduously collected data concerning it, and caused experiments bearing thereon to be made and carefully recorded at his own estate at Montbard. The results of these inquiries, which he reports in the chapter " On

---

[34] Just a year earlier we have found Buffon using the most exaggerated language possible about the changelessness of species.
[35] *Hist. Nat.*, Vol. xiv, 1766, p. 335; italics mine.

Mules" (in the third supplementary volume, 1776), led him to the conclusion that hybrids are not necessarily without hope of posterity. On the testimony of an affidavit from a gentleman in San Domingo, Buffon declares that in hot climates mules have been known to beget offspring of mares, and females of their kind to breed with horses. "One was therefore wrong formerly in maintaining that mules are absolutely infertile." Other experiments in the crossing of goats and sheep, dogs and wolves, canaries and goldfinches, are recited; they all go to show that sterility is merely a question of degree.

> All hybrids (*mulets*), *says prejudice*, are vitiated animals which can not produce offspring. No animal, *say reason and experience*, is absolutely infertile, even though its parents were of separate species. On the contrary, all are capable of reproduction, and the only difference is a difference of more or less.[36]

That hybrids are *relatively* infertile, and probably incapable of breeding with one another, Buffon still maintains; "their infecundity, without being absolute, may still be regarded as a positive fact." Something, therefore, is still left of his test of unity of species. But now that it seemed to be reduced to a mere difference in degree, it was no longer the sharp-cut, decisive, impressive thing that it had at first appeared. And, feeling that his criterion of species had a good deal weakened, Buffon was led—not, indeed, even now to an altogether unequivocal affirmation of the descent of real species from one another—but to a confused, half-agnostic utterance, in which he seems to take at least the possibility of such descent for granted:

> In general, the kinship of species is one of those profound mysteries of nature which man will be able to fathom only by means of long and repeated and difficult experiments. How, save by a thousand attempts at the cross-breeding of animals of different species, can we ever determine their degree of kinship? Is the ass nearer to the horse than to the zebra? Is the dog nearer to the wolf than to the fox or the jackal? At what distance from man shall we place the great apes, which resemble him so perfectly in bodily conformation? Were all the species of animals formerly what they are today? Has their number not increased, or rather, diminished (*sic*)? ... What relations can we establish between this

[36] *Ibid.*, Supp., Vol. III, 1776, p. 20; the italics are Buffon's.

kinship of species and that better known kinship of races within the same species? Do not races in general arise, like mixed species, from an incapacity in the individuals from which the race originated for mating with the pure species? There is perhaps to be found in the dog species some race so rare that it is more difficult to breed than is the mixed species produced by the ass and the mare. How many other questions there are to ask upon this matter alone—and how few of them there are that we can answer! How many more facts we shall need to know before we can pronounce—or even conjecture—upon these points! How many experiments must be undertaken in order to discover these facts, to spy them out, or even to anticipate them by well-grounded conjectures! [37]

This passage certainly indicates a strong inclination towards an acceptance of a thorough-going doctrine of descent; yet in Butler's lengthy compilation of the evidences of Buffon's evolutionism it is not cited at all! The volume containing it, says Butler, offers " little which throws additional light upon Buffon's opinion concerning the mutability of species "! [38] In truth, it offers one of the best of the extremely few passages which give some plausibility at least to the theory that Buffon was continuously working towards an unqualified transformism and actually arrived at that position in his later life. But if he reached it (which his language just quoted does not quite justify us in declaring) he did so only in a transient mood. For, as we have already seen, in 1779, in the *Époques de la Nature*, we once more find him asserting—though no longer upon the ground of the sterility of hybrids—that the " constitutive form " of each separate species is the same today as in " the earliest ages."

5. It is more important, and it is commonly easier, to determine what opinions a man's writings tended to encourage than to determine what opinions he actually held. Mind-reading is perhaps no essential part of the history of science. If, then, in conclusion, we raise the more important question with respect to Buffon, it is evident that his work both fostered and hindered the propagation of evolutionary ideas in biology. Earlier than any other except Maupertuis, he put the hypothesis of organic evolution before his contemporaries in clear and definite form. He called to their attention, also, the facts of comparative anatomy which constitute one of the

---

[37] *Ibid.*, pp. 32-33.   [38] *Evolution Old and New*, p. 165.

principal evidences for that hypothesis. Throughout the rest of the century we never cease to hear about the wonderful "unity of type" characteristic of the vertebrates and perhaps of all living things. It was this consideration which led Kant as near to evolutionism as he ever came; Herder and Goethe are full of it, though the former never admitted its full evolutionary consequences; and all, it is evident, learned it directly from Buffon. He, says Goethe, was the first to recognize *eine ursprüngliche und allgemeine Vorzeichnung der Tiere*. Buffon, moreover, once and for all inscribed upon the program of natural history, as its primary problems, the reduction of the number of separate species to a minimum, the derivation of highly divergent forms from a common origin through natural descent, and the discovery of the causes and methods of modification. He, finally, did more than any one else to habituate the mind of his time to a vastly (though not yet sufficiently) enlarged time-scale in connection with the history of organic nature, a necessary prerequisite to the establishment of transformism.

These were great steps in the progress of evolutionism. But it is equally true that Buffon probably did more than any other eighteenth century writer to check the progress of evolutionism. He did so partly by the authority which, for his contemporaries, attached to those personal utterances of his favorable to the doctrine of immutability. These utterances were far more numerous and more categorical than those which could be quoted on the other side; and they certainly were not taken as ironical by the average reader of the period. But, what is still more important, Buffon put into currency what passed for a scientific and serious argument against any wholesale theory of descent. In the eyes of many learned men of his own and later generations, perhaps his chief single contribution to science was his definition of species. This, as we shall see, was regarded as of immense importance by Kant, and was, indeed, the starting point and the controlling principle of that philosopher's biological speculations. "It is Buffon," wrote Flourens as late as 1844, " who has given us the positive character of a species." Now before the Buffonian criterion of species was propounded, there already existed a tendency toward evolutionism, fostered by the principle of continuity and by such embryological conceptions as those of Maupertuis—a tendency to disregard species altogether and to infer from the variability of individuals to an unlimited and rather promiscuous mutability of the successive generations of living things. If it had not been for Buffon, transformism would probably have developed at first through the increase and diffusion of this tendency; and its development might well, in that case, have been more rapid. But when species came to be regarded as real " entities of nature," determined by the objective criterion of the sterility of hybrids, this somewhat too facile evolutionism received a check, and a certain presumption of the constancy of true species seemed to be created. This presumption had all the more force because it left room for a large measure of mutability in the case of varieties, and thus gave a sort of appeasement to the strong impulse towards genetic modes of thought which was already active in the mid-eighteenth century. But more than all this, Buffon, as we have seen, from the first managed to associate with his definition of species the assumption that the sterility of one animal when crossed with another was a character that (unlike almost all others) could not have been produced in the course of descent with modification. And this supposition that the sterility of hybrids was incapable of an evolutionary explanation long remained a serious obstacle to the acceptance of the theory of descent, even with those little influenced by theological prejudices against the theory. We find even Huxley in 1862 troubled over the difficulty. In his Edinburgh lectures of that year " he warned his hearers of the one missing link in the chain of evidence—the fact that selective breeding has not yet produced species sterile to one another." The doctrine of descent was merely to be " adopted as a working hypothesis, . . . subject to the production of proof that physiological species may be produced by selective breeding." [39] Since Buffon appears to have been the first to emphasize the notion of physiological species, and to give currency to the supposition that the sterility of hybrids affords a presumption against any thorough-going transformism, he must be regarded as having done more than almost any man of his time to counteract the tendency which he also, perhaps, did more than any other to promote.

[39] Huxley's *Life and Letters*, I, 1900, p. 193.

FIVE

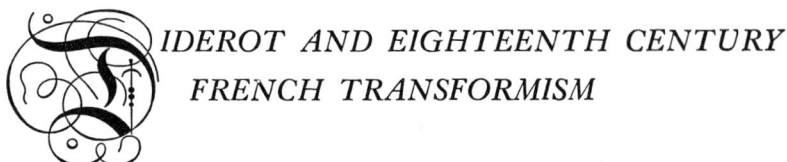IDEROT AND EIGHTEENTH CENTURY
FRENCH TRANSFORMISM

LESTER G. CROCKER

Diderot was bound to come to the idea of evolution. It was not only that his searching mind drank in the newest scientific notions and imaginatively grasped their furthermost implications. More specifically, he needed such a theory in order to complete the materialistic interpretation he was building of man and the universe.

It frequently happens that a concept which to a later generation has become a commonplace—often accepted without sufficient critical questioning, like current coin whose face we no longer examine—simply did not exist in the minds of an earlier period, and was conceivable to those who gave it birth only by dint of extraordinary intellectual courage, imagination, and travail.

> Yet naught there is
> So easy that it standeth not at first
> More hard to credit than it after is;
> And naught soe'er that's great to such degree,
> Nor wonderful so far, but all mankind
> Little by little abandon their surprise.
> (Lucretius, Bk. II)

The concept of biological transformism, which was to revolutionize man's thinking in manifold ways, is surely an idea of this sort. There were many obstacles to hinder its coming into being. Not the least of these was a version of the world and of life which bridled men's thinking all the more effectively by combining with the strength of universal acceptance the power of sacredness and the might of

114

institutions committed to it. Scientific evidence of an incontrovertible nature, that might embolden a few courageous thinkers to risk martyrdom, was, to put it at the best, scanty. A dawning notion of geological history, the elements of a science of comparative anatomy —and even these hotly challenged—here was insufficient basis for a scientist to venture very deep into the dangerous waters of unorthodox speculation. This was for the philosophers to do, especially for those with an unreined poetic imagination.

But even a philosopher could not step unimpeded into the brightly illumined truth of organic evolution. He was separated from it by intellectual barriers. Were not animate and inanimate beings so totally distinct from each other, that life could not be conceived of except as an act of divine creation? Were not matter and mind— *res extensa, res cogitans*—so completely incompatible, that the presence of the latter could never be accounted for by anything that might happen to the former? Did not all living organisms, and all their organs, present such clear evidence of design, that a prearranged plan could alone explain them? Outside of God, there was only nature. Nature was matter and its laws. Of matter, Wollaston expressed the common opinion when he wrote, " Matter is incapable of acting, passive only, and stupid."[1] As for its laws, they were, in the prevailing Newtonian view, the geometrical regulators of a stable, equilibrated machine. Everything was in its proper place in the great chain of being.

All of these obstacles were to be vanquished in Diderot's mind. We shall try to follow, in his writings, the birth and growth of his evolutionary thinking.

In the late 1740's, Diderot began to formulate his concept of nature. Whereas both to Christians, and to Newtonian deists (like Voltaire), nature was God's completed and immutable creation, Diderot began to think instead of dynamic forces which, self-existent and self-directing, have evolved the universe we know from a primeval chaos, to which it might again return in the course of its necessary unfolding. In 1746, in his first original work, the *Pensées philosophiques*, Diderot is still a deist, but is clearly tempted by atheism. Two factors retard his becoming an atheist: the ethical and the cosmic. Only the second of these need concern us here. If Diderot clings to his deism, it is partly because he has not yet reached this concept of nature as a self-sufficient All, containing

---

[1] *The Religion of Nature Delineated*, London, 1726, p. 74.

within itself the origin and the explanation of all its phenomena. His position may be termed anti-evolutionary. Spontaneous generation is rejected as the source of higher forms of life. Only an act of creation can explain the existence of the germ cells from which we come. The wing of a butterfly, the eye of a mite—these are sufficient to prove intelligent design. Yet, at the same time, Diderot is strongly drawn to the atheist's argument: if we concede matter to be eternal, and if we grant that movement is inherent in matter, an infinite number of "throws of the dice" was bound to bring about the combination which is our universe. It should be noted that there is nothing evolutionary in this thought; it is a quantitative mechanism that offers no perspective of qualitative change. The inclusion of motion as essential to matter is, however, a first step in that direction. It was an idea that Diderot had probably encountered in the "radical" irreligious manuscripts that were freely though illegally circulated; he conceivably could have met it in Toland's *Letters to Serena*; and he had undoubtedly read it in the *Metaphysics* of Aristotle.

In 1749, Diderot published his next important work, the *Lettre sur les aveugles*. We are immediately aware that there has been a significant revision in his thinking. In the three preceding years, he had been swept away by the strong current of interest that turned his contemporaries towards the startling discoveries in the biological sciences. He had read Nieuwentijdt, Nollet, William Derham; Bonnet's and Réaumur's works on insects. Needham's *Account of Some New Microscopical Discoveries* seemed to him a convincing demonstration of spontaneous generation. But it is unlikely that he as yet knew Buffon's theory of "organic molecules," for he seems to have become acquainted with the newly published *Histoire naturelle* during his imprisonment in the dungeon of Vincennes, which followed publication of the *Lettre sur les aveugles*. A particularly important influence was that of Trembley's memoir on fresh-water polyps (1744).[2] Trembley's discovery of the regenerative powers of the polyp threw the biological world of France and England into a ferment; his conclusions were taken up by Bonnet, Réaumur, and others. Maupertuis, as early as 1745, expressed his amazement and saw new perspectives opening up. The

---

[2] *Mémoires pour servir à l'histoire d'un genre de polypes d'eau douce*, Leiden (and Paris), 1744. For an important discussion of its influence, see Aram Vartanian: "Trembley's Polyp, La Mettrie, and eighteenth century French materialism," *J. Hist. Ideas*, 11:259-86 (1950).

philosophical repercussions were even more startling. The animal soul was clearly shown to be material. The heterogenetic origin of life seemed to be confirmed. Bonnet suggested that the polyp bridged the animal and vegetable kingdoms, nullifying their supposedly complete separation. Later he declared the polyp to be the key to the interpretation of nature.[3] The most radical conclusions were drawn by La Mettrie, in his notorious *L'Homme machine*, published towards the end of 1747. In this work, La Mettrie excludes any spiritual principle from the universe and rejects the biological finalism of Newton. Trembley's polyp, he states specifically, proves that nature (read: *matter*) contains within itself the power that produces its activity and—most important—its organization. (One might consider this a rebirth, on a new scientific basis, of the Aristotelian view that there is no matter without directing purpose or form.) The mechanical atoms-plus-chance materialism of Lucretius and Gassendi is declared to be insufficient, for nature involves design—design which springs from the dynamics of matter in action, free from the finalism of an extra-natural intelligence.[4] The vital point in this concept of nature, which makes it of particular concern here, is that a theory of transformism could scarcely have been conceived without the belief that matter, and living organisms in particular, possess a self-organizing power or impulse.

The importance of Leibniz in the formulation of this new outlook has, at least in recent years, been largely overlooked. The biological developments must be given their full due. They were crucial, in imparting reality and concrete importance to a concept which, however, already existed in a different frame of reference. Thus, in the *Monadology*, Leibniz writes "that the natural changes of the Monads come from an *internal principle*, since an external cause can have no influence upon their inner being." (Par. 11.) All simple substances " have a certain self-sufficiency which makes

[3] *Oeuvres*, Neufchâtel, 1779, IV, 340.
[4] "We do not know Nature at all; hidden, inner causes may have produced everything. Just look at Trembley's polyp! Doesn't it contain within itself the causes which give rise to its regeneration? What absurdity, then, would there be in thinking that there are physical causes for [by?] which all has been brought about, and to which the entire chain of this vast Universe is so necessarily tied and subjected, that nothing that happens in it could not have happened? . . . Thus, to destroy chance is not to prove the existence of a Supreme Being since there may be something else which would be neither chance, nor God, I mean Nature. . . ." (*L'Homme machine*, éd. Maurice Solovine, Paris, 1921, p. 108-9).

them the sources of their internal activities and, so to speak, incorporeal automata." (Par. 18). Matter, too, is undergoing constant modification, perpetual change, within the integrity of its substance (*Théodicée*, par. 396). Finally, "every present state of a simple substance is naturally a consequence of its preceding state, in such a way that its present is big with its future." (*Monadology*, par. 22). The biological discoveries, then, and the speculations they aroused in several thinkers (whom we know to have been familiar with Leibniz) added little new to a concept which already existed full-blown, though only in the abstract. It was the startling and unexpected confirmation of abstract speculation by empirical observation that brought the former to life, and into an actively operational status, in a direct application to the problems of biology. I suggest, then, that this decisive reorientation of eighteenth century thought came about through the fusion of an idea and a discovery, a fusion of the thought of Leibniz with the observations of Trembley. This fusion was made by the men we are now discussing.

(Leibniz's contribution to the rise of the new biological thinking has generally been limited to his reformulation of the idea of a "chain of beings," or "principle of plenitude." Certainly this was a notion that permeated the thought of the eighteenth century. It was of consequence for the development of transformism, inasmuch as it tended to create a continuous series of links between all species and "kingdoms." For this static concept to become evolutionary, however, it had, in the first place, to be transmuted into the concept of a process of dynamic change, occurring through a vast period of time—a transformation Arthur O. Lovejoy has studied in his *Great Chain of Being*. But even this would not have been enough to produce a biological theory of transformism. A still more significant change in thinking was required. Nature had to be conceived of as a self-creating, self-patterning force, as an experimenting —and a blindly experimenting—force. Furthermore, in the actual history of the idea, the second step really preceded the first. Consequently, the new concept of transformism, though it was aided and abetted by that of the chain of beings, was so distant from it, that it cannot be said to have derived directly or mainly from it. And, curiously, one concept of Leibniz may thus have had a share in the later transforming of another of his concepts.)

Although La Mettrie is often given credit for having been the first to set forth the new concept of nature, Maupertuis had, in fact, preceded him by two years. In his revolutionary work, the *Vénus physique* (1745), Maupertuis attributed to nature such a self-organizing and self-patterning power, based on motion and gravitational attraction (which, he emphasizes, is a non-mechanical principle). In other words, he made development inherent in matter. We shall not go into his theories at this point, since they are concerned essentially with the processes of reproduction and variation; but inasmuch as he does seek analogies in non-living matter, the principle is adumbrated in general form.[5] La Mettrie was almost certainly familiar with Maupertuis' work; but his thinking seems to have evolved independently. He was inspired by the polyp, and Maupertuis by his observations on genetic variation. Both reached the same conclusion.

In *L'Homme plante*, published in 1748, La Mettrie takes one more step in building the foundations of the eighteenth century theory of transformism. Like Bonnet, he emphasizes the "chain of beings" concept, and places the polyp in the key position of bridge between vegetable and animal. If nature is dynamic and self-determining; if nature consists of an unbroken gradation of beings from simplest to most complex, with the specific difference inhering in their material organization—if these two propositions be established, then the third and concluding step becomes inevitable.

With this background, it is now time to return to Diderot. Since we are interested in studying the genesis and development of his idea of evolution, as well as that idea itself, there is an essential point to note at this stage. It appears probable that Diderot did not know La Mettrie's work until after the publication of the *Lettre sur les aveugles*.[6] What we find in the *Lettre* must not be attributed, as is sometimes done, to the influence of La Mettrie.

At the base of the *Lettre sur les aveugles* lies a firm belief in the unity of nature, and in its dynamism. The psychological phase of Diderot's discussion rejects psycho-physical dualism in favor of a physical explanation of mental activity. Similarly, as he views the formation of the universe, he sees no final separation between the organic world and the inorganic. Such a view of the unity of nature is another essential preliminary to a concept of organic evolution. How far do these ideas carry Diderot in 1749? Has he yet reached such a concept?

[5] See notes 33, 34 *infra*, and *Vénus physique* (*Oeuvres*, Lyon, 1756), II, 85 ff.
[6] *Lettre sur les aveugles*, éd. Robert Niklaus, Genève, 1951, p. xxi.

Significantly, he speaks first of animals. Breaking loose from the teleology which, in the *Pensées philosophiques*, struck him as a necessary conclusion from the obvious adaptive design of all living organisms, he proclaims this design to be independent of a divine intelligence, plan, or guiding hand. As the blind Saunderson explains it to his would-be converter, the minister Holmes, the error—and Diderot is aware that it had been his, too—lies in the assumption that this order always existed as we now see it in successful forms of life. But if we postulate, instead of a single and final act of Creation, an original chaos from which nature gradually formed itself, under the impulsion of its intrinsic dynamism, we get a picture that is in much closer agreement with the known facts. The principal facts that Diderot has in mind are the complete failure of some natural productions to embody a successful design, and the imperfection of others. Saunderson's blindness is evidence of the latter, and the production of nonviable "monsters" proof of the former. These facts tend to exclude God from the natural processes, since he could create only fit beings; and they simultaneously point to a persistent creative process in the initial formation of the world, a self-determining but blind impulsion to trial and error.[7] Such a process is equally alien to the workings of a Perfect Being. Let us now read Diderot's own words:

> Imagine, then, if you will, that the order which strikes you has always subsisted; but allow me to believe that it is not at all so; and that, if we went back to the birth of time and of things, and if we perceived matter stirring and chaos disentangling itself, we should encounter a multitude of shapeless beings for a few well-organized.... I may ask you, for instance, who has told you, Leibniz, Clarke and Newton, that in the very beginning of animal formation, some were not without a head and others without feet? I can assure you that some had no stomach, and others no intestines... that all the monsters were successively annihilated; all the imperfect combinations of matter have disappeared; only those remain in whose mechanism there was no imperfect contradiction and which could exist by themselves and perpetuate themselves. On this supposition, if the first man had had his larynx blocked, lacked proper food or properly formed genera-

---

[7] Then Nature, delivered from every haughty lord,
And forthwith free, is seen to do all things
Herself and through herself of own accord,
Rid of all gods.   (*De rerum natura*, Bk. II)

tive organs ... what would have become of the human species? It would have been enveloped in the general cleansing of the universe, and that proud being who calls himself man, dissolved and dispersed among the molecules of matter, would have perhaps forever remained in the number of possibles. ... But order is not so perfect, that there do not still appear, from time to time, monstrous productions ... I therefore conjecture that in the beginning, when matter in fermentation produced the universe, my fellow men were quite common.[8]

Diderot then extends his concepts of trial and error and fitness to embrace the physical evolution of the cosmos.

But why should I not apply to worlds what I believe of animals? How many misshapen worlds have dissolved, are reformed and dissolved again, perhaps at this very moment, in the immensity of space, where I cannot touch, and you cannot see, but where movement continues and will continue to combine accumulations of matter, until they have found an arrangement in which they can persist? Oh philosophers! transport yourselves with me to the end of this universe, beyond the point where I can touch and where you see organized beings; walk over this new ocean, and seek through its irregular agitation for vestiges of that intelligent Being whose wisdom you admire here! ... What is this world, Mr. Holmes? a composite, subject to revolutions that all indicate a continual tendency to destruction; a rapid succession of beings that follow each other, push each other and disappear; a passing symmetry, a momentary order.

The influences that have shaped Diderot's thinking are obvious. The disentangling chaos and the importance of monsters hark back to Empedocles, Epicurus, and Lucretius, whose philosophy he knew well, and to a later echo in Shaftesbury's *The Moralists*. There are also evidences, in the *Lettre*, of the impact of Cartesian philosophy.[9]

---

[8] Diderot, op. cit. p. 42-3.
[9] It is not probable that Diderot knew of Hooke's speculations which, according to Louise D. Patterson, enlarged the Cartesian physics to include the organic world in a sort of evolutionary theory. See "Hooke's gravitation theory and its influence on Newton," *Isis*, 40: 327-41 (1949); 41: 32-45 (1950). While this statement is perhaps too sweeping, there are several points of interest in Hooke's *Micrographia* (London, 1665). He finds nothing to make him "positive" that animal and plant are "altogether *heterogeneous*, and of quite differing kinds of nature," some species being, on the contrary, transitional or undecided (p. 124). He sees no fixity in species. One kind of creature "may produce several kinds" (p. 193-4). A plant, though produced

Doubtless Descartes' concept of the physical universe as necessarily evolving under fixed law (in contrast to the static Newtonian universe) was a factor of some importance. If we add to these influences that of Trembley, we can understand how Diderot was led to express a view that links the organic and inorganic worlds in a process of creation, patterned according to necessary laws that impel matter to find temporary stability of physical organization and viability of organic form. The motion inherent in matter lies at the base; as Diderot later wrote, thinking of Leibniz, "In general there is no force whatever that is not a principle of change." Time and becoming, the continuity of process (once more we cannot help thinking of Aristotle), and the selection of the fit—all these have brought him to the threshold of a theory of organic evolution.

To the threshold—but not across it. Some critics have erroneously seen in Diderot's statements the first announcement of such a theory. Actually, his concept applies only to the fundamental dynamism of nature. But he has no notion, as yet, of *transformism*, that is, of a species having itself a history, of becoming what it is out of something else that it was. Matter spawns its productions as fully formed species, whose "fitness" or "monstrosity" consists essentially of internal viability. Thus men were produced, in the original disentangling process, along with all the rest, even as Lucretius describes it in his fifth book. The very process of the universe is that of creation, growth and destruction.

> Geburt und Grab,
> Ein ewiges Meer,
> Ein wechselnd Weben,
> Ein glühend Leben.
> (*Faust*, Part I)

It must further be noted that chance persists (even as it does in modern biology) alongside inherent laws of viable structuring; and that there are only quantitative combinations, with no concept of qualitative change. Diderot has not yet escaped (as it is often

---

by spontaneous generation ("casual putrefactive production") may germinate and thus "propagate its own, that is, a new species" (p. 124). However, he sees new species produced by putrefaction as inevitably of "inferior" or degenerate rank (p. 122-3), reminding one somewhat of Buffon's later theories. Although Hooke's influence in France has not been studied, his description of salts taking on shapes resembling living things (p. 129-130) is too close to Maupertuis' description (*Système de la nature*, *Oeuvres*, and *Vénus physique*, I, 85-86) not to suspect a direct influence.

asserted) from mechanical materialism. The persistent, creative dynamism of his nature concept rests perhaps on a broader base than the Lucretian; but it still terminates in stability, which is dissolved in a cycle of formlessness and reformation. An evolutionary or a dialectical materialism, on the other hand, requires a dynamism that drives through stability, in a process of advancing *neo*-formation. In fact, there is not a single idea, in all we have quoted, that does not seem directly and immediately inspired by the second and fifth books of *De Rerum Natura*.[10] There is also no doubt that the notion of nature's self-creative dynamism was becoming widespread, and Diderot had assuredly read these lines in Pope's popular *Essay on Man:*

> See, thro this air, this ocean, and this earth,
> All matter quick, and bursting into birth.
> (I, 233-4)

The definite exposition of a theory of transformism comes at last in 1753, in the *Pensées sur l'interprétation de la nature*. Many influences, accumulated during the intervening five years, had been working on Diderot's mind.

Mention must first be made of Benoît de Maillet's *Telliamed* (which, despite its 1748 date, had no apparent influence on the *Lettre sur les aveugles*). Historians agree that this work was the first to contrive a genuine notion of transformism; but it is a very limited concept. *Telliamed*, a remarkable fantasy composed of a mélange of shrewd intuition and absurd credulity, reminds one of the imaginary voyages of the seventeenth century sceptic, Cyrano de Bergerac (to whom it is dedicated). De Maillet develops a theory that the seas once covered the earth, and are still receding. This accounts for the presence of what we now call fossils. In the last section of his book, he takes up the origin of animals and man. He mocks both spontaneous generation and the materialist hypothesis of atoms combining, comparing such theories to primitive myths of man formed from the earth. His own theory is that the universe is filled with seed ("semences") of all species, and that the sea acted as the female matrix which allowed their development. From the sea, species "transmigrated" to the land. Thus every species of land animal has its marine equivalent. Flying fish became birds.

---

[10] For a contrary view, see A. Vartanian: "From Deist to Atheist," in *Diderot Studies*, ed. Torrey and Fellows, Syracuse, 1949, p. 46-63.

There are "sea-dogs," "sea-monkeys," and of course, a "sea-man." He quotes many instances of "sea-men" having been observed, and assures us that if we do not come upon them more frequently, it is only because they dwell by preference near the poles. There is no notion of transformism in all this. What impressed others, however, was his idea that there is a limited adaptive change in the process of accommodation to the new milieu, which may in some few cases be abetted by the crossing of species. Thus one traveler had observed that a Hollander who, by a not unusual reverse process, had become a "sea-man," was covered by scales and had hands like fins.[11] De Maillet here foreshadows the theory of transmission of acquired characteristics.

A more important influence on Diderot was surely the great impact of La Mettrie's writings, as we have described them above, and as they were completed by the publication, in 1750, of the same writer's *Système d'Epicure*. It has in fact been claimed that La Mettrie was the first eighteenth century writer to suggest, in his *Système d'Epicure*, the fundamentals of transformism.[12] Actually, the most that can be said is that he skirts the edge of such a theory. He never reaches it. What he gives us, first of all is a variation of de Maillet's theory. The seeds of all the species are in the air. When the oceans receded, they left the "human egg" on the shore, "where it incubated in the sun's heat." Thus "man and all other animals left their shells, as certain ones still do in warm countries."[13] The hatched human nourished himself, helped by kindly animals. When La Mettrie says that men have not always "existed as they

[11] *Telliamed, ou Entretiens d'un philosophe indien avec un missionnaire français*, Basle, 1749. *Telliamed* went through four editions between 1748 and 1755. La Mettrie specifically refers to it in his *Système d'Epicure*, and Bonnet, in his use of terminology such as "lion-marin," "veau-marin," and in his insistence on the orang-outang as an "homme sauvage," also reveals de Maillet's influence. (See *Contemplation de la nature* (1764), *Oeuvres*, IV, 105, 116). The latter notion will again appear in Monboddo's theories. Even Buffon's "organic molecules" may owe something to him. It has also been said that de Maillet suggested the origination of men from mermaids or sirens; on the contrary, he rejects this as an illusion due to posture (p. 336-7). Among the fantastic incidents recounted by de Maillet is the capture of a "sea-girl" who felt inevitably drawn back by the sea; it is curious that this is precisely the case of the heroine of a recent Spanish play, Casona's *La sirena varada*.

[12] Cf. Raymond Boissier: *La Mettrie* (Paris, 1931, p. 45-64); A. Vartanian: "Trembley's Polyp," p. 273.

[13] *Oeuvres philosophiques*, Amsterdam, 1764, II, 141-2, 152-3. He also says the "seeds" develop now in the matrix of the spermatic vessels.

are," he means only that "the earth must have served as uterus to man, that she must have opened her bosom to human germs, already prepared." Beyond this, La Mettrie, who followed Diderot's writings closely, echoed the ideas expressed in the *Lettre sur les aveugles* about monsters and viability. "The first generations must have been very imperfect," some being without esophagus, generative organs, etc. Only the viable survived. Nature had to essay many combinations before coming up with a perfect animal. At this point La Mettrie expresses an idea which might be construed as containing a hint of transformism. Through how many more combinations of matter did animals pass, he exclaims, "before generations became as perfect as they are today!"[14] It is probable, however, that La Mettrie is here referring only to the elimination of "monsters" in reproduction, or "generation." The context of the passage and the surrounding ideas justify only this interpretation. Similarly, when he speaks of animals being produced before men, "because it requires more time to make a man" than an imperfect being, he is referring to his idea that nature reworked the same matter into various combinations.[15] The essential element of transformism, that present forms *depend* on earlier forms, is lacking in La Mettrie. The closest we come to it is in this reference to a hypothesis he does not firmly embrace: "The animals which hatched from an eternal germ, whatever it may have been, the first to come into the world, by mingling among each other have, according to some philosophers, produced this handsome monster called man; and the latter, in turn, by his mixture with animals, may have brought about the different peoples of the universe."[16]

An important role in the development of Diderot's thinking was undoubtedly played by the writings of Maupertuis, especially the *Système de la nature* (1751, 1754). Curiously, Diderot failed to see the tremendous significance of Maupertuis' greatest anticipation—his work on a genetic theory of heredity.[17] Maupertuis clearly per-

---

[14] Ibid., p. 144-6.
[15] Ibid., p. 153, 140. Cf. *L'Homme machine* (p. 100): "animals formed from the same matter which lacked only a degree of fermentation . . ."
[16] *Système d'Epicure*, p. 157.
[17] See Bentley Glass: "Maupertuis and the beginnings of genetics," *The Quarterly Review of Biology*, 22:196-210 (1947). The *Système de la nature* was first published in 1751 as the *Dissertatio inauguralis metaphysica*, and translated into French in 1754. Diderot had read it in the Latin. The essentials of the theory are already outlined in *Vénus physique* (1745), but there is no evidence to indicate that Diderot was familiar with that work.

ceived in genetic mutation and recombination the possibility of an explanation of biological variation, and in this respect he stands, a unique figure, enouncing a truth others of his time were unable fully to grasp. Diderot did not realize until much later how remarkably Maupertuis had accounted for the monsters which had struck him as evidences of nature's dynamic trial and error process.[18]

Why did Diderot, always so avid for new ideas, fail on this occasion? A good part of the explanation probably lies in the language used by Maupertuis, in the attribution to his genetic particles of a low form of intelligence, a property "akin to what in us we term desire, aversion, memory."[19] Diderot could not accept this panpsychism, even as he was not able definitely to accept the quality of Buffon's "organic molecules" which had been adopted by Maupertuis in his theory. Both concepts implied a hylozoic view of ultimate particles; and the "organic molecules" perhaps also involved a final separation between the inanimate and animate world. Furthermore, Maupertuis specifically denies the possibility that matter plus motion could, ultimately, as a result of organization, produce sensation and then intelligence, a belief which is at the heart of Diderot's materialist evolutionism.[20] Maupertuis avoids this obstacle by redefining matter to include "intelligence." Now this hypothesis is unacceptable to the materialist, even though it does lead Maupertuis to an important coincidence of thought with La Mettrie. Maupertuis, influenced, like Diderot, by *L'Homme machine*, is emboldened to develop the idea of nature he had sketched in *Vénus physique*. He declares that the old atoms-plus-chance materialism does not explain how matter combines into living forms possessing purposeful design. There had to be properties and laws of matter, and a selective power in the genetic particles that were actively self-regulating. Maupertuis believed his redefini-

---

[18] Ibid., p. 206. See Arthur O. Lovejoy: "Some eighteenth century evolutionists," *The Scientific Monthly*, Sept. 1950 (reprinted from *The Popular Science Monthly*, 1904), p. 167. For the "production of new species," and its relation to recombination of genetic materials by chance or by human design, and for an explanation of the disappearance and reappearance of variations, see *Vénus physique*, p. 108 ff., and 119 ff. In the *Système de la nature*, Maupertuis emphasizes his new concept of matter, and the possibility of "mutations"; however the latter is not conceived of as resulting from any change in the inner structure of genetic particles themselves, but only in their accidental or spontaneous displacement and recombinations.

[19] *Système de la nature*, II, 157-8; also p. 149 ff.

[20] Ibid., p. 139-41, 146-7, 153-6, 166-7.

tion of matter (to include "memory") solved the difficulty.[21] At the same time, he rejected atheism and led the ultimate explanation back to the Divine Intelligence, a view which could only grate upon Diderot's atheism.[22] Still a third reason for Diderot's failure to grasp the significance of Maupertuis' discovery was the influence exercised on his mind by the Leibnizian principle of plenitude. *Natura non facit saltus.* This may be said to have been a preconception that blocked his view.

If Maupertuis' brilliant apprehension of the principles and problems of heredity and genetic variation did not inspire Diderot, he did find another passage in the *Système de la nature* which he could enthusiastically embrace. We shall shortly see how it left its mark on him. Granting his theory of variation, continues Maupertuis,

> Couldn't we explain in this way how the multiplication of the most dissimilar species might have resulted from two single individuals? Their origin would be owing only to a few accidental productions in the embryo, in which the elementary particles would not have retained the arrangement which they had in the father and mother animals: each degree of error would have created a new species; and by dint of repeated divergences there would have come about the infinite diversity of species which we see today, which will perhaps increase with time, but to which the passing of the centuries perhaps brings only imperceptible additions.[23]

---

[21] Ibid., p. 146-7 (xiv), 183. See A. O. Lovejoy, loc. cit. Maupertuis also applies this theory to inanimate or inorganic matter (p. 150-5); for the earlier version, in the *Vénus physique*, see pp. 85 ff.

[22] Ibid., p. 156-8 (xxix, xxxi). Of particular interest is Maupertuis' separation of the moral intellect and his attribution of it, along with our knowledge of God, to the distinctive human soul (p. 159-61, xvii).

[23] Ibid., p. 148-9 (xlv). Dr. Glass (op. cit., p. 206), correctly points out that Maupertuis largely limits variation to genetic recombination and mutation. In other passages of the *Vénus physique*, however (p. 121 ff.), and in the *Système* (p. 159), Maupertuis seems to accept the possibility of transmission of acquired characteristics, as by repeated mutilation.

Note should be taken of a passage in an earlier work by Maupertuis, the *Essai de cosmologie* (Oeuvres, I, 11-12), in which he argues against the proof of God from biological design. "But could we not say that in the chance combination of Nature's productions, as only those could subsist which embodied relationships of fitness, it is not astonishing that this fitness is found in all species now existing?" Only a few animals, he continues, were so constructed by the workings of chance; "animals without a mouth could not live, others lacking generative organs could not perpetuate themselves: the only ones which have endured are those which possessed order and fitness; and those species, which we see today, are only the smallest part of what a blind

Buffon's influence is more uncertain. Diderot toyed with his theory of organic molecules, which possessed the merit of explaining development by the inherent power of an internal design. But Buffon, precisely because transformism stood in contradiction to this favorite theory of organic molecules as the direct source of all creatures, was opposed to the idea of evolution. In the fourth volume of the *Histoire naturelle* (1753—the year of Diderot's *Pensées*) he maintained that no new species have come into existence, despite infinite combinations of matter and of living things.[24] Buffon makes this denial despite the fact that he is impressed by anatomical similarities among animals. At the same time he denies categorically the possibility that some species may be degenerate forms of a prototype. Buffon includes in his speculation a curious passage. If it were true that a single species had been produced by degeneration (such as the ass from the horse), "there would no longer be any limits to Nature's power, and one would not be wrong in supposing that from a single being she has been able to draw, with the passage of time, all other organized beings. But no, it is certain, through Revelation, that all animals have participated equally in the grace of creation, that the first two of each and every species came fully formed from the Creator's hands."[25] Scholars have disagreed in reading this passage as a sincere or an ironical asseveration. Expression of heretical theories *via* ironical disclaimer were common at the time; and yet, Buffon provides no solid evidence elsewhere of his accepting a notion of transformism until much later.[26] In view of the fact that in this same passage he goes

fate had produced." Some have seen in this passage an early expression of such evolutionary processes as adaptation and survival of the fit (cf. Lovejoy: op. cit., p. 164-5). The date of this work (1750) is, however, significant. This passage is obviously an echo of Diderot's statement, which we have quoted, in the *Lettre sur les aveugles* (1749). Again, there is no relation to environment, or gradual selection, but only an immediate inherent viability dependent on chance. This Lucretian position, as we know, was soon to be surpassed by both men; although Maupertuis (and probably Diderot, too) kept a greater role for chance, in his theory of variation, than some scholars allow (cf. Vartanian, "Trembley's Polyp," p. 281).

[24] *Histoire naturelle*, Paris, 1753, IV, 182-3.
[25] Ibid.
[26] In a much later work (1766), he admitted the origination of certain new species by a process of "degeneration," but denied this was a general process. It was not until 1774 that he propounded his geological theory of "les époques de la terre." Buffon's greatest contribution was his insistence on homologies of structure in different species. For further discussion, see Buffon: *Oeuvres philosophiques*, éd. Jean Piveteau, Paris, 1954, p. xxxii-xxxv. The question,

on to affirm, in specific terms, that animals are not related to each other and do not have a common origin, and the fact that he insists on the fixity of species, it seems to me that Buffon is sincerely protesting against Linnaeus' classification and its consequences, and perhaps also against the passage we have previously quoted, on the same subject, from Maupertuis. On the other hand, we know from Diderot's own words that although he, too, accepted Buffon's sincerity, he preferred to read Buffon's lines in the way that suited him.[27] The reader may judge for himself the relation between Diderot's theory, to be quoted below, and the statements of Maupertuis and Buffon.

It is not possible to give an account of all of Diderot's readings during this period. There is ample evidence, in his *Pensées sur l'interprétation de la nature* and elsewhere, to show that he seriously followed the latest biological developments, including embryology, that he studied physiology and chemistry, and absorbed Daubenton's findings in comparative anatomy. He was also to make fuller use, here and in future writings, of Bonnet's development, in his *Traité d'insectologie* (1745) of the popular scale-of-beings concept, in which the polyp served as bridge between the plant and animal worlds.

Bringing to a focus all his studies, and his own earlier ideas on the inherent force in nature, on change, cycles, and viability, Diderot, in 1753, in the *Pensées sur l'interprétation de la nature*, reaches a concept of evolutionism which is the most complete and brilliant speculative exposition of that doctrine in his time. He refers to transformism three times in the course of this brief work.

There is a general statement, based partly on Leibniz' assertion that no two blades of grass or grains of sand can ever be identical:

> If beings change successively, passing through the most imperceptible nuances, time, which does not stop, must eventually put the greatest difference between forms that existed in ancient times, those which exist today and those that will exist in far-off centuries; and the *nil sub sole novum* is only a prejudice based on the weakness of our organs, the imperfection of our instruments and the shortness of our lives.[28]

---

however, is more complicated than Piveteau realizes, because of Buffon's extremely broad concept of species.
[27] *Oeuvres complètes*, éd. Assézat et Tourneux, Paris, 1875, II, 15nb. 2, 16.
[28] Ibid., p. 55.

There is a more specific declaration:

> Just as in the animal and vegetable kingdoms, an individual begins, so to speak, grows, exists, declines and dies, is it not the same with entire species? ... Would not a philosopher without religious faith suspect that the necessary elements for animal life have existed from the beginning of eternity; that it chanced that these elements united, because such a thing was possible; that the embryo formed by these elements has passed through an infinity of forms and developments; that it had, successively, motion, sensation, ideas; ... that millions of years passed between each one of these developments; that there are others yet to come, unknown to us; that there has been or will be a stationary period ... that it [animal life] departs or will depart from this state by an eternal decay ... ; that it will disappear forever from nature, or rather that it will continue to exist in it, but in a different form and with different properties than we see in it in this instant of its duration? Religion spares us many wanderings and much labor.[29]

Here, finally, is a still more detailed outline of Diderot's conjectures:

> It seems that Nature has taken pleasure in varying the same mechanism in an infinity of different ways. She abandons one kind of production only after having multiplied its individuals in every possible respect. When we consider the animal kingdom and perceive that among quadrupeds, there is not a single one whose functions and anatomy, especially internal, are not entirely like those of another quadruped, would we not naturally conclude that there has never really been but one first animal, prototype of all animals, some of whose organs nature has only lengthened, shortened, transformed, multiplied, obliterated? Imagine the fingers of the hand drawn together, and the material of the nails so abundant that, spreading and swelling it envelops and covers the whole; instead of a man's hand you will have the hoof of a horse. When we see the metamorphoses of the embryo bringing closer the different kingdoms by imperceptible degrees ... who would not be impelled to believe that there was only one prototype of all beings? But, whether we accept this philosophical conjecture as true, as does Dr. Baumann [Maupertuis], or reject it as false, like M. de Buffon, it cannot be denied that we must entertain it as a hypothesis essential to the progress of experimental physics, to that of rational philosophy, to the dis-

---

[29] Ibid., p. 57-8.

covery and explanation of phenomena which depend on organization.[30]

The essential ideas presented by Diderot are these: (1) each species has had a history; (2) it has evolved over a long period of time; (3) new species appear through a process of variation, but maintain a relation to each other. Despite his regrettable failure to utilize Maupertuis' genetic findings, it is noteworthy that he does introduce, in addition to comparative anatomy, a reference to embryology, to the recapitulation theory. No one yet could call upon fossil evidence. But it is clear that Diderot, working on the most advanced findings available, has united them with his general philosophy of nature to produce this animated vision of biological history. Another important shortcoming in his theory is the absence of what Julian Huxley has termed the quintessence of Darwinism— the selective action of external conditions. The history of species is accounted for entirely on the grounds of internal dynamism and viability.

The concept of transformism takes its place, in Diderot's mind, in an expanding philosophy of materialism. Several points must be noted briefly, as strictly complementary to the theory itself. His materialistic monism seeks the fundamental unity of nature. All of nature's manifestations, he asserts, must be reducible to one basic law, and all phenomena must be interdependent. Nature "has perhaps produced only a single act." The absolute independence of any single phenomenon would disrupt the whole and vitiate nature's order. Eventually, he predicts, such phenomena as matter, weight, electricity, gravitation, and magnetism will be unified in a single law. Nature's dynamism, in its constant flux, is not submitted entirely to chance. "Chaos is an impossibility; for there is an order inherently consistent with the primary qualities of matter." Molecules, because of their shape and attraction, have definite interrelationships.

It would be similarly impossible to admit a definitive separation between inert and organic matter. On the one hand, Diderot is dubious about Buffon's "organic molecules." But he is undecided,

---

[30] Ibid., p. 15-16. For the comparison of the horse's hoof and a man's hand, Diderot refers in a note to Daubenton's description in Buffon. This description points out "a general, primitive pattern" underlying animal variations from fish to mammals, but attributes this to God's wishing "to use only one idea and vary it at the same time in as many ways as possible, so that man could admire equally the magnificence of the execution of the simplicity of the design." (Op. cit., IV, 379.)

speaks himself of "organic molecules," and even says, "it is evident that matter in general is divided into dead and living matter." Yet he immediately inquires, "How is it possible that matter is not one, either all inert or all living? Is inert matter really inert? Doesn't living matter die? Doesn't inert matter ever begin to live?" On the other hand, he cannot go all the way with Maupertuis' materialization of Leibniz' monad (or spiritualization of Epicurus' atom), although he is attracted by it. Maupertuis, he says, should have kept his system within proper limits. It is enough to include in the concept of matter, in addition to mass and movement, the notion of sensitivity—a dull, blunted sensitivity, much inferior to that of the lowest living things. However, Diderot does not as yet clearly extend the idea of sensitivity beyond "organic molecules," to "dead matter." As it will appear in his later writings, he is not certain whether sensitivity is an attribute of all matter, or whether it is an emergent from organization. He will alternate between these two positions. As a result of this sensitivity, there would be for any living molecule "only one situation more comfortable than others," which it would tend to assume through a kind of "automatic restlessness." [31]

Here, then, is a key to a question which, philosophically at least, is inseparable from that of organic evolution, that of the origin of life. Life is materialized, but matter is partially vitalized. "Is there any other difference," queries Diderot, "between dead matter and living matter, except organization, and real or apparent spontaneity of movement?" [32] This, of course, implies spontaneous generation; but we will be less shocked by it if we remember that this theory, in a more sophisticated form, is the accepted doctrine of modern biology.

Clearly, an important change in the thinking of Diderot, as well as of Maupertuis, had come about shortly after 1750. A glance back at the chronology will be fruitful. The two crucial events had been the work on the polyp and Maupertuis' discovery of genetic variation. In exposing the latter discovery, in the *Vénus physique*

---

[31] Ibid., p. 49. The idea of sensitivity appears in La Mettrie's *Traité de l'âme* (La Haye, 1745, p. 29-36), where he is equally hesitant about its inherence in matter as an actuality or a potentiality. It was also suggested to Diderot by the discovery of muscular irritability (contractility), first mentioned in La Mettrie's *L'Homme machine* (1747, p. 113 ff.), and then in Albrecht von Haller's *Elementa physiologiae* (1757). Haller may have derived the notion from the seventeenth century English doctor, Francis Glisson.

[32] Ibid., p. 58-9.

(1745), Maupertuis had simultaneously brought out the central idea of the new materialism: the self-patterning force of nature, containing in itself the laws of its organization and development. This view was to enable materialism to advance from the mechanical atoms-plus-chance concept, to that of the emergence of novelty from the increasingly complex organization imposed by natural process. At the time, Maupertuis based this force on an "attraction," or selective affinity, between chemical particles, essentially similar to the gravitational attraction between celestial bodies.[33] He also extends this principle to non-organic chemical substances.[34]

In 1747, La Mettrie had taken the next step of formulating this concept of nature in a more general and, simultaneously, a more precise fashion.[35] Diderot's *Lettre sur les aveugles* in 1749, and in 1750, Maupertuis' *Essai de cosmologie* both express a theory of universal change and cosmic evolution, of organic trial and error based on internal viability. Meanwhile, in 1749 Buffon had given out his important notion of organic molecules. Again in 1750 (*Système d'Epicure*) La Mettrie insists on nature's creative dynamism, and brings together de Maillet's *Telliamed* and Bonnet's "chain of being." In 1751, Maupertuis revises his notion of the self-organizing power of organic and inorganic matter, by the attribution of sensitivity and elementary intelligence and memory. At the same time, he hints at the possibility of evolution from a common ancestor. Buffon, in 1753, again expresses the latter idea, but apparently rejects it. Finally, towards the end of 1753, Diderot brings together all these ideas in a synthesis that is sharper, firmer, more cogently developed, more clearly evolutionary.[36]

---

[33] "Why, if this force exists in Nature, should it not act in the formation of animal bodies? Let there be in each of the seminal fluids particles intended to form the heart, head, entrails, arms, legs; and let each of these particles have a greater uniting ratio (relation) with that one which, in order to form the animal, has to be its neighbor, than with any other; the foetus will form itself: and were it a thousand times more complex than it is, it would form itself." (*Oeuvres*, II, 89.) After Maupertuis, La Mettrie also proposed to explain the formation of specialized organs by "necessary movements" of the foetal matter (*L'Homme machine* [1748], p. 110; *Système d'Epicure* [1750], p. 149-50). For Bonnet's interesting absorption of this idea into the preformation theory, see Glass: op. cit., p. 207-8.

[34] Ibid., p. 85-7.

[35] See Note 4, *supra*. "Granted the least principle of motion, animated bodies will have all they need to move, feel, think, feel remorse and behave, in a word, in the physical realm and in the moral realm which depends on it." (*L'Homme machine*, p. 113.)

[36] In passing, some note should be taken of Rousseau's awareness of the

After 1754, Diderot turned to other fields of activity. He was not again to take up scientific questions until the late 1760's. His later work, however, was to remain unpublished until well into the nineteenth century. Despite the lapse of fifteen years, his writings present evidence of continuous reading in the field, and greater maturation of thought. Aside from von Haller, whose *Elementa physiologiae* (1757-1766) he studied intensively, the principal new influences were those of Bordeu, Robinet, and Bonnet. Bonnet, in his *Contemplation de la nature* (1764), develops further the idea that all is linked in nature, there being, in consequence, no essential difference between animal and vegetable. He also expands his " scale of beings " hypothesis, which he bases on Leibniz; since there are no jumps in nature, there is a gradual gradation of beings. Nevertheless, he rejects the idea of organic evolution, and—a metaphysician at least as much as a naturalist—is concerned with proving universal harmony.[37]

Robinet, in *De la Nature* (1761), also developed the idea of nature having varied a single prototype, from stones to men, forming a natural gradation of beings. The polyp is the proof of this chain of beings. The " prototype " is " a germ which tends naturally to develop itself. . . . Its energy cannot be repressed. . . . The

notion of transformism. In the *Discours sur l'origine de l'inégalité* (1755), he comments very explicitly on the possibility that man has originated, by a process of change and adaptation, from more primitive, non-human forms. He dismisses the hypothesis, not out of hand, but as unprovable in the present state of biological knowledge, and as superfluous to his particular theme. The close friendship between Rousseau and Diderot at the time, and the acknowledged influence of the latter on the writing of the *Discours*, seem to account for Rousseau's entertaining of the hypothesis. His references to it, however, are a precious indication of the intellectual ferment. (See Vaughan: *Rousseau's Political Writings*, Cambridge, 1915, I, 142, p. 198, note C. In note j on p. 208, Rousseau does accept the transmission of acquired characteristics.)

[37] *Contemplation de la nature*, Oeuvres, IV, 38-115. He declares species to be invariable. Some importance has been attached to Bonnet's insistence on a possible close relationship between man and higher apes. He states that the orang-outang is probably only a variety of our species, an " homme sauvage " (p. 115-6 n.). It would once again be a rash (and I believe, erroneous) judgment to conclude from this that Bonnet entertained a belief in transformism as the origin of species. Aside from his own contrary statements, we must remember that the orang-outang-man *rapprochement* had been made earlier by Rousseau—from whom it seems that Bonnet borrowed it (*Discours sur l'inégalité*, note j, p. 208-9); and that Rousseau accepts the " homme sauvage " as a fact that exists in his mind quite outside of the hypothesis of transformism (which he had excluded as unproven), and not necessarily related to it. These are connections that we now inevitably make, in our intellectual frame of reference; but in the eighteenth century, they were not at all obvious.

germ develops, then, and each degree of development gives a variation of the prototype, a new combination of the primitive universal plan." The character of the prototype remains in all beings, and the only difference between stone, plant, and animal is "the measure in which they participate in that essence." Nature is then dynamic in its original formation of species; but they are formed by a series of trials, out of the same unformed material, which contains the germs of all beings. In 1768, in his *Considérations philosophiques sur la gradation naturelle des formes de l'Etre, ou les Essais de la nature, qui apprend à former l'homme*, Robinet expressed his theory even more strongly. "A stone, an oak, a horse, a monkey, a man are graduated variations of the prototype which began to form itself (*se réaliser*) with the least possible number of elements. A stone, an oak, a horse are not men; but they can be regarded as more or less rough types in their relation to a single primitive design, and in that they are all the product of a single idea, more or less developed." We must consider the succession of individuals "as so many steps of being towards humanity." [38]

The significance of Robinet's hypothesis should not be underestimated. Buffon had previously expressed the idea of the "prototype" in the fourth volume of the *Histoire naturelle*, but only within a single species. In enlarging it to cover all species, Robinet is drawing closer to a concept of organic evolution. He envisions nature as dynamic and continuously experimenting, and all species as related. In Robinet, the static concept of the chain of beings has become dynamic. Yet it would be too much to say that he had any clear idea of transformism. All that he postulates is a series of trials ("*essais*") in an ascending scale of complexity. But each trial is, so to speak, an entirely fresh start from the relatively unorganized stage of the *original* prototype. Now this concept certainly goes beyond the earlier one of Diderot, in the *Lettre sur les aveugles*, in which all of nature's trials took place within a single period of original disentanglement. It is close to the theory of La Mettrie which we have examined. But species are not conceived of by Robinet as having themselves had a history, nor as derivative in a directly unfolding process. Consequently, we are not surprised to see Robinet affirming the absolute fixity of species.

[38] P. 7. This idea is related to Daubenton's work. We may perhaps see in it an anticipation of the "unity of plan" concept that later formed our evidence of evolution.

He considers anatomical similarities to be deceptive, and insists that each species has a specific organization; else they would be variable, "could change, mingle and disappear." [39]

What may have been Robinet's most significant contribution to Diderot's transformism was an offshoot of his metaphysical theory of evil. This was his picture of a struggle for existence, which he considered the necessary mechanism, established by God, for maintaining the equilibrium of the species. This process of mutual destruction involves the sacrifice of the individual, whose only function is reproduction; it provides a balance among living things which, if destroyed by the supremacy of any one species, would lead to its own extinction by hunger.[40] It is possible that these ideas had an important influence on Herder. In Robinet's mind, however, there is no relation whatsoever of this struggle to any concept of natural selection, variation of species, or transformism.

Théophile de Bordeu, finally, one of the brilliant physicians of the age, was a friend of Diderot's and a collaborator in his *Encyclopédie*. He was a member of the Montpellier school of vitalists, and carried forward the thinking of Stahl and Boerhaave. His influence is found in several details of the *Rêve de d'Alembert* (1769), the great dialogue in which Diderot developed the most complete system of materialism in the eighteenth century. Diderot was also impressed by some of his more general ideas. Bordeu conceived the organs of the body to have an independent sensitivity and life of their own. Like Haller, he held the "fibre" to be the basic element of all animal structure, and described its properties as irritability, sensitivity, and elasticity.[41]

---

[39] *De la nature*, Amsterdam, 1761, I, 70; IV, 1-28 (1767).

[40] *Ibid.*, Première Partie. In an article shortly to appear in the *Proceedings of the American Philosophical Society* (Dec. 1958), Prof. Gilbert Chinard has brought to light the little-known development of Robinet's theory by Rétif de la Bretonne, in his *Ecole des peres* (1776). Rétif also sees a struggle for existence limiting population, and preventing undue domination by any one species. The harmony of the animal system is thus maintained. War, luxury and disease produce the same effect among men. Rétif also claims that all terrestrial animals descend from marine amphibians. He emphasizes nature's creative force, resulting in constant variation and the appearance and disappearance of "races" and "species." Unfortunately, his concept of "species" is so vague, contradictory and ambiguous that we cannot be sure of his real meaning. Once again the concepts of struggle and of evolution are not related.

[41] For a thorough study of Bordeu and Diderot, see Herbert Dieckmann: "Théophile Bordeau und Diderots *Rêve de d'Alembert*," *Romanische Forschungen*, 52:55-122 (1938).

In the *Rêve de d'Alembert*, Diderot envisions the universe, life, and man, under the light of a radical materialism derived from Spinoza, Descartes, and the science of his time. We shall here pass over his theory of a living universe, self-sufficient, perpetually reorganizing and renewing itself, forming an immense, unitary Whole. Diderot, in this development, is inspired by the Stoics and by the eighteenth century misinterpretation of Spinoza. But he further applies the concept in its conventional "chain of beings" form, to biology. "All beings circulate in each other. All is in perpetual flux. Every animal is more or less man; every mineral is more or less plant; every plant is more or less animal." [42]

At repeated intervals in the dialogue, the problems of life are taken up, one by one. How did life come to be? All matter, Diderot is now prepared to assert, involves sensitivity. There is only one substance, diversely organized. Change in organization converts "inert sensitivity" into "living sensitivity." This phenomenon occurs in the mineral-plant-animal cycle of nature, in the process of growth and nourishment.[43] Further, the act of fertilization is a mechanical process, changing the non-living germ cells into life, under the stimulus of heat produced from motion, in an embryological process that Diderot describes in some detail. Within the inert germ itself is the latent force or element ready to impose its organization. "And he who would expound to the Academy the progress of the formation of a man or an animal, would use only material agents whose successive states would be an inert being, a feeling being, a thinking being, a being solving the problem of the precession of the equinoxes, ... an aging being, decaying, dying, dissolved and returned to vegetative earth." [44] But where did the germ first come from? queries d'Alembert. Diderot's reply relates the problem of the origination of life to the theory of evolution. "If the question of the priority of the egg over the chicken or of the chicken over the egg embarrasses you, it is because you suppose that animals were at first what they are now. What madness! We no more know what they were than what they will become. The imperceptible earthworm that wriggles in the mud is

---

[42] The reader will recognize La Mettrie, Bonnet, and Robinet as the sources of this idea.

[43] This idea Diderot may well have taken from Toland. See L. G. Crocker: "John Toland et le matérialisme de Diderot," *Revue d'histoire littéraire*, 53: 289-95 (1953).

[44] *Rêve de d'Alembert*, éd. Vernière, Paris, 1951, p. 13-14.

perhaps on his way to becoming a large animal. . . ."[45] If the sun were to be extinguished, and then rekindled, life would eventually start up again, but the results would be quite different; all in nature is connected, and to suppose a new phenomenon is to create a new world.

In Diderot's materialistic explanation of life and its variations, there are, in the order in which he takes them up, five interrelated steps: universal sensitivity (at least in virtuality); its transformation in the life-cycle (epigenesis); transformism; fertilization; and a repeated adherence to the theory of spontaneous generation.

We must neglect Diderot's brilliant psycho-physical explanation of thought, memory, and the self, a discussion which includes dreams, obsessions, illusions, and amnesia. Again the source is in La Mettrie, but Diderot goes far beyond him. A theory of moral determinism follows. Returning then to questions of biology, he discusses the nature of individual organs and their submergence in the unity or whole of the living body, by their submission to a collective life and consciousness—an idea whose sources are in Bordeu, Maupertuis, and the second book of Lucretius. It is this phenomenon that is peculiar to life. Abiogenesis is firmly adhered to. Once more the discussion leads back to problems of evolution. What is the mechanism of change and variation? Unfortunately, Diderot had not been impressed by the genetic theories of Maupertuis. He now suggests the alternative hypothesis of the transmission of acquired characteristics. Rejecting, on the one hand, Christian finalism, and on the other, the Lucretian theory that organs are prior to any use made of them, he coins an oft-quoted phrase: " organs produce needs, and reciprocally, needs produce organs." [46] Bordeu, speaking in the dialogue, avers that he has seen two shoulder-blades lengthen into stumps of arms—a purposefully fantastic exaggeration of regeneration in crustaceans, which Réaumur had studied long before.

---

[45] Here we can see the great difference between Diderot's theory—a true transformism—and those of Bonnet and Robinet, whom Vernière mistakenly considers as true evolutionists (p. LI-LII). In the minds of Bonnet and Robinet, the different species are separate " trials " by Nature, as she devises ever higher combinations. The species themselves are invariable. When Bonnet speaks of " passage " from one species to another, it is always contiguity in the chain that he has in mind.

[46] Ibid., p. 67. In the later *Réfutation d'Helvétius*, Diderot was to put the proposition in a more acceptable form, stating that exercise strengthens organs, and disuse atrophies them. (*Oeuvres*, II, 410.)

Suppose a long sequence of armless generations, suppose continuous efforts, and you will see the two sides of this pincers extend, extend more and more, cross behind the back, return forward, and perhaps form digits at their extremities, and refashion arms and hands. The original conformation is altered or perfected by necessity and habitual function. We walk so little, work so little and think so much, that I should not think it strange if man ended up by being only a head. (p. 69)

It is to be observed that the stimulus to adaptation is not a changing environment, but internal dynamism, or an internal teleology. There are really three distinct theories of which we have given an account in this study, and they should be carefully distinguished. Benoît de Maillet is the true predecessor of Lamarck, in his rather fantastic notion of acquisition of new characteristics in direct response to environmental change, and their transmission. Maupertuis explains variation, in an entirely opposite fashion, as the result of random genetic change. Diderot, on the other hand, is impressed by the dynamism of life, and by its purposeful response to the self-contained needs of the organism.

In a remarkable passage which follows, Diderot describes the differentiation of organs in the embryo, which he attributes to the action of certain "threads," each carrying its own design and finality. Extract from the foetus one of the "threads," the corresponding organs will never develop; mutilate it, and a monster will result (Diderot has finally explained his monsters!); double the thread, and two heads may result, or six fingers; change their position, and the lungs or heart will be reversed. By a mechanical process of extracting "threads," a genius could theoretically be converted into an unorganized mass of flesh; then, by their appropriate restoration, this mass returned to its proper state. At last Maupertuis' genetic theory seems to have entered the picture; but Diderot fails to apply it directly to evolution. He develops one aspect of it, but only to explain the defects of his "monsters." Commenting on their irregular reappearance in different generations, "Bordeu" says,

> It takes two to make a child, as you know. Perhaps one of the agents repairs the defects of the other, and the defective network is reborn only at a time when the descendant of the monstrous race predominates and determines the formation of the network. The bundle of threads constitutes the original and primary dif-

ference between all animal species. The variations in the bundle of any one species produce all the monstrous varieties in that species. (p. 91)

It is curious that Maupertuis, in his notion of "particles," foreshadows the concept of genes, while Diderot, with his "threads," anticipates the discovery of chromosomes. In making this claim, however, we must not fail to include an important reservation. Diderot's notion of the "threads" is developed in a highly ambiguous fashion. His first description of them would indicate that they are not related to our idea of chromosomes, but are materials that are formed epigenetically, as a *result* of fertilization, in the embryo. He seems to be thinking of the "threads" he may have observed in a fertilized chicken's egg.[47] As the discussion proceeds, however, the concept seems to change, and as we have seen in the passage quoted, the "threads" become themselves the constituents of heredity and the genetic constitution of species. Furthermore, we may even (if we wish to be so bold) see in Diderot's words an intuition of the theory of recessive genes.

Much of the remaining part of the dialogue completes Diderot's theory of materialism by its extension to ethical questions. Virtue and vice have no reality, except as social impositions; all actions, even the so-called abnormal, if they produce pleasure without hurt, are equally natural, and morally indifferent. We are reminded that Diderot's beloved Horace had said something like this in his third satire, "Nec natura potest iusto secernere iniquum." Implicit in this discussion is a theory previously developed in the *Neveu de Rameau*, and closely tied to Diderot's view of nature and evolution: in the competition for survival, all that matters is success. "There is nothing solid but drinking, eating, living, loving and sleeping," says d'Alembert in his dream. Life, and persistence in being are the only realities nature knows.

There is one more work in which Diderot touched for a last time on the questions of life and evolution. His *Eléments de physiologie* (1768-1780) are a series of notes, often daring or fantastic, stemming from his prolonged study of Haller. Organs, he states again, are differentiated by an inherent patterning power within the germ cell ("molecule"). The undifferentiated "molecule"

---

[47] "You were, in the beginning, an imperceptible point, formed by smaller molecules, scattered in the blood and lymph of your father or your mother; this point became a long thread, then a bundle of threads." (*Oeuvres*, II, 144.)

produces an eye "by an inherent tendency (*une disposition première*) which can, with nutritive material, produce no other effect."[48] Diderot is now somewhat doubtful about life arising from a complex organization that changes inert into active sensitivity. Irritability and sensitivity are different from all other forces, are life, and they do not exist without life.[49] What else can account, he inquires, for the difference between the living man and the corpse? Is there not, in living bodies, "*un principe vital*," the extinction of which is death, the presence of which animates, creating the unity of the Whole, and the continuity of the life process? He foresees the modern view that all life is organized energy engaged in a struggle, destined to failure, against entropy. "What is an animal, a plant? A coordination of infinitely active molecules, a linkage of small, living forces, against which all things are working together to dissolve. Is it then astonishing that these beings end so soon?"[50]

Diderot's *principe vital* is very close to Bergson's *élan vital*. Were there any reason to believe that he knew the work of Caspar Wolff, we should think it a reminiscence of the latter's *vis essentialis*.

Of the phenomenon of organic evolution he now expresses his firm conviction. "We must not believe that [animals] have always been and will always remain such as we see them.... Why could not the long series of animals be varying developments of a single one?... One of these [absurd] suppositions is that there is on the surface of the earth a single being, an animal which has always been what it now is."[51] His remarks on the mechanism of evolution fall under two heads. He repeats his belief in the inheritance of acquired characteristics. "Organization determines functions and needs; and occasionally needs revert upon the organization, and this influence can go sometimes so far as to produce organs, always so far as to modify them.... I am not averse to believing that the prolonged eradication of an arm would bring about a one-armed race."[52] Of greater interest is a new viewpoint that appears for the first time. The significant shift is now made from inner viability to successful relationship with environment. He definitely glimpses, though the

---

[48] *Oeuvres*, IX, 411. As previously stated, the sources are in Maupertuis and La Mettrie.
[49] Ibid., p. 269, 274.
[50] Ibid., p. 255.
[51] Ibid., 264-5.
[52] Ibid., p. 336, 419.

thought is still vague, the essential elements that had been lacking in his theory of transformism—adaptation to environment, and selection by environment. There is no question, he writes, of a tendency towards perfection as life evolves (an idea he conceivably might have encountered in Leibniz);[53] there is only a purge of ill-adapted beings. "Why cannot man and all animals be considered species of monsters that are a little more enduring? ... What is a monster? A being whose survival is incompatible with the existing order. But the general order is ceaselessly changing; in the midst of this mutability, how can the continuance of the species remain the same? ... The vices and virtues of a preceding order have brought about the existing order, whose vices and virtues will bring about an order to come, without there being any question of the whole getting better or worse." Diderot clearly implies that if the environment suddenly changed, the most "perfect" and "beautiful" beings would become imperfect, and perish unless they were able to re-adapt themselves. In this context of existence, he concludes, spanning nature and ethics, "the world is the house of the strong."[54]

To attempt a summary of Diderot's "theory of transformism" would be futile, since his thought was itself a process of ideological transformation. But we may recall some basic assumptions. Diderot could not believe that matter had to be defined solely in terms of human sense impressions (extension, impenetrability, etc.), and that it could have no other mode of existence. It also seemed incredible that matter could have two distinct ontological states, inert and living. But whereas modern biology, analytic and quantified, tends to study life in terms of its elementary components, or the non-living, Diderot took the other approach. To explain life (being a philosopher he could not merely accept it), he postulated "sensitivity" (and motion) in matter. When he doubted sentience, he doubted, too, the possibility of a materialistic explanation. He also required a theory of organism. The organism "controls" the behavior of its parts, in view of a goal. For this reason life possesses behavior, and ultimately, consciousness, thought and will. They cannot be explained through the activity of the components, but only through the organization. What matters is the dynamism of life, imposing design and creating change, ever seeking stability, never resting in it, always in process, always becoming. All this is

[53] See A. O. Lovejoy: *The Great Chain of Being*, Cambridge, 1948, p. 256-7.
[54] *Oeuvres*, IX, p. 418-9, 428.

far from the Darwinian combination of chance variations of undirected particles and environmental selection (which Diderot glimpsed too late).

Diderot's theories of transformism, of which only the earlier group was known to his contemporaries, had little direct influence on the growth of that hypothesis. The notion itself, however, was disseminated among the encyclopedists, and eventually scientists in France, Germany, and England became absorbed in its development. It should be recalled that Lamarck was a student of Buffon's and a contributor to the later *Encyclopédie méthodique*.

Throughout the course of his writings, Diderot's theorizing reveals two principal characteristics. One is his absorption of all available suggestions put out by contemporary investigation and speculation. The other is the power of animation and synthesis which enables him to complete the idea, to vivify it, and—most distinctively of all—to perceive its connection with related problems. In Diderot's mind, the hypothesis of transformism is a vital link in a coherent philosophy of materialism that extends from inanimate matter to the decisions and ethical judgments of the being who embodies matter's most complex organization. In a real sense, the theory sprang from a materialistic view of nature, and in turn, it became an important contribution to that view, making possible its completion in a universal form.

Furthermore, the development of a theory of transformism is intimately related to the entrance of historicism into the European intellectual outlook. The rise of relativism in ethics and social thought, as evidenced in the writings of Montesquieu, Diderot, and others, is a complementary part of a general evolutionist view of the universe, which embraces the cosmos, life, and societies. The new historicism will counter the Greek and Christian vision of a universe stabilized by eternal laws. From it will grow some of the deepest conflicts of the contemporary world, as the evolutionist philosophy leads both to the relativism of social Darwinism (or to that of Existentialism), and to the absolutism of Hegel and Marx.

Finally, to anyone interested in the history of ideas, who cares to see how a vital concept germinates, comes to life, and grows, summoned forth by the discoveries and nourished by the speculative necessities of an age, the history of Diderot's theory of transformism and its relations with the thought of his time offers a revealing, unforgettable page from man's intellectual history.

SIX

HEREDITY AND VARIATION IN THE EIGHTEENTH CENTURY CONCEPT OF THE SPECIES

BENTLEY GLASS

I

From the standpoint of evolutionary theory the great question during the eighteenth century was how to deal with variation within the species. The inner struggles of Linnaeus with the problem sharply illuminate the matter, for Linnaeus symbolizes the fixity of species in contradistinction to Buffon, just as Bonnet represents preformation in opposition to Maupertuis, and Spallanzani represents biogenesis over against Needham. Yet it would be a gross oversimplification of eighteenth century biological thinking to say simply that Buffon, Maupertuis, and Needham were evolutionists, whereas Linnaeus, Bonnet, and Spallanzani were not. There are shades of opinion and changes of mind through the years, and in the end the conception of species and their modifications through time which were arrived at by Linnaeus and Buffon were not very different.

The rise to fame of Carl von Linné was not only meteoric but precocious. When the first edition of the *Systema Naturae* appeared from the press in 1735 and made him overnight the most famous biologist of his time, he was 28 years of age; and he had already written several deserving books before that. Linnaeus was born with a natural gift, almost a mania, for classifying everything, but he combined with it a deep love of nature and masterly keenness of observation. His interest in plants made him undertake to develop a better system of classification than any in existence at the time, since he saw clearly the imperfections of those of Tournefort and Ray. Through a treatise of Vaillant, *De sexu plantarum* (1717), Linnaeus became aware of Camerarius' discovery in 1694 of the male and female organs of flowers and the nature of fertilization by pollination. He may already, and quite independently, have reached the same conclusion regarding the significance of the stamens and pistils. At all events, he soon convinced himself that a workable classification could be based upon the differences and resemblances in these organs from group to group. The classification was extended to the animal and mineral kingdoms in the *Systema Naturae*.

Probably at this time Linnaeus, in the great pressure of his creative activity, had little time to think about the problem of variation. He was working out the details of his great scheme, and in four years published no fewer than eight significant books. He would be tempted to set aside vexatious complications, in the knowledge that he could return to them later. At least, it is clear that at this time he fully accepted the fixity of species. The *Fundamenta botanica* No. 157 (1736) contains already the famous dictum: *Species tot sunt diversae quot diversas formas ab initio creavit infinitum Ens*; and over and over through the works of these early years, up to 1751, there recurs the expression *nullae species novae*. This belief on the part of Linnaeus may have sprung partly, as Perrier has suggested, from his devotion to the idea of Leibniz that Nature has no gaps.[1] Every species must then be strictly intermediate between two other species, or the plenum would not exist. The belief perhaps sprang also from Linnaeus' rejection of spontaneous generation and his alliance with the ovists. In any case, it led him to make a great reduction in the number of species (7,300) he would admit as true, the remainder of Ray's 18,000 entries being reduced to the rank of varieties.[2] One must then ask what in Linnaeus' mind these varieties were. The answer is clear. In the *Philosophia botanica* (1751), the definition of "varieties" immediately follows the statement quoted above that "there are as many species as the Creator produced diverse forms in the beginning," and further that "these forms multiply and produce, according to

---

[1] E. Perrier, *La philosophie zoologique avant Darwin*, F. Alcan, Libr. Germer Baillère & Cie., Paris, 1884, p. 34.

[2] Ray himself did not regard them all as certainly deserving the denomination of species.

the laws of generation, forms always like themselves. This is why there are just so many species as there exist today diverse forms or structures." Linnaeus therefore yielded no consideration to hereditary variants. Varieties were simply " plants of the same species modified by whatsoever occasional cause . . ." such as " the climate, sun, warmth, the winds, etc." They relate only to " stature, color, taste, odor." Furthermore, upon the return of the plant to its former environment it reverts, so he thought, to its original type. Botanists need take no account of such slight and evanescent variations.

This was Linnaeus' original view. Yet even while he was still reiterating words in this tenor, his mind was troubled and undergoing a change. It seems to have begun in 1742, when a student named Zioberg brought him, from an island near Uppsala, a collection of plants of the common toadflax, or butter-and-eggs, *Linaria vulgaris*, among which were specimens of a quite unknown plant. This new form, in all respects like the common *Linaria* except in its flowers, instead of the expected snapdragon-like blooms had tubular flowers expanding into five uniform petals at the mouth, and having five spurs at the base instead of a single one. Linnaeus called it *Peloria*, and in the *Amoenitates academicae* of 1749 (through a dissertation by his student D. Rudberg) he speaks of it with admiration and excitement. The ancients, he said, imagined that rye could become transformed (*transmutari*) into barley, barley into oats, and oats into tares. The mutation of *Linaria* into *Peloria* is a metamorphosis even greater than such transformations ("*mutationibus illis multo adhuc majorem esse metamorphosin quam mutata in Peloriam Linaria subierit*"). Yet *Peloria* must be considered, he thought, a species, albeit a new one, since it formed perfect seeds and bred true to type. It thus becomes clear that this hereditary variation was to Linnaeus no mere variety unworthy of a name, but in fact a new species. To distinguish the difference he even uses the word *mutation* rather than *variety*.

Now such a phenomenon was not unheard of. According to Guyénot,[3] an apothecary of Heidelberg named Sprenger had in 1590 discovered in his garden, where he grew medicinal herbs, and among them celandine (*Chelidonium majus*), a new form of the latter with very greatly divided leaves. He named it *Chelidonia major foliis*

[3] E. Guyénot, *Les sciences de la vie aux XVIIe et XVIIIe siécles: l'idée d'évolution*, Albin Michel, Paris, 1941, p. 374.

et floribus incisis, and sent it to Bauhin and other botanists of the time, all of whom recognized it as a new and previously unknown species. It is still cultivated and is known as *Chelidonia laciniatum*. A more famous and fully documented case was the discovery in 1715 by J. Marchant of two new species of *Mercurialis* growing in his garden; one of them he named *Mercurialis foliis capillaceis*, and the other *Mercurialis foliis in varias et inaequales lacinias quasi dilaceratis*. Both were cultivated until 1719, when Marchant communicated his discovery to the Académie des Sciences.[4] "There would be reason to suppose," wrote Marchant, "that the Omnipotent, having once for all created individual plants, as models for each genus . . . these models, in perpetuating themselves, would finally have produced varieties, among which those that remained constant and permanent have constituted species, which, through the succession of time, and in the same manner, have made other different products." It seems that in the mind of the discoverer, and likewise of the Historian of the Académie, who commented upon the communication, there was no question: these transmutations constituted new species.

Linnaeus, faced with his *Peloria*, recognized that it could obviously not be one of those temporary varieties produced by the action of the environment. Not only did it breed true, but it had occurred in nature more than once. Gmelin in 1749 reported that a *Peloria* had been found in another species of *Linaria* near Nuremberg; Linnaeus mentioned several places in Sweden; and a M. Ludolfe, a locality near Berlin. Gmelin had moreover transplanted the only two Siberian species of *Delphinium* to his garden in St. Petersburg and thereafter found five or six new fertile forms. He concluded that besides the species originally created, the number had since the beginning " doubled, tripled, or multiplied to infinity." This speculation pleased Linnaeus. It permitted one to adhere to the principle of the immutability of original species and at the same time admit new species produced through hybridization. He set his students to work upon the problem of looking for hybrids between well-known species and even hybridizing species to originate new ones.

It may be pointed out that the theory was certainly extreme. Neither *Linaria* nor *Antirrhinum*, the snapdragon, nor other plants

[4] J. Marchant, Observations sur la nature des plantes, *Mém. Acad. Roy. Sci.*, 1719, pp. 59-66.

of the Scrophulariaceae normally have peloric flowers. Either one must assume that a hybrid between two similar forms could bring forth something completely new; or else the hybridization must have been between a *Linaria* and a plant completely outside its family. Yet *Peloria* was in all respects, except for its flower structure (and fruit), a typical *Linaria*. Why did no other characters of the presumed widely different parent show up? Today, since the peloric mutation has recurred many times in these species, we know that it may be produced by a single gene alteration, and that it breeds true upon self-fertilization because it is a recessive type and consequently homozygous. What a pity that Linnaeus did not undertake to hybridize the *Peloria* with the common *Linaria*. He might in that case have anticipated Mendelism! Instead, he and his students Dahlberg, Gråberg, and Haartman began looking for new species that might have arisen by hybridization whenever parent species from distant realms had been brought together in a garden.

A large number of genuine and purported hybrids were discovered in the ten or fifteen years following the first finding of the *Peloria*.[5] Only a few of them can be mentioned, to serve as examples. One was a hybrid thistle (*Carduus*), its putative parents the common hairy thistle and the thistle of the Pyrenees. Several were *Verbena* hybrids, one being presumably from a cross between *V. americana* and *V. humilior*. *Pimpinella agrimonoides* was regarded as a hybrid between the common pimpernel (*P. sanguisorba minor laevis*) and agrimony (*Agrimonia officinarum*), a truly startling combination of a member of the rose family with one of the primrose family, had it been authentic. Another, somewhat less dubious, was found in the Uppsala garden in 1750, and described in fullest detail by Haartman. It was named *Veronica spuria* by Linnaeus, and was held to have come from a cross of *Veronica maritima* female and *Verbena officinalis* male. This hybrid represents a cross between species of different families belonging to two distinct subgroups of the Tubiflorae. It was a perennial, multiplying only from the roots, since it was quite sterile. Like the other reported hybrids, its flower structure most resembled the female parent, but the stem and leaves were very like those of a *Verbena*. It seems that Linnaeus, faithful to an old idea of Cesalpino, con-

---

[5] A quite detailed account, including bibliography, is given in H. F. Roberts, *Plant Hybridization before Mendel*, Princeton University Press, Princeton, 1929, pp. 15-33; cf. Guyénot, op. cit., pp. 368-374.

sidered the stamens of a flower to correspond to the wood and the pistil to the pith. "It is impossible to doubt," Linnaeus wrote, "that there are new species produced by hybrid generation. From all these things, we learn that the hybrid is brought forth, as to the medullary substance or the internal plant or fructification as the exact image of the mother, but as to its leaves and other external parts it is as that of the father. These considerations, therefore, lay down a new foundation for the students of nature, to which many things contribute. For thence it appears to follow, that the many species of plants in the same genus in the beginning could not have been otherwise than one plant, and have arisen from this hybrid generation."[6] Where the pollen parent was unknown, as in the case of many of these purported hybrids, Linnaeus believed that he could deduce it from the nature of the leaves and stem of the hybrid, and the general availability of possible pollen-providers. The description therefore signifies only that the purported hybrid had vegetative features somewhat resembling those of a *Verbena*. No one appears to have repeated this cross, or even to have regarded it as worth the attempt. Did Linnaeus deceive himself?

There was, notwithstanding, one hybrid which Linnaeus himself produced in his garden in 1757 through an artificial cross of *Tragopogon pratense* female by *T. porrifolius* male. It was reported in the *Disquisitio de Sexu Plantarum* (1760), which was submitted to the Imperial Academy of Sciences in St. Petersburg along with some of the seed of the *Tragopogon* hybrid. It thus chanced that Koelreuter had an opportunity to grow the hybrid and examine it when it bloomed in the spring of 1761. His comment that it was a mere "half-hybrid" was based on an erroneous theory of his own, and there seems no particular reason to doubt the authenticity of this first artificial interspecific plant hybrid.

If any further evidence were needed to demonstrate how far Linnaeus had shifted from his early view of the fixity of species than the clear statement just quoted a paragraph above, it is to be found in his numerous references in the *Species plantarum* (edition of 1764) which express a conjecture that some particular species in question has been derived from some other. Green has compiled many of these:

---

[6] C. Linnaeus, *Disquisitio de sexu plantarum*, 1760, p. 127-128; the translation is that of Roberts, op. cit.

Thalictrum lucidum . . . The plant is possibly not so very distinct from *T. flavum*. It seems to me to be the product of its environment.

Clematis maritima . . . I should rather think it is derived from *C. recta* under altered conditions.

Achillea alpina . . . May not the Siberian mountain soil and climate have moulded this out of *A. ptarmica*? [7]

There are numerous others, all of which imply that Linnaeus not only had accepted the origin of new species through hybridization, but was wondering about the origin of good species—no mere temporary varieties—through the action of the environment over long periods of time. The *Metamorphosis plantarum* of 1755 deals not only with fixed species but also with monstrosities and varieties. If, as Linnaeus had heard, Réaumur had produced a hybrid between a rabbit and a hen, was it not more likely that hybrids within the order of anthropoids, which Linnaeus had proposed as a fancy, might indeed have occurred? And was man by any significant character not of the order of anthropoids?

In 1762, Linnaeus conjectures that the work of God in the Creation stopped at providing the common source of each genus, or maybe only of each order, and that diversification within the limits of these original groups has made the species. In 1766, he omits from the final edition of the *Systema Naturae* his famous dictum that there are no new species. Finally, in 1779, he expresses the view that "species are the work of time," and that one of the great scientific enterprises of the future will consist in demonstrating this truth.[8]

Was Linnaeus, the master proponent of the fixity of species, then transformed by the hard pressure of facts into an evolutionist in his later years? That would be saying too much, perhaps; for the old man hardly thought his way clearly through his puzzlement. How

---

[7] E. L. Greene, *Carolus Linnaeus*, Christopher Sower Co., Philadelphia, 1912, pp. 79, 81, 82.

[8] Quoted by J. Rostand, *L'évolution des espèces*, Libr. Hachette, Paris, 1932, p. 24, from *Amoenitates Academicae*, 1762; and again J. Rostand, *Esquisse d'une histoire de la biologie*, 12th edition, Gallimard, Paris, 1945, from *Amoenitates Academicae*, 1779. [I have not been able to consult the original.] Cf. also K. Hagberg, *Carl Linnaeus*, E. P. Dutton and Co., New York, 1953, chap. XII, "The Origin of the Species," for interesting sidelights in spite of a sketchy treatment of the subject.

clear and sharp scientific distinctions had been in his youth! How turbid the waters he now beheld! Nevertheless, the view here set forth is agreed to be correct by every modern student of Linnaeus' work and thought.[9] One can put it this way. In the end he believed in the evolution of the smaller systematic categories, of the species as he knew species, and maybe of the genera. But the original Creation was still that of a multitude of forms, distinct then and forever. In holding to this conviction his Lutheran conscience was salved.

II

Were further evidence necessary of Linnaeus' ultimate conversion to belief in the origin of new species, it would be sufficient to cite a contribution to the *Histoire de l'Académie des Sciences* made by Adanson in 1772. Entitled "Examen de la question, si les espèces changent parmi les plantes; Nouvelles Expériences tentées à ce sujet," [10] this remarkable paper is concerned wholly with the refutation of Linnaeus' views. Quoting Marchant's opinion about the significance of his mutant *Mercurialis*, Adanson says: " This opinion remained forgotten until the year 1744, when M. Linnaeus after having regarded species as constant in 1740, commenced to doubt it, and even to believe in the production of new species when speaking of the plant he called *Peloria*; and after having cited the two examples of M. Marchant, he enumerated six others upon which we will report." Adanson thereupon proceeds to describe his own experimental tests of these reputed new species, and to demolish their claims to such status. Before considering this highly significant, though almost unknown, piece of eighteenth century evolutionary controversy, a sketch of the earlier career of Michel Adanson, the botanist involved, will be relevant.

Adanson is a figure of disputed stature in the history of science. To Julius Sachs, whose *History of Botany* (1875) has become a classic, it seemed clear that the " first great advance in the natural system " was due to Antoine Laurent de Jussieu (1748-1836), prin-

---

[9] Including Loren Eiseley, in his recent book, *Darwin's Century*, Doubleday & Co., Garden City, 1958, pp. 16-26.

[10] *Histoire de l'Académie Royale des Sciences*, 1769 (publ. 1772), pp. 71-77 (commentary of the Historian of the Académie) and *Mémoires*, pp. 31-48.

cipally by providing the families of plants with distinctive characters.[11] The work of the uncle, Bernard de Jussieu, marked little advance over the effort of Linnaeus, who, in his *Classes plantarum*, had formulated the basis of a natural classification of plants that would take account of all characters and show true affinities, instead of being admittedly arbitrary and artificial like his empirical reliance on the sex organs of the flowers in the *Species plantarum*. Sachs remarks:

> Adanson's claims of priority over Bernard de Jussieu may be passed over as unimportant. The natural system was not advanced by Adanson to any noticeable extent; how little he saw into its real nature and into the true method of research in this department of botany is sufficiently shown by the fact, that he framed no less than sixty-five different artificial systems founded on single marks, supposing that natural affinities would come out of themselves as an ultimate product,—an effort all the more superfluous, because a consideration of the systems proposed since Cesalpino's time would have been enough to show the uselessness of such a proceeding.[12]

On the contrary, a more modern historian of science, Émile Guyénot, hails Adanson as the true founder of the natural method of classification and the greatest botanist France ever produced—a man at least the equal of Linnaeus in genius.[13] The reasons for Adanson's eclipse and the neglect of his work appear to lie, like those relating to Maupertuis, in the machinations of an evil genius, in this instance his rival Antoine Laurent de Jussieu, who through nepotism succeeded his uncle as director of the Jardin des Plantes and successfully devoted himself throughout his life to the derogation of Adanson and the enhancement of the reputation of his uncle. The facts seem to be as follows.

A precocius youth with a passion for the natural sciences, Adanson at 21 lacked the financial resources and influence needful to obtain a post in the limited scientific world of the time; and in 1748 he went as a business agent to Senegal, where he spent four years and a half. Two years later he wrote to his onetime teacher, the elder

---

[11] J. von Sachs, *History of Botany (1530-1860)*, trans. by H. E. F. Garnsey, rev. by I. B. Balfour, Oxford, at the Clarendon Press, 1890, p. 116. Original German edition, 1875.
[12] Ibid.
[13] Guyénot, op. cit., pp. 29-38; 366-367; 377-379.

de Jussieu, of the inadequacies he had found in the systems of Tournefort and Linnaeus in classifying tropical plants, and of his conviction that such systems were faulty because they selected a single part of the plant as a basis for classification. One must compare all parts minutely, he thought, and thus there will arise the character of the whole organism which can lead to a natural classification. On this basis, he at once met with a startling success, by divining the status of the great tropical baobab tree as a member of the mallow family, the Malvaceae. After returning to France, he published in 1757 his mémoire on the baobab, and immediately set to work on the plan of the great *Familles de plantes*, the outline of which he read to the Académie in 1759, while the entire work was completed in 1763. It was based on his natural method, which, he declared, is one wherein the classes contain "no plants which do not agree together" in the *ensemble* of their characters. Reliance upon this method was necessary, in Adanson's mind, because of his complete and implicit belief in the Great Chain of Being and the Principle of Continuity. From a belief that nature makes no leaps, that there are no real voids between organic forms, he, like Buffon in the beginning, held that in nature there are really only individuals, and neither genera, nor classes, nor species. The latter, he thought, are conveniences of the classifier's mind. The reality was the individual and the variability of those individuals. Nevertheless, upon close examination one can discern fissures in the apparent continuity, "lines of separation." These may be the result of our ignorance of intermediate forms, or of their disappearance, the latter being especially evident from "the bones of monstrous quadrupeds, the skeletons and impressions of fishes and of plants, and a prodigious number of shells, so different from those which live today in the seas." [Adanson, while in Senegal, had made huge collections of plants and animals, especially mollusks, and was the first to recognize, in a monograph of 1752, that the shipworm *Teredo* is a mollusk. One must not judge by the shell, but by the nature of the animal itself, he emphasized.][14]

A corollary of Adanson's view of variability was that it is very difficult to determine what is a variety and what is a species. Hence he deplored Linnaeus' exclusion from his system of all varieties and

[14] M. Adanson, *Histoire naturelle de Sénégal: Coquillages*, Paris, 1752. Cf. Guyénot, op. cit., p. 84.

his great reduction in the number of admitted species. "Seeds," said Adanson, "are the source of a prodigious number of varieties, often so changed that they could pass for new species, especially when they reproduce [again] by means of seeds. . . ." He did not think that these supposed new species—and he mentions specifically as examples the two mercurialis of Marchant, Linnaeus' *Peloria*, and Gmelin's delphinium—came from hybridization between species, as Linnaeus proposed. Among seedplants similar changes may be procured "either by the reciprocal fecundation of two individuals differing in some character, although of the same species, or by culture, soil, climate, drought, humidity, shadow, sunlight, etc. . . . These changes are more or less prompt, more or less enduring, disappear after one generation or perpetuate themselves for several generations, according to the number, the force, the duration of the forces which unite to form them." [15]

That was in 1763. Three years later a remarkable new strawberry was discovered by the Duchesnes, father and son, horticulturists of Versailles. It had a simple, entire leaf instead of the three distinct leaflets characteristic of the ordinary strawberry. The younger Duchesne concluded it to be a new species, and proceeded to a hypothesis that all existing kinds of strawberries had probably been derived by variation from one original type, by the influence of climate or unknown causes.[16]

At the Jardin des Plantes Adanson proceeded to investigate very thoroughly and scientifically the nature of this monophyllous strawberry, together with Linnaeus' *Peloria* species, the *Mercurialis* species of Marchant, Vaillant's Miracle, or Smyrna, wheat, and a variety of barley Adanson had bred himself, with 4-ranked instead of 2-ranked ears but otherwise possessing the characteristics of the 2-ranked sugar-barley (*orge sucrion*). Stating that to start with he had himself been inclined to agree with the suppositions of Linnaeus regarding new species, Adanson was convinced to the contrary only by repeated and painstaking experiments, continued over the space of seven years.

The *Mercurialis* of Marchant which Adanson had at his disposal had lower leaves like those of the common mignonette (*Reseda*). Hence Adanson not only crossed the *Mercurialis*, which was a male

---

[15] Adanson, *Familles des plantes*, Paris, 1763. Cf. Guyénot, op. cit., p. 377.
[16] Duchesne fils, *Histoire naturelle des Fraisiers*, Paris, 1766. Cf. Guyénot, op. cit., pp. 375-377.

plant, with two ordinary female *Mercurialis* plants, but also attempted to pollinate ordinary *Mercurialis* by mignonette and flax plants, respectively, since according to Linnaeus' conjectures the new species might have arisen as a hybrid between *Mercurialis* and one of the two latter species. The seeds of the cross of the abnormal *Mercurialis* by the normal *Mercurialis* germinated poorly (hardly 10 per cent). All of the plants which grew from these seeds or from a second, third, fourth, or fifth generation were ordinary *Mercurialis*. The attempted crosses with the mignonette and flax plants were wholly sterile, except for a few ordinary *Mercurialis* progeny that were produced—Adanson concluded—by contamination with pollen from neighboring gardens in spite of all his care, or perhaps by the presence of some male flowers on the female *Mercurialis* plant, a phenomenon Bernard de Jussieu had reported as occurring. Adanson concluded from these experiments that the *Mercurialis* with dissected leaves was no new species, but a monstrosity—as we would say today, a mutant—and this seemed to be confirmed by its numerous defective features not only of leaves but of vessels, veins, and stamens. Today we know that Adanson was unquestionably correct. The laciniated *Mercurialis* is a simple mutant variety.

Adanson's next experiments were done with *Peloria*. Contrary to Linnaeus' report, this form failed to breed true, either in the original variety which Linnaeus had sent to the Jardin des Plantes, or in a new peloric type which had arisen there in a Spanish species of *Linaria*. In both, reported Adanson, peloric flowers and typical flowers were borne on the same stalk, or plants might have only typical irregular flowers, or only peloric flowers; but in every case the peloric blooms were sterile. The only seeds came from the ordinary flowers, and these seeds yielded now ordinary *Linaria*, now *Peloria*. Once again Adanson was forced to conclude that the supposed new species was a monstrosity, with certain organs developed to excess (*monstre par excès*) and others developed weakly or not at all (*monstre par défaut*).

As for the strawberry with simple leaves, Adanson bred this too, and found that the leaves were in actuality quite variable in the basic defect of the veins of the leaves, so that it was apparent that the single-lobed leaf was an imperfect fusion of the three lobes of an ordinary leaf. The flowers, too, had more petals and sepals than usual, while the number of stamens was reduced; and both stamens and fruit were malformed. Here, then, was another example of a mutant, "monstre par excès dans le calice et la corolle; monstre par défaut dans les feuilles, les étamines, et la fruit." Every botanist was familiar with monstrosities: "one knows sufficiently what a difference there is between a species and a monstrosity; a species is comparable to a new species, but a monster can be compared only to an individual of the species from which it originates."

The barley studied by Adanson had a 2-ranked ear, but he noticed a slight tendency in some plants for the ear to become quadrate. Through selection continued over five or six years (10-12 generations), he raised the frequency of 4-ranked ears until all were of that sort, although within the ears there was still variability and a tendency to revert to a 2-ranked type. As for the famous Miracle wheat, which Linnaeus for some reason had ignored in the *Species plantarum*, this too proved to be a "monstre par excès." Among its branched heads, when carefully examined, there proved to be two to four per cent of simple heads; and by sowing the wheat too late or in spring, or in an earth too poor or too dry, virtually all branching of the heads would disappear. [Premonition of Lysenko!]

From these repeated results on every variety of supposed new species he could obtain, Adanson came to the conclusion that Linnaeus was quite wrong. ". . . The transmutation of species has not taken place among the plants, no more than in animals. . . ." Finally, a most pregnant conclusion: "To see the harmony which reigns in all parts of the Universe, every reasonable philosopher is at first moved to believe that these deviations also have their laws and their limits: in fact, the more one observes, the more one convinces himself that these monstrosities and variations have a certain latitude, necessary without doubt for the equilibrium of things, after which they return into the harmonious order preestablished by the wisdom of the Creator."

It is surprising that the erudite author of *Les sciences de la vie aux XVIIe et XVIIIe siècles* failed to discover this final mémoire of Adanson's career, for it has been referred to by Rostand [17] and others. Guyénot consequently arrived at a rather dubious conclusion, in holding that Adanson had not conceived clearly the idea of evolution. On the contrary, he furnished crucial experimental evidence to destroy the principal form in which it existed in his

---

[17] Rostand, *L'évolution des espèces*, p. 25.

time. Nevertheless, we are scarcely in error if we grant that his achievement of a truly natural classification based on real affinities, his emphasis on the variability of species and the analysis of mutant forms, his suggestion of the causes of hereditary variation, and his emphasis on fossils as indicating the demise of species which once filled the plenum of Nature, were all more in the true direction of the ultimate evolutionary theory than Linnaeus' theory of the origin of new species by hybridization. What irony in the fact that the tested "new species" of Linnaeus were all demonstrably mere mutants, although in principle he was right, since new species of plants can and do arise by hybridization upon occasion! The fate of Linnaeus' theory of the origin of new species within the bounds of the originally created forms, call them what you will, reminds us forcibly of the similar fate of Hugo de Vries' mutation theory, which was likewise wrong in the particular instances on which it was based, yet right in principle.

The last years of Adanson were lamentable. Even Lamarck may be considered to have been blest in comparison. Thirty-seven years were to pass after the publication of Adanson's last great mémoire before he died—years during which he was not only excluded from every official post and opportunity, but was first vilified and then forgotten. When still only 62 years of age, at the outbreak of the Revolution, he was deprived even of his two pensions, the one royal, the other from the Académie. During ten years of penury he lived on the sugar and coffee provided by a loyal servant, a peasant woman who worked at night, unknown to him, to make the extra pittance necessary for that poor fare. Crippled with rheumatism, without light and without fire, he tasted the final despair of inability to continue his work. Only in 1799, when his pension from the Académie was restored, was there a measure of relief; but it came much too late to revive his efforts. He died in 1806, when Lamarck was in the ascendant—Lamarck who wrote the article on "Famille" in the *Encyclopédie methodique* without so much as mentioning Adanson's name, even when describing the plants brought by Adanson from Senegal.[18]

---

[18] Guyénot, op. cit. I have relied almost exclusively upon this work in the present treatment of Adanson, except for a careful study of the Mémoire of 1769. The facts of Adanson's life are recounted in A. Chevalier, *Michel Adanson*, Larose, Paris, 1934.

III

One further crushing blow was dealt Linnaeus' theory of the origin of new species by hybridization. This was through the analysis of numerous interspecific hybridizations performed by the first great plant hybridizer, Joseph Gottlieb Koelreuter, who reported them in detail in a "Preliminary Report" (1761) and three "Continuations" (1763; 1764; 1765).[19] Koelreuter's experiments were models until the time of Mendel. Applying the discovery made by Camerarius of the sexual organs of flowers, and relying on their exclusive pollination by insects which might be readily excluded from access to the flowers, he performed numerous artificial pollinations with success. In 1761 he obtained the first well-authenticated interspecific plant hybrid, from a pollination of *Nicotiana rustica* by *N. paniculata*. Perfect seeds were obtained, they germinated well, and the young hybrid plants, extremely vigorous and fast-growing, were intermediate in type between the parent species. However, when they came into bloom, the anthers of the hybrid's flowers were not filled with pollen, and the pollen grains present appeared under the microscope to be shrivelled and empty. The hybrid, in fact, proved to be completely self-sterile, and Koelreuter therefore called it a "mule-plant." However, when the hybrid was pollinated with pollen from either of the parent species, it set some fertile seed. In these back-crosses Koelreuter observed that the progeny tended most to resemble the maternal parent; but in successive crosses gradually to go back to the type of original species used in the backcross. Thus, according to his findings, Linnaeus' new species could not be true-breeding hybrids, as Linnaeus thought, for hybrids are sterile except in backcrosses, and then they tend to return to the original forms.

As to Linnaeus' one artificial hybrid, the *Tragopogon*, Koelreuter was dubious of that too, although for the wrong reason.

> For the hybrid goat's-beard, which the celebrated Linnaeus considers in his new prize essay, is not a hybrid plant in the real sense, but at most only a half hybrid, and indeed in different degrees,

---

[19] J. G. Koelreuter, *Vorläufige Nachricht von einigen das Geschlecht der Pflanzen betreffenden Versuchen und Beobachtungen, nebst Fortsetzungen 1, 2, und 3*, Leipzig, 1761-1766. Reprinted in *Ostwald's Klassiker der exakten Wissenschaften*, No. 41, Engelmann, 1893. Cf. Roberts, op. cit., chap. II, for a relatively extended analysis.

as I will clearly and plainly demonstrate at another opportunity, with many reasons which appear in part from the nature and peculiarity of the composite flowers, and from certain experiments instituted upon the time of fertilization of the same; in part from the structure of the above-mentioned presumed hybrid itself, which had been raised by me from seeds which Linnaeus had sent, together with his prize essay, to the Honorable Russian Imperial Academy of Sciences, and which have bloomed the past spring in the Academy's garden at St. Petersburg.[20]

What Koelreuter meant by a half hybrid is apparent from the context. He assumed that always in hybridizations there would be a "tincture" of self-fertilization, in proportion to the amount of pollen from the anthers of the maternal parent to the amount of foreign pollen placed upon the stigma. [This was before the time when it became common practice in cross-pollination to remove the anthers of the flower selected as the female parent before their dehiscence.] Kolreuter thought, in other words, that a given plant of the progeny could share inheritance from two male parents, in proportion to the relative amounts of pollen contributed by them. This belief was grounded in his observation that it required 50 to 60 pollen grains to fertilize all 30 ovules in an ovary of *Ketmia* or *Hibiscus*. The actual nature of the fertilization process was not to be discovered until a century later; but Koelreuter himself soon showed that when the stigmas of a flower are pollinated by their own pollen and foreign pollen at the same time, fertilization is through the former. His theory of half hybrids was thus unfounded, and the objection to the validity of Linnaeus' hybrid *Tragopogon* may be rejected.

A more serious objection to Linnaeus' theory of the formation of new species by hybridization was, however, raised in this same communication (Erste Fortsetzung). Koelreuter showed that in his numerous crosses involving species of *Nicotiana* (tobacco), *Dianthus* (pink), *Ketmia* (a composite), *Leucojum, Hyoscyamus* (dogbane), *Verbascum* (mullein), and *Matthiola* (stock), in general only closely related species may be crossed at all, and not always even these. The wide hybridizations conjectured by Linnaeus on the basis of the supposed combination of maternal and paternal characteristics in the "new species" thus fell, and along with them were discredited

---

[20] Koelreuter, op. cit. Roberts' translation, p. 43.

his more reliable instances of species hybrids, among them, for example, the alsike clover and the artificially produced *Tragopogon*. In fact, it became common practice to say that Linnaeus had made no effort at all to test his theory experimentally (Duchesne, 1766; Sachs, 1875; Guyénot, 1941).[21]

The most significant bearing of Koelreuter's hybridizations upon the evolutionary theory lay in the results of the self-fertilizations of the hybrids and crosses of them to the original parent species. Many of these were performed, of which the *Dianthus* crosses may serve as example. There were 7 of these reported in the Erste Fortsetzung, 8 in the Zweite Fortsetzung, and no less than 29 in the Dritte Fortsetzung. Among salient observations were the following: (1) the F_1 hybrid was always very uniform in type but the progeny of selfings of hybrids and of backcrosses showed considerable variation; (2) at least three selfings were fertile; and (3) in a cross of *D. chinensis* by *D. hortensis* in which the latter had double flowers, all the progeny had double flowers.

One may emphasize first of all the great tendency to variation when the hybrids were bred. This observation was totally destructive of Linnaeus' theory of the formation of true-breeding new species from hybrids.

Of at least equal interest, however, were the self-fertile hybrids, even though the backcrosses had already demonstrated that many hybrids are not totally sterile, at least on the female side. This discovery of fertile hybrids between species rendered utterly untenable the sole criterion of species which Buffon, and after him Kant, were utilizing to maintain their views of the constancy of species.[22] Of course, it might be said that the test proved the parent types to be no true species, but mere varieties. Yet they had been well recognized by all systematic criteria as distinct species. The logical conclusion had either to be that the forms were not true species, or else that true species are able in some instances to cross and yield fertile hybrids. To accept the former view would lead to the expansion of the species until it engulfed the genus or even the entire family, and to admit that within its thus enlarged limits variation, hereditary or acquired, would have full scope. To accept

---

[21] Duchesne, op. cit., cited from Guyénot, op. cit., p. 373. Sachs, op. cit., p. 89; 400. Guyénot, op. cit., p. 373: "Il est à peine besoin de dire que les interprétations de Linné ne reposait sur aucune détermination expérimentale."

[22] See Professor Lovejoy's chapters on Buffon and Kant in this volume.

the latter view was more frankly to admit the probability that species may become altered by intermingling with other species. But then what keeps the species in nature breeding true to type?

Buffon chose the first way out of the dilemma: reduce the number of true species to a minimum—all quadrupeds to 38 species, for instance—and the criterion of hybrid sterility holds fast. But within each such "species" enormous modifications can take place. Linnaeus, so late in life, could not forsake his well-ordered system of species. He therefore abandoned the idea of the constancy of species, and resorted to the constancy of originally created "stocks" from which groups of related species have been derived. In reality, the difference between this conception and Buffon's is purely semantic. Koelreuter—and this is most curious—seems to have evaded altogether the issue of the nature of species and the light thrown upon it by his own experiments. It seemed enough that he had demolished Linnaeus' theory of new species through hybridization. He had saved the dogma of the constancy of species from the hands of its faithless creator. He did not see that in doing so he had in truth opened Pandora's box, since henceforth no one would ever be able to distinguish species from races and varieties by any simple criterion. Why did some, but only some, of Koelreuter's species hybrids manifest complete sterility, as among the mulleins (*Verbascum*) he intercrossed? Koelreuter remarks on the singular absence of such hybrids in nature, although some of the species involved in the crosses had lived in proximity for thousands of years. Was it not, he thought, teleologically, because of their very sterility? But in that case, why were the more fertile interspecific hybrids not formed either? The question indeed might lead one by minute gradations back to mere racial and varietal distinctions. Koelreuter failed to pursue it beyond the demonstrably greater potency for fertilization of a flower's own pollen than of foreign pollen.

The dominance of double flowers in the *Dianthus* cross might have provided a clue to the particulate nature of heredity; but these hybrids do not appear to have been backcrossed on the recessive parent. Koelreuter was not looking for purity of genetic character, since he had made many other crosses in which, even for simple traits, the hybrid had been intermediate in type. He had, for example, crossed *Mirabilis jalapa* red-flowered by yellow-flowered in reciprocal directions, and had done the same for *Leucojum* red-flowered by white-flowered, *Datura stramonium* white-flowered by

D. tatula violet-flowered, and the like. In each of these cases the flower color of the hybrid was intermediate, or at least not strictly like either parent. The *Mirabilis jalapa* hybrid had mixed red and yellow, or orange-yellow flowers; the *Leucojum* hybrid had whitish-violet blooms; the *Datura* flowers showed " a whitish color playing a little into the violet; the flower-tubes marked with five violet stripes, and the others [?] sky-blue." Undoubtedly results such as these and many others of like nature found in the succeeding century blinded Nägeli to the general significance of Mendel's discovery.

Finally, it should be apparent that Koelreuter's experimental findings, properly understood, constituted a body-blow to the doctrine of preformation in the form of emboîtement. Sachs has compared his influence in this respect with that of Caspar Friedrich Wolff.

> . . . Wolff is usually said to be the writer who refuted the theory of evolution [as Bonnet called his doctrine of preformation]. It is certainly true that in his dissertation for his doctor's degree in 1759, the well-known ' Theoria generationis,' he appeared as the decided opponent of evolution; but the weight of his arguments was not great, and the hybridization in plants which was discovered at the same time by Koelreuter supplied much more convincing proof against every form of evolution. Wolff conceived of the act of fertilization as simply another form of nutrition. Relying on the observation, which is only partly true, that starved plants are the first to bloom, he regarded the formation of flowers generally as the expression of feeble nutrition (vegetatio languescens). On the other hand the formation of fruit in the flower was due to the fact, that the pistil found more perfect nourishment in the pollen. In this Wolff was going back to an idea which had received some support from Aristotle, and is the most barren that can be imagined, for it appears to be utterly incapable of giving any explanation of the phenomena connected with sexuality, and especially of accounting for the results of hybridisation. Wolff may have rejected the theory of evolution on such grounds as these, but he failed to perceive what it is that is essential and peculiar in the sexual act.[23]

While this may be less than fair to Wolff, it certainly places the general significance of the studies of hybridization made by Koelreuter in a proper light. As Sachs well says: " It is impossible to rate too highly the general speculative value of Koelreuter's artificial

---

[23] Sachs, op. cit., p. 405.

hybridisation. The mingling of the characters of the two parents was the best refutation of the theory of evolution [read *preformation*], and supplied at the same time profound views of the true nature of the sexual union." [24] It is only fair, however, to point out also that Maupertuis' study of polydactyly and his views of the formation of the foetus had done the same, ten years earlier, and that Maupertuis' speculative extension of his views into the realm of evolution far exceeded Koelreuter's reluctance to grapple with the problem of species.

At all events, Koelreuter's experiments, though largely neglected in his own lifetime by the Linnaean school and most other systematists, bore fruit especially in the work of Konrad Sprengel, who took up Koelreuter's observations of insect pollination, demontrated the function of nectaries in flowers, and revealed the ingenious devices whereby, in dioecious plants, the fertilization of a flower by its own pollen is commonly prevented. In the next century the hybridizers Knight, Herbert, and Gärtner continued the analysis of species crosses; but—in Sachs' words again—" it was reserved for Darwin's wonderful talent for combination to sum up the product of the investigations of a hundred years, and to blend Koelreuter's, Knight's, Herbert's, and Gärtner's results with Sprengel's theory of flowers into a living whole. . . ." [25]

Lacking in the picture, of course, is Mendel, for Sachs wrote in ignorance of the revolutionary experiments with peas in the little monastery garden in Brünn. Yet Mendel, too, owed his inspiration to Koelreuter, whose work he knew so well. And if Nägeli and the other plant hybridizers of the time were so preoccupied with species hybrids and the seeming insolubility of the problem of species in relation to hereditary variation that they failed to understand Mendel, that preoccupation too stemmed from Koelreuter. How could one suppose that the study of simple variations of hereditary nature would one day throw more light on the nature of the origin of species than all the ten thousand crosses and three hundred and fifty different hybrids Gärtner had produced and described so painstakingly? Only Darwin might have understood, and fate kept Darwin and Mendel apart.

[24] Ibid., p. 413.
[25] Ibid., p. 431.

IV

In the eighteenth century, the theory of "evolution" was the theory of Charles Bonnet, and it had a far different sense than the theory of evolution of the nineteenth century. The older theory had reference to the evolution of the individual from the germ, which Bonnet supposed must always exist prior to fecundation, else there would be no heredity; and this led inevitably to emboîtement. The new evolution is the evolution of the species, presupposing hereditary change and variation, whereas the old evolution theory presupposed an unchanging species, based on the eternal conservation of its hereditary nature. In the new evolution, as we now see it and as Maupertuis saw it in the mid-eighteenth century, the ultimate event is abrupt change, mutation; as Bonnet saw it in his evolutionary scheme, the ultimate event was original creation. In the one conception, hereditary variation is ever new; in the other, it has been present from the beginning, requiring only the right conditions for its release and expansion. "Nature is assuredly admirable in the conservation of individuals;" said Bonnet, "but she is especially so in the conservation of species. . . . No changes, no alteration, perfect identity. Species maintain themselves victoriously over the elements, over time, over death, and the term of their duration is unknown." [26] From this position, as C. O. Whitman has ably set forth, Bonnet never wavered.

After wrestling with all the perplexing questions presented in Hydra; after accounting for sex as a means of diversifying the unity of the *beau physique,* and sexual reproduction as a device for expanding the germ and preserving regularity of specific form; after reconciling the existence of varieties with the permanence of species; after contending that a mule is a disguised horse and a hinny a disguised ass, and that the sterility of hybrids is to be regarded as fertility kept dormant by lack of adequate means to unfold; after reducing all heredity to likeness of original, contemporaneous, and independent creations, unfolding under similar conditions; after elaborating a scheme of "natural evolution" broad enough to take in any number of cosmic revolutions, and

---

[26] C. Bonnet, *Considérations sur les corps organisés* (1762), in *Oeuvres d'histoire naturelle et de philosophie,* Neuchatel, 1779, T. III, chap. VIII, sect. CXL, p. 90. Translation by C. O. Whitman, "Bonnet's Theory of Evolution— a System of Negation," in *Biological Lectures, Woods Holl,* 1894, p. 237.

CHARLES BONNET.
né à Genève le 13 Mars 1720.
FUTURI SPES VIRTUTEM ALIT.

provide for the ultimate perfection of every organism as an immortal being;—in a word, after setting "Ferien" to all creative activity, Bonnet resolutely undertook to devise a scheme that would keep the holiday repose forever inviolable. With a zeal never daunted, and an ingenuity seldom baffled, never defeated, he piled mountain upon mountain of negation, rolling Ossa on Olympus and Pelion upon Ossa, until the whole organic world seemed to be completely buried under a stupendous mass of negations, blending in one infinite negation—No CHANGE.[27]

It is then truly of interest to see how Bonnet treated the problem of variation in relation to species.

Whitman points out that the passage immediately preceding (sect. CXXXIX) the words of Bonnet quoted above, is commonly taken to indicate that Bonnet had an evolutionary trend to his thought. The passage is as follows:

One cannot doubt that the species which existed at the commencement of the world were no less numerous than those which exist today. The diversity and the multitude of combinations (*conjonctions*); perhaps also even the diversity of climates and nourishments have given rise to new species or to intermediate individuals. These individuals having mated in their turn the intergrades (*nuances*) have become multiplied, and in multiplying have become less detectable (*sensibles*). The *pear-tree* among *plants*, the *chicken* among *birds*, the *dog* among *quadrupeds* furnish us with striking examples of this truth. And in this respect we would not at all have to speak of the varieties which are to be observed among men derived originally from two individuals!

One is not surprised that Guyénot (op. cit., pp. 384-385) sees in this an expression of the multiplication of species through the action of the environment. But, as Whitman says, such passages must be read in context. Bonnet frequently indulges in seeming contradictions that, properly understood in his frame of thought, are no contradiction at all. Says Whitman: "... there is not the least inconsistency in it. Bonnet describes *appearances*, and he expects the reader to remember, what he has so often repeated, that appearances are deceptive. In many instances he uses the language of modern evolutionary doctrines without having any conception of them, and carrying always ideas that contradict them."

[27] Whitman, op. cit., pp. 239-240.

Nevertheless, so flat a contradiction as these two sections pose is very disconcerting. From the examples cited by Bonnet, it seems probable that he has here used the term "espèces" not in the Linnaean sense, but as signifying simply sorts or varieties. If so, he intended to say in sect. CXXXIX that various breeds and races, or "espèces," can be produced by the action of different conditions on the germ; and thereafter in section CXL to insist that the germ itself is unchanged, the real species unaltered. The germ of the barnyard chicken is the same as that of the wildfowl, the germ of the dog no different from that of the wolf, and all the races of mankind are brothers of the same species regardless of their environmentally induced distinctions.

Three aspects of the thought of Bonnet must be carefully distinguished. The over-riding conception is that of preformation. To this idea the theory of emboîtement was secondary, and indeed, as Bonnet found difficulties in applying it in every instance, it could conveniently be surrendered when necessary. What was essential to the major idea was not a practically infinite series of models of the adult each successively enclosed in turn within the preceding generation, but simply the existence of some organic preformation by means of which development could proceed under control. It was in particular the consideration of the formation of buds by the *Hydra*, the little "polype à bras" made so famous by his cousin Abraham Trembley, that forced Bonnet to these conclusions. The bud clearly can form anywhere on the body of the parent *Hydra* and does not contain parts within it, like the bud of a plant, all ready to expand and unfold. It is a mere bump, an excrescence on the parent's body. Yet, as it grows in size, it puts forth tentacles, develops a mouth in their center, becomes a fully formed *Hydra*. There must then be something, reasoned Bonnet, to make this happen, something present from the beginning—"certain particles which have been preorganized in such a way that a little polyp results from their development."[28] Since, too, the polyp can regenerate from any portion of its body when this is cut into small pieces, the particles must exist in every part of the whole. This consideration led Bonnet to the third concept of his theory, the concept of the germ.

[28] Bonnet, "Tableau des considérations," Art. xv, p. 68, *La Palingénésie philosophique*, 1769, in *Oeuvres*, T. VII.

" The germ need not be an actual miniature of the organism, but ... merely an 'original preformation' capable of producing the latter," was Thomas Henry Huxley's rendering of Bonnet's thought.[29] The actual statement reads:

> I add here that I understand in general by the word germ every preordination, every preformation of parts capable by itself of determining the existence of a Plant or of an Animal.[30]

These words, written first in the *Contemplation de la Nature* in 1764 and repeated without change in the *Palingénésie* (1769; 1783), stand as the ultimate, mature reflection of Bonnet's natural philosophy. Bonnet's maturing concept of the germ is recapitulated as follows in the later work:

> *I at first assumed, as a fundamental principle, that nothing was generated*; that everything was originally preformed. ...
> I postulated that all organized bodies derived their origin from a germ which contained *très en petit* the elements of all the organic parts.
> I conceived the elements of the germ as the *primordial foundation*, on which the nutritive molecules went to work to increase in every direction the dimensions of the parts.
> I pictured the germ as a network, the elements of which formed the meshes. The nutritive molecules, incorporating themselves into these meshes, tended to enlarge them. ...
> Strictly speaking, I said (Art. 83, pp. 47, 48), the elements [inorganic] do not form organic bodies; they only develop them, and this is accomplished by nutrition. The primitive organization of the germs determines the arrangement which the nourishing atoms must take in order to become parts of the organic whole.[31]

Thus conceived, Bonnet's ingenious scheme is far different than the ludicrous view commonly attributed to him. The essence of the difference between his *germ* and the reproductive particles postulated by Maupertuis lies simply in the distinction that in Bonnet's thinking they could not, as Maupertuis supposed, be disorganized and dependent upon forces of mysterious nature, con-

---

[29] See Whitman, " The Palingenesia and the Germ Doctrine," in *Biological Lectures, Woods Holl*, 1894, p. 260.
[30] Bonnet, ibid.
[31] Bonnet, *Palingénésie*, Part VII, chap. IV, p. 205. Translation of C. O. Whitman, ibid., p. 246.

jectured chemical affinities resembling "what in us we term desire, aversion, memory," to bring each to its proper place in the body. The wholeness of the organism was something too intrinsic to be built up of inorganic matter. The organic fabric must exist, thought Bonnet, from the beginning. Hence preformation.

Hereditary differences—variations—thus existed for Bonnet as for other observers of Nature, but they were the consequence of the differences between germs, primordial, created once for all in the beginning, destined successively to enlarge and transform through varieties of nutritive condition, capable of metamorphosis to seemingly totally new forms under altered conditions, but in death reduced again to the undying germ. Even progress was envisaged.

> Perhaps there will be a continued progress, more or less slow, of all species towards a higher perfection, such that all degrees of the scale will be continually changing in a constant and determined order: I mean the mutability of each degree will always have its reason in the degree that shall have immediately preceded it.[32]

Now this passage has been taken by some writers to indicate that Bonnet too had conceived of evolution in the form of a transformation of species. But the entire context of his work shows this to be untrue. The germs, he believed, existed from the beginning of Creation. If there was progress, it took place *within* each species because the Creator had made the germs of each succeeding generation superior to those of the generation before. Plants might rise to the state of animals, oysters and polyps to that of birds and quadrupeds, monkeys to that of men, and men to that of angels; but for each it would be a metamorphosis, not a transmutation of species. Each germ had already within it, from the beginning, its potentiality to become something different, to take on a higher form, like the "evolution" of the butterfly from egg into caterpillar, caterpillar into chrysalis, and chrysalis into winged beauty.

Bonnet, of course, was profoundly right in seeing that there must be some organic entity persistent throughout all growth and change of form, some specific pattern or primordial foundation existing not alone at the origination of the individual but descending through the filiated generations. Today we call it the genotype. But he was just as profoundly in error in supposing that the genotype must

---

[32] Bonnet, *Palingénésie*, Pt. III, chap. III, p. 149-150. Cf. Whitman, ibid., pp. 255-257.

be static, or that it could not be generated in each new individual through the contributions of both parents. Having as a young man become famous for his discovery of parthenogenesis in plant-lice, he might be excused for naturally holding to the doctrine of ovism; but how could he blind himself to the clear implications of the biparental transmission of polydactyly in human families, or the presence in species hybrids of the features of both parent species?

In the *Considérations sur les corps organisés* (Part II, Chap. VIII, sect. CCCLV) Bonnet undertakes to consider the inheritance of 'sex-digitisme" in the famous Kelleia family, and in the final revision of this work (1779) he appended a note regarding Maupertuis' study of the Ruhe family of Berlin, of which he had been ignorant when the work was first written (1762). He adds comments regarding another report of inherited polydactyly in the *Journal de Physique*, November 1774, and concludes: "It seems, then, that one must recognize that polydactyly is transmitted or appears to be transmitted by the one and the other sex." [33] But does this conclusion force Bonnet to admit biparental heredity or even the possible inheritance of a monstrous germ? Not in the least. Just as earlier he had supposed that in the cross between a mare and a jackass the seminal liquor of the latter modifies the growth of the germ (which remains that of a true horse) so that the ears become longer, the natural fertility is impaired, and the modified larynx gives forth a bray, so now he suggests that the seminal fluid of a six-fingered male modifies the growth of the germ derived from a normal five-fingered female, but in no way changes the germ itself. As for the transmission through the mother, if abnormal conditions introduced by the semen of the male can modify the growth of the germ, why could not abnormal conditions in the intra-uterine development of the foetus convert a normal germ into a polydactylous one? Monsters are the accidents of nature. They arise through interference with the normal "evolution" of the germ.

No new generation! How apparent the link here between the faith in preformation and the rejection of spontaneous generation! For it was new generation of every sort, the origin of any living organism from separate organic particles rather than from a prior germ containing the essence of the whole, that was repugnant. The

---

[33] Bonnet, *Considérations sur les corps organisés*, Pt. II, chap. VIII, sect. CCCLV, p. 534 fn.

spontaneous generation of bacteria claimed by John Turberville Needham, the organic particles of Buffon, and the hereditary particles of Maupertuis were just as odious as the epigenesis of Caspar Friedrich Wolff. These views Bonnet shared with his friend and mutual admirer, the Abbé Spallanzani.

Lazaro Spallanzani was probably the greatest and most versatile biological experimentalist of the eighteenth century. His earliest studies, conducted at Modena, made him quickly famous. Needham, who had published a series of experimental studies on bacteria, was a close friend of Buffon, and an ardent believer in the reality of spontaneous generation, a view demanded by Buffon's theory of organic particles supposed to be disseminated throughout nature and to constitute a source of living organisms. In these experiments of 1745-50, Needham had placed meat broth in vessels corked and then sealed with mastic, after which he had set them by the fire and heated them to a temperature he supposed sufficient to kill all life in the flasks. After setting them aside for a few days, he invariably found the broth swarming with bacteria. Spallanzani (1765-76) repeated the experiments, but with greater care to assume perfect sealing. He used glass flasks with slender necks which could be fused in a flame, and thus sealed hermetically after the nutrient broth had been introduced. Thereafter the flasks were immersed in boiling water for three-quarters of an hour. (Needham had failed to record either the durations or the temperatures of the heat treatments he applied.) Spallanzani's flasks of broth remained clear and free of bacteria. When Needham objected that such severe heating had destroyed the capacity of the infusions to support life, Spallanzani triumphantly showed that when air was readmitted to the sealed vessels they were soon teeming with microbes. Just a century after the work of his compatriot Redi, Spallanzani had, at least temporarily, banished spontaneous generation from biology.

Spallanzani's conviction in the correctness of the preformation theory grew steadily during the succeeding years of experiment and communication with Bonnet. He turned to studies of the regenerative powers and reproduction of vertebrates, especially frogs and salamanders. Undismayed by the failure of so eminent an investigator as Réaumur, he succeeded in determining that the male frog emits seminal fluid at the very time when the female with which he is coupled lays her eggs. This he did by making for the male frogs little pantalettes of waxed taffeta, with suspenders to go over the shoulders. The pantalettes in no way interfered with the clasping of the female by the male, and when she laid her eggs Spallanzani was able to collect a glistening drop or two of the semen in the recesses of the garments. Not only did he then show that such eggs uniformly failed to develop, unlike those laid by females mated normally, but he also demonstrated that when the batch of eggs was divided and the diluted semen was added to one batch and not to the other, the eggs of the inseminated batch would develop normally into tadpoles, while the others failed to develop at all.

Notwithstanding this first successful artificial fertilization of animal eggs, Spallanzani failed to draw the correct conclusions. He had already determined, he thought, the existence of the germ in the frog's egg; and he held, in agreement with the views of Haller and Vallisneri, that the semen enters the egg by tiny pores and gently stimulates the heart of the foetus and sets it to beating. He tried many artificial agents—acids, poisons, blood, urine—nothing could replace the semen in this mysterious power. He heated the semen and found that it lost its fertilizing power after 2 minutes at 35°. He filtered it through fine cloth or blotting-paper. The power to fertilize was lacking in the filtrate; but pure water in which the filter was rinsed acquired the power to fertilize the eggs. Thus Spallanzani did every imaginable experiment needed to show that the spermatozoa were the active, fertilizing agent of the semen and that the fluid portion itself was impotent. Yet he failed to draw any such conclusion. No generation! The germ must be in the egg. The little animalcules of the semen, he knew them well; but they were neither the mediate nor immediate authors of generation. "Those of the man are not little men, those of the horse are not little colts." [34]

He tried one more experiment. Let a drop of semen evaporate, and watch it through the microscope as the spermatozoa recede from the edge of the drop and congregate in the center. Now immerse a needle in the edge of the drop and withdraw a droplet seemingly devoid of spermatozoa and apply it to an unfertilized egg. Success! it develops—the animalcules are after all not the source of the fertilizing power of the semen. Thus he performed what

[34] Quoted by J. Rostand, *La formation de l'être*, Libr. Hachette, Paris, 1930, chap. XII, p. 148. A valuable and graphic account of Spallanzani's work on fertilization. See also Rostand, *Esquisse d'une histoire de la biologie*, chap. VII; and Guyénot, op. cit., pp. 283-286.

seemed to be the crucial experiment, and thus he who had scorned Needham's subterfuges trapped himself in the delusion fostered by his own preconceptions.

For the time, preformation seemed firmly and permanently established. Wolff had been set down by the magisterial von Haller, who brushed away the young man's observations as those of a novice. Bonnet had met every argument of the epigenesists with subtle, untestable reasoning; and Maupertuis was dead. Spallanzani had overthrown the claims of Needham and Buffon regarding the spontaneous generation of bacteria and infusoria. The doctrine of the primordial, preorganized germs, created in the beginning, enduring unchanged until the end, reigned supreme. Perhaps in far-off Uppsala the old man whose brilliant youthful achievement had created a new biology pondered with failing mind the denial of his transformist views through the experiments of Adanson and Koelreuter, and thought: *nullae species novae*. Perhaps it was so after all. It would take a new generation to find out.

It would indeed take a new generation, the generation of Lamarck, and after that yet another, the generation of von Baer, before the transformation of species could be seriously reconsidered. And still beyond that time, another century would lapse ere the germ of truth embodied alike in the preformation theory and the epigenetic view of development would become reconciled in a higher synthesis, the genetic theory of development. And it would take that long, too, for the genetic analysis of variation to make clear the problem of hybrid sterility, to establish genetic mutation as the source of hereditary variation, and fully to characterize the species in dynamic, evolutionary terms.

SEVEN

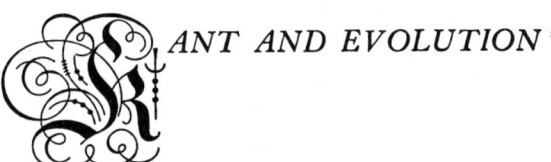ANT AND EVOLUTION *

ARTHUR O. LOVEJOY

It has long been one of the generally accepted legends of the history of science that the author of the *Kritik der reinen Vernunft* was also a pioneer of evolutionism. In the anthropological essays of the Koenigsberger, for example—we are assured by the writer of a German treatise on Kant's philosophy of nature—[1] " we already find the most essential conceptions of the modern theory of descent indicated, at least in germ—and, indeed, in a way that marks Kant out as a direct precursor of Darwin." The same expositor says:

> Throughout these writings the idea of evolution plays everywhere the same role as in contemporary science. . . . The series of organisms is for Kant in a constant flux, in which the seemingly so stable differentiae of genera and species have in reality only a relative and subsidiary significance.

In a famous passage of the *Kritik der Urteilskraft*, says another writer, " the present-day doctrine of descent is clearly expressed in its fundamental features." [2] Haeckel, who is in the main followed

---

* Reprinted in part from an article in *Popular Science Monthly*, January, 1911; it has been extensively revised, and Section 4 has been added.

[1] Drews, *Kants Naturphilosophie*, 1894, pp. 44, 48.

[2] Schultze, *Kant und Darwin*, 1875, p. 217. Schultze's monograph, perhaps the earliest, and hitherto the most comprehensive, on the subject, seems to be responsible for much of the error into which subsequent writers have fallen. It consists, indeed, chiefly of reprints of the greater part of each of the writings in which Kant approaches the topic in question; but it is accompanied by a commentary and notes in which Schultze gives a highly misleading impression of Kant's actual utterances.

by Osborn, goes even farther in his ascription of Darwinian and "monistic" ideas to Kant's earlier works, though he thinks that in later life Kant fell from grace. Haeckel says:

> In various works of Kant, especially in those written in his earlier years (between 1755 and 1775) are scattered a number of very important passages which would justify our placing him by the side of Lamarck and Goethe as the principal and most interesting of Darwin's precursors. . . . He maintains the derivation of the various organisms from common primary forms, . . . and was the first to discover the principle of the 'struggle for existence' and the theory of selection. For these reasons we should unconditionally have to assign a place of honor in the history of the theory of development to our mighty Koenigsberg philosopher, were it not that, unfortunately, these remarkable monistic ideas of young Kant were at a subsequent period wholly suppressed by the overwhelming influence of the dualistic, Christian conception of the universe.[3]

Yet even at the last, though Kant's nature-philosophy became less "monistic," Haeckel finds his biology scarcely less evolutionistic. In the *Critique of Judgment* Kant, according to Haeckel, still "asserts the necessity of a genealogical conception of the series of organisms, if we at all wish to understand it scientifically." In the supposition of a marked "change of view from Kant's earlier to his later years" with respect to the applicability of the principles of natural causation in the realm of the organic, Osborn concurs with Haeckel. The writer of the historical article in the volume issued by English biologists in commemoration of the Darwin semi-centenary, says that Kant may be "best regarded as the culmination of the evolutionist philosophers" of the eighteenth century.[4] Rádl in his *Geschichte der biologischen Theorien* apparently accepted Schultze's description of Kant as a *Vorläufer Darwins*.[5] And during the last half-century the same legend seems to have passed for one of the established facts about the history of evolutionism. The late Ernst Cassirer, a distinguished member of the Marburg school of Kantian idealists, wrote in his *Rousseau, Kant, Goethe* (1945), that Kant "clearly envisaged the task and the goal of a general theory of evolution," and cited as evidence of this a passage, to be quoted

[3] *History of Creation*, Lankester's translation, 1892, p. 103. Cf. Osborn, *From the Greeks to Darwin*, 1894, pp. 98-99.
[4] J. Arthur Thomson in *Darwin and Modern Science*, p. 6.
[5] Rádl, op. cit., 1905, I, pp. 255-6.

later in this paper, in which Kant declared that, although "*a priori*, in the judgment of Reason alone there is no contradiction here, . . . experience gives no example of it; according to experience, all generation that we know is *generatio homonyma*. This is not merely *univoca* in contrast with a generation out of unorganized matter, but in the organization the product is of the same kind as that which produced it; and *generatio heteronyma* [production from a different species], so far as our empirical knowledge of nature extends, is nowhere found."[6] Even the author of a recent (1955) and generally admirable brief exposition of Kant's philosophy writes that . . . "Kant does not reject the main ideas of the theory later developed by Darwin."[7]

These accounts of Kant's historic position in relation to transformism would be interesting, and important for the historian of modern science, if they were correct. But they are all very far from correct.[8] Kant wrote for the most part at a time when the conception of organic evolution had been made familiar to naturalists by two of the most celebrated and most influential men of science of the time, Maupertuis and Buffon. He was himself in his later life especially interested in two distinct scientific problems, both of which made a consideration of the hypothesis of the transformation of species inevitable, and an acceptance of it natural. He therefore frequently refers to it. But in no case does he unequivocally express belief in it, and in most cases he vehemently rejects it. Least of all can he be correctly called a "forerunner of Darwin." For Darwin was the originator, not of the theory of organic evolution, which by 1859 was an old, though controverted, doctrine, but of the theory of natural selection as an explanation of the origin of new species. This was what Kant would have termed a purely "mechanistic" in contrast with a "teleological" explanation. And he was most of all hostile to the supposition that any of the phenomena of organic life can be completely explained mechanistically.

[6] The passage referred to by Cassirer is in *K. d. U.*, Sect. 80. In sect. 81, to which Cassirer does not refer, it will presently be seen that Kant asserts that the Creator "would, in the original products of his wisdom, have supplied merely the predispositions by which an organic being produces *another of like kind and the species perpetually maintains itself*."
[7] S. Körner, *Kant*, in Pelican Philosophy Series, p. 210.
[8] I had myself, before coming to close quarters with the subject, fallen into the prevalent error of classifying Kant among the early evolutionists. (*Popular Science Monthly*, November, 1909). Yet for the past seventy years a substantially correct account of the matter has been accessible in a brief article by J. Brock, *Biologisches Centralblatt*, Bd. VIII, 1888-9, pp. 641-8.

It must, however, be remembered that in so far as Kant's denial of the possibility of the transformation of species rested upon the "Buffonian rule" (that, since the offspring of organisms of different species, e. g. mules, are incapable of reproducing their kind, they cannot be descended from any common ancestors) he was in accord not only with theological orthodoxy but also with the biological orthodoxy of his time. For the fact of the infertility of hybrids continued to be generally regarded as a decisive argument against the theory of organic evolution until the 1850's. Kant, therefore, cannot be much blamed for accepting this current assumption and using it as a weapon against the transformists of the eighteenth century. Yet it was a completely invalid argument, and there were reasons why he, at least, should have seen that it was invalid. There are few more curious aberrations in the history of scientific opinions than the acceptance of it by most naturalists for more than a century as if it were axiomatic. For it was not self-evident, and there were no rational grounds for supposing it probable, that two species of organisms which, when mated, can engender only sterile hybrids, can themselves have had no ancestors in common; and we now know that it is not true in fact. It has been generally agreed among paleontologists and zoologists since the 1870's that the ass and the horse, parents of our useful but sterile mule, are both descended, through divergent lines, from the extinct fossil species *Eohippus*. This fact could not, of course, have been known to Kant; but he should have seen that his own doctrine in the *Kritik der reinen Vernunft* implied that the proposition "parents of sterile hybrids had no ancestors in common" was incapable of proof. For it was obviously not an "analytic" proposition, like "all equiangular triangles are triangles"; it was (in his terminology) a "synthetic" proposition, but *not*, a "synthetic proposition *a priori*." It was an empirical proposition which, if it were true, could, on Kantian principles, be known to be so only *a posteriori*, i.e., by observation of instances—and, since it was a general proposition, of all instances—of the generation of the species in question (i.e., of those which, in mating, produce sterile hybrids.) But it was also a negative proposition about all such species: none of them have had any common ancestors. This negative general proposition, even if it happened to be true (which it was not), could not possibly have been proved by any empirical evidence available in the eighteenth century; such proof would have required a complete genealogy, a "family tree," of all extant species of this kind, showing that the same forebears could be discovered nowhere in either of the two lines of descent.

We shall now review in approximate chronological order Kant's writings pertinent to the subject.

1. *The Review of Moscati on Man's Upright Posture.* In 1771 Kant wrote a review of a disquisition by an Italian anatomist, Moscati,[9] on the difference between the structure of man and that of the lower animals. Moscati's principal contention was that the upright posture is not "natural" to man, and was not his primitive attitude. Upon this Kant remarks in part as follows:

> Here we have once more the natural man upon all fours—an acute anatomist having traced him back to that condition. Dr. Moscati shows that the upright gait of man is forced and contrary to nature, and that his structure is such that this position, when it has become necessary and habitual, entails upon man various disorders and diseases—clear proof enough that he has been led by reason and imitation to depart from his primitive animal posture. In his inner constitution man is not formed otherwise than as are all the quadrupeds.... Paradoxical as this conclusion of our Italian physician may seem, yet in the hands of so acute and philosophical an anatomist it attains to almost complete certainty (*erhält er beinahe eine völlige Gewissheit*). We see from this that Nature's first care was for the preservation of man as an animal, in his own interest and that of the species; and for this purpose the posture which was best adapted to his internal structure, to the position of the foetus, and to protection against dangers, was the four-footed one; but we see also that there lay in man a germ of reason, through the development of which he was to become fitted for society. He consequently assumed the posture most suitable to this, that of a biped. By virtue of this, man, on the one hand infinitely surpasses the animals; but, on the other hand, he is obliged to endure certain disorders that afflict him in consequence of his having raised his head so proudly above his former comrades.

Here, then, Kant readily accepts the doctrine that man was originally a four-footed animal, which, *pari passu* with its unique development of rationality and of the social instincts, assumed the

---

[9] Moscati was professor of anatomy at the University of Pavia. His book appeared in 1770; a German translation by Beckmann, professor in Göttingen, was published in 1771.

upright attitude. His promptness in making the views of Moscati his own certainly indicates a general predisposition to evolutionary ways of thinking; and, if we had no other expressions of Kant's dealing with the subject more directly, it would be not unnatural to construe this assertion of the descent of civilized man from quadrupedal ancestors as equivalent to an assertion of the mutability of species. Yet the latter doctrine, it must be noted, is nowhere expressed or directly implied in the review of Moscati; and it will presently become clear that Kant would not have regarded it as a legitimate inference from any of his admissions about the earlier condition of humanity. From the time of publication of this review to the end of his life Kant seems to have remained what may be called an anthropological evolutionist; but he deliberately refused to make the transition from this position to a general biological evolutionism.

2. *The Two Essays on the Conception of "Race," 1775, 1785.* The review of Moscati was the earliest indication among Kant's writings of a growing interest in a group of scientific problems which always thereafter much occupied his attention: namely, the genetic problems of physical anthropology. The beginnings of that science, in its systematic form, are usually credited to the treatise of Blumenbach, *De generis humani variatione nativa*, 1775. Blumenbach, says the historian [10] of eighteenth century anthropology, "derived his zoological facts chiefly from Buffon. His philosophy, and in particular his fundamental conception of man's place in nature, were founded on the system of Leibniz. The opening sections of his book at once show his principal preoccupations in the inquiry— viz., to establish the limits, on the one hand, between man and the animals, and, on the other, between the different races of men. These two remained the chief themes of anthropology throughout the succeeding period." It was to the second of these themes that Kant especially addressed himself. His first discussion of it appeared in the same year as Blumenbach's treatise. In the "preliminary announcement" to his *Lectures on Physical Geography*, delivered in the summer semester of 1775, Kant took for his topic "The Different Races of Men"; [11] he reverted to the subject in an article

[10] Günther, *Die Wissenschaft vom Menschen im 18ten Jahrhundert*, p. 287.
[11] "Von den verschiedenen Racen der Menschen." This writing will here be referred to as the "Physical Geography." It is to be found in Hartenstein's edition, 1867, II, p. 433.

in the *Berliner Monatsschrift* for November, 1785, entitled "Elucidation of the Conception of a Race of Men."[12] These two essays do not significantly differ in doctrine, and they may most conveniently be dealt with here as slightly variant expressions of the same arguments and conclusions. They are among the most important documents for the determination of Kant's position with respect to the theory of evolution.

Kant derived not only most of his zoological facts, but also some of his ideas of scientific method, from Buffon. The latter, like Maupertuis, had ridiculed the "systems" and "methods" of the great systematists, Linnaeus and Tournefort, and had looked with a good deal of contempt upon their absorption in purely descriptive and classificatory science. Schemes of classification were convenient, no doubt, and accurate description essential; but there was a higher stage of scientific inquiry to which these were merely vestibulary. Buffon wrote:

> We ought to try to rise to something greater and still more worthy of occupying us—that is to say, to *combine* observations, to generalize the facts, to link them together by the force of analogy, and to endeavor to arrive at that high degree of knowledge in which one can recognize particular effects as dependent upon more general effects, can compare nature with herself in her larger processes.[13]

This spirit Kant had in some degree caught; and in the "Physical Geography" he proposes a modification in the nomenclature of the sciences which should express the distinction between two types of scientific inquiry. He observes:

> We are accustomed to use the words "*Naturbeschreibung*" (description of nature) and "*Naturgeschichte*"[14] (natural history) as synonymous. But it is manifest that the knowledge of the things of nature as they now are still leaves to be desired a knowledge of what they previously have been, and of the changes through which they have passed in order to arrive at their present condition. A "history of nature"—such as is still almost com-

---

[12] "Bestimmung des Begriffs einer Menschenrace," here referred to as the "Conception of Race." V. Hartenstein edition, IV, p. 215.

[13] "Discours de la manière d'étudier et de traiter l'histoire naturelle." In *Oeuvres*, Lanessan ed., I, p. 6.

[14] Later (in the "Use of Teleological Principles") Kant proposed to express this distinction by the words "physiography" and "physiogony."

pletely lacking—would make known to us the alterations of the form of the earth and those which the terrestrial creatures (plants and animals) have undergone in the course of their natural migrations, and their consequent divergences from the primitive type of their ancestral species (*Stammgattung*). Such a science would probably reduce a great number of seemingly distinct species (*Arten*) to mere races of a single genus (*Gattung*), and would transform the now current artificial system (*Schulsystem*) of nature-description into a physical system for the understanding.

In this, manifestly, Kant shows a lively sense of the nature and importance of genetic problems in the investigations of the naturalist. It is true that he somewhat naïvely makes the distinction between the genetic and the descriptive equivalent merely to the distinction between past and present. It need hardly be said that genetic inquiries in science are not necessarily purely historical or archeological inquiries, since phenomena of genesis may be recurrent phenomena, taking place in accordance with the same laws in past or present. But, though he blurred the idea somewhat, it remains true that, in his contrast between two types of scientific research, Kant exhibited his inclination to what, in the vaguer sense, may properly be described as an evolutionary habit of mind. It still remains, however, to determine just how far this carried him, when he came to the consideration of definite problems.

His problem of predilection, as I have said, was that of the nature of a "race," the relations of different races, and the causes of their diversity in physical characters. And this made necessary, at the very outset, a consideration of the nature of a "species." Here, once more, Kant follows Buffon: " Animals, however different they may be in form, belong to the same physical species if, when mated with one another, they produce fertile offspring."

> This Buffonian rule gives a definition of natural species as such (*die Definition einer Naturgattung der Tiere überhaupt*), in contrast with all artificial species (*Schulgattungen*). The artificial classification deals with *classes*, which are grouped together upon the basis of similarity, the natural classification deals with lines of descent, grouping animals according to blood-kinship. The one provides an artificial scheme to aid the memory, the other a natural system for the understanding. The purpose of the former is merely to bring animals under labels, that of the latter is to bring them under laws.

These references to *Naturgattungen*, determined by the criterion of fertility of offspring, are themselves hardly in the language of transformism. Yet one who employed such language might still regard these "true species" as eventual results of divergent descent from common ancestors. But when we examine Kant's way of further defining these species, we find that his notion of them expressly precludes the possibility of any transformation of one into another through descent. By the Buffonian test, he says:

All human beings belong to one and the same natural species, since in mating they always beget fertile offspring, however, dissimilar the parents may be in appearance. For this unity of natural species there can be but one natural cause, viz., that all men belong to a single stock (*Stamm*), from which they have originated or at least could have originated. In the former case [i.e., of actual descent from common ancestors], they belong not only to one and the same species, but also to one family; in the latter case they would be similar to one another but not related, and it would be necessary to assume a number of separate local creations: an opinion which multiplies causes beyond necessity.[15]

This argument, by which Kant reasons that all men are of one *Stamm*, directly implies that men and other animals are *not* of one *Stamm*, i. e., are not related through any lines of natural descent. For he makes identity of species synonymous with community of descent, and diversity of species synonymous with separateness of descent. In other words, his manner of distinguishing a species from a race rests upon wholly anti-evolutionary presuppositions.

Within the limits of a species, however, Kant holds that very considerable modifications of physical character may be brought about in the course of successive generations. Now (apart from individual variations not transmitted to offspring), there seem to Kant to be two significantly different types of heritable peculiarities: those which are *invariably* inherited, and those which are only *alter-*

[15] The same ideas are perhaps still more clearly expressed in the article "On the Use of Teleological Principles in Philosophy," 1788: "There could be no more certain test of diversity of stock (des ursprünglichen Stammes) than the inability of two different hereditary branches of mankind to engender fertile offspring. But where the generation of such offspring is possible, the utmost diversity of external appearance is no obstacle to regarding the parents as having a common descent. For if they can, in spite of this diversity, produce offspring that exhibit the characters of both parents, then they may be classified as belonging to two *races* of a single stock, which originally had latent within itself the characters that were to be developed in each separately."

natively inherited. Thus the colors of a negro and a white who marry are both manifested in the offspring; children of such marriages are always mulattoes. But the complexions of the children of a dark man and a blonde woman are not necessarily a compromise between the complexions of their parents. Some or all of the children may resemble one parent only, and show (with respect to any given character) no marks of their descent from the other. By means of this distinction Kant differentiates a "race" from a "variety." Those members of a single species which also possess in common characters of the invariably hereditary sort belong to the same race; those which possess in common (and, so long as they mate with their own like, transmit to their offspring) characters that, upon cross-breeding with other types, are only alternatively hereditary, constitute only "varieties."

These definitions of "species" and "race," it is true, involve—as Kant recognizes—some revision of the classifications of the systematists.

> Originally, when only similarity and dissimilarity were taken into consideration, it was customary to group classes of creatures under *genera* (*Gattungen*). But if it is their descent we are considering, it is necessary to ask whether these classes are species (*Arten*) or only races. The wolf, the fox, the jackal, the hyena and the domesticated dog, are so many classes of quadrupeds. If one assumes that each of them has a special descent (*Abstammung*), they constitute so many species; if one grants that they may have sprung from a single stock, they are simply races of that stock. In "natural history" (*Naturgeschichte*), which has to do only with generation and descent, the words *Art* and *Gattung* mean the same;[16] only in "nature-description," where it is merely a question of the comparison of characters, does a distinction between them find place. What in the latter is called a species must in the former often be designated as a race.

Kant's elaboration of an ethnological scheme upon the basis of these definitions does not here concern us. But it is worth noting that he believes that the only character which is "invariably inherited" from both parents—and therefore the only mark of a true or "natural" race—is skin-color; and that, using this criterion, he

---

[16] It is for this reason that, in translating Kant's expositions of his own doctrines, I have, so far as possible, rendered both *Art* and *Gattung* by "species." The citation is from the "Conception of Race," Sect. 6, *n*.

finds that there are just four races of men, the white, the negro, the Mongolian or "hunnish," and the Hindu. From these four originals Kant was prepared to explain all the hereditary shadings of the various peoples of the earth as the results of diverse hybridizations. The question of greatest interest of all, from the standpoint of biological theory, still remained to be asked. Within the limits of a "natural species," as we have seen, Kant recognized that profound modifications of physical characters took place, and became permanent and transmissible through heredity. Thus, he thinks it at least a probable conjecture that the original type of man was white. But from white ancestors black and yellow and brown races have been developed. How did this come about? What, in Kant's words, are "the immediate causes of the origination of these different races"? He has his own entirely confident answer to the question. A natural answer for a late eighteenth-century biologist would have been to say that these differentiated racial characters are the results of environmental modifications of individuals, which gradually have become hereditary. But such an explanation Kant emphatically rejects. It would hardly do to call him an eighteenth-century Weismannist; but he was (though not without serious but unrecognized inconsistencies) a vigorous opponent of the supposition that acquired characters can be inherited, and an unqualified partisan of the doctrine of the continuity and unmodifiability of the germ-plasm. His reasons for taking this position betray once more his entire inability to conceive of the transformation of "real" species into other species.

There are current, he admits,[17] many, though poorly authenticated, stories of cases in which acquired characters have been inherited: tales of the "influence of the imagination of pregnant women" upon the foetus; of "the plucking out of the beard of entire peoples, and of the docking of the tails of English horses, by which nature was compelled to eliminate from the processes of reproduction in these organisms a product for which those processes were originally organized"; accounts of "the artificial flattening of the noses of new-born infants, which peculiarity nature is supposed finally to have taken up into the reproductive faculty." Kant rightly regards all such stories with a sceptical eye; but his theoretical reasons for doing so are significant. These accounts are to

[17] "Conception of Race," Sect. 5, and *Anmerkung*.

be rejected because they conflict with a general principle or presumption of science which must be adhered to at any cost, namely:

> that throughout organic nature, amid all changes of individual creatures, the species maintain themselves unaltered (*die Species derselben sich unverändert erhalten*)—according to the formula of the schools, *quaelibet natura est conservatrix sui*. Now it is clear that if some magical power of the imagination, or the artifice of men, were capable of modifying in the bodies of animals the reproductive faculty itself, of transforming Nature's original model or of making additions to it, which changes should then become permanent in subsequent generations, we should no longer know from what original Nature had begun, nor how far the alteration of that original may proceed, nor—since man's imagination knows no bounds—into what grotesqueries of form species might eventually be transmogrified (*in welche Fratzengestalt die Gattungen und Arten zuletzt noch verwildern dürften*). In view of this consideration, I for my part adopt it as a fundamental principle to recognize no power in the imagination to meddle with the reproductive work of Nature, and no possibility that men, through external, artificial modifications, should effect changes in the ancient original of a species in any such way as to implant those changes in the reproductive process and make them hereditary. For if I admit a single instance of this sort, it is as if I admitted the truth of a single ghost-story or tale of magic. The boundaries of reason are then once for all broken through, and errors rush in by thousands through that opening. There is, meanwhile, no danger that, in adopting this conclusion, I may take a position of blind or stubborn incredulity towards real facts of experience. For all these romantic (*abenteuerlich*) occurrences have without exception one peculiarity, namely, that they can not be subjected to experiment, but are supposed to be proved merely by casual observations. But whatever, though capable, indeed, of experimental testing, offers no experimental evidence, or employs all sorts of excuses to avoid such a test, is mere fiction and illusion.[18]

---

[18] The "Physical Geography" is equally emphatic in repudiating both inheritance of acquired characters and mutation of species: "External things may, indeed, provide the occasions, but they can not be the efficient causes, of the appearance of characters that are necessarily transmitted and inherited. As little as chance or physico-mechanical causes can bring an organic body into existence, just so little can they imprint anything upon the reproductive faculty, that is, produce any effect that is itself reproduced, either as a special form or as a relation of the parts. Air, light and nutrition can modify the growth of an animal body, but they can not furnish this change with a power

Nothing could better exhibit Kant's characteristic state of mind on biological questions than this passage. There are occasional bits of sound sense in it and of discriminating judgment about scientific method; and there is a power of at least seeing where the significant problems lie. Yet, though he had come under the influence of evolutionistic conceptions, and is in these very writings endeavoring to apply genetic methods to certain biological inquiries, he recoils in horror before the idea of admitting that real species are capable of transformation. It is primarily in the name of a pseudo-axiom of Scholasticism that he pronounces for the fixity of species! But in reality, as his expressions show, it was because of certain temperamental peculiarities of his mind—a mind with a deep scholastic strain of its own, one that could not quite endure the notion of a nature all fluent and promiscuous and confused, in which series of organisms are to an indefinite degree capable of losing one set of characters and assuming another set. He craved, after all, a universe sharply categorized and classified and tied up in orderly parcels. And thus, though he had learned from the newer scientific tendencies of his time that the business of science is with processes, and especially with genetic processes, this scholastic side of his mind prevented him from making any thorough application of the principle to biology. He was prepared to go a considerable distance upon the path of evolutionism—but to admit that organisms (always to Kant, because of their "teleological" character, forming in nature a realm apart) were so far plastic that the very archetypal traits of species could, under the play of ordinary, environmental agencies, be altered past recognition—that was too much!

Meanwhile, it must be remembered that he was already committed to the admission of a large measure of modification *within* the species. But if it were so incredible a thing that the "original form" of a species should be radically altered, why was it not equally incredible that black men should be descendants of white men? Why did not the arguments against the transformation of one species into another species apply equally to the transformation of one race into another race? Why should one who supposed—as Kant supposed—that the wolf or hyena may have developed into

of reproducing itself after its original causes are no longer operative. . . . For it is not possible that anything should so penetrate into the reproductive faculty as to be capable of gradually removing the creature from its original determination and bringing about a real and self-perpetuating departure from the specific type (*Ausartung*)."

the extraordinarily diversified breeds of our domestic dogs, have found it an intolerable paradox to suppose that the horse may have developed into the donkey, or both from a common ancestor? To such questions as these Kant's theory concerning the causes of the origination of races was called upon to provide an answer. The answer has an appearance of great simplicity: Kant merely said that in reality races had no characters which were not present, *but latent*, in their species from the start. In other words, he escapes the difficulties of his position by the easy artifice of a hypothesis of preformations, borrowed from Leibniz. Nothing has been added to or taken from the germ-plasm of the species "man" since the beginning; the reproductive faculty merely contained in itself always certain alternative potencies—especially with respect to the production of skin-color—one or another of which was called into play in accordance with variations of external circumstances.

> Any character that was to be transmissible (*was sich fortpflanzen soll*) must have already lain beforehand in the reproductive faculty, predetermined to develop at the proper occasion, in conformity with the circumstances amid which the animal might find itself and in which it would be obliged to maintain itself.... This precaution of Nature to equip all her creatures for all kinds of future conditions by means of hidden inner predispositions, by the help of which they may maintain themselves and be adapted to diversities of climate or soil, is truly marvelous. It gives rise, in the course of the migration and change of environment of animals and plants, to what seem to be new species; but these are nothing more than races of the same species, the germs and natural predispositions for which (*deren Keime und natürliche Anlagen*) have developed themselves in different ways as occasion arose in the course of long ages.[19]

Kant's conception of the "grounds" for the existence of these *Anlagen* is manifestly teleological in the most *naive* way; the species was fitted out beforehand with distinct elements in its germ-plasm in order to prepare its later representatives for specific contingencies that had not yet arisen, and in some cases never would arise. This idea Kant elaborates in detail in the case of the skin-color of the negro; the passage is so delightful a combination of teleological "explanation" and phlogistic chemistry [20] that it deserves to be quoted:

---

[19] Cited from the "Physical Geography."
[20] Kant was, of course, by no means abreast of the best chemistry of his

The presence of purposiveness in an organism is the general ground from which we infer an original preparation in the nature of a living being, having this [purpose] in view, and—if the purpose is only later fulfilled—infer the existence of duly furnished germs. Now, this purposiveness can be in no race so clearly shown as in the negro. . . . It is already known that human blood turns black simply through becoming overcharged with phlogiston (as may be seen from the under side of a cake of blood). Now the strong odor of the negro, which cannot be removed by any degree of cleanliness, already leads us to surmise that his skin eliminates a great deal of phlogiston from the blood, and that Nature must have so organized his skin that it is capable, in much greater degree than is ours, of dephlogisticating the blood —this being, with us, accomplished chiefly by the lungs. But the true negroes live in lands where the air, because of the thickness of the trees and the marshiness of the surroundings, is so heavily phlogisticated that, according to Lind's account, English sailors run the risk of death from this cause when they ascend the river Gambia even for a single day, for the purpose of procuring meat. It was, therefore, a very wise arrangement of Nature so to organize the skin of the negroes that their blood, even if the lungs do not sufficiently eliminate phlogiston, is yet far more thoroughly dephlogisticated than ours. Their blood must therefore deposit a great deal of phlogiston in the ends of the arteries, so that at this place—that is to say, just under the skin—it shows through as black, though in the interior of the body it is red enough.

Such, then, are reasons why our African brother is black and has a distinctive odor.

Kant's principles of the fixity of the specific type and the essential unmodifiability of the "reproductive faculty" imply that the diverse heritable and adaptive characters of what he calls "varieties," no less than those of races, preexist in the species ready-made from the outset, in the form of special "germs" or *Anlagen.* In writing the "Physical Geography" and the "Conception of Race" Kant does not seem to have clearly perceived this implication; but in his essay "On the Use of Teleological Principles in Philosophy" (1788), he expressly draws the inference.

As for what are called varieties in the human species, I remark

---

time. The passage cited was published two years after Lavoisier's direct and decisive refutation of the phlogiston theory.

only that in respect to these, as well as to the racial characters, nature must be conceived, not as producing forms with entire freedom, but as merely *developing* forms in a way predetermined by original predispositions (*Anlagen*). For varieties (as well as races) show purposiveness and adaptation, and therefore can not be the work of chance. . . . The varieties among men of the same race were in all probability no less purposively implanted in the original stock (*Stamm*), in order to make possible the utmost diversity for the sake of endlessly various ends, than were the differences of race, in order to assure adaptation to fewer but more important ends. . . . There is, however, this difference, that the racial *Anlagen*, once they had developed—which must have already happened in the earliest period—no longer produced any new forms, nor yet permitted the old ones to become extinguished; while the *Anlagen* of varieties—at least so far as our knowledge goes—seem to indicate a nature inexhaustibly productive of new characters, both inner and outer.

It is a conventional practice, especially among German writers on philosophy, to speak in a tone of reverent admiration of Kant's profound insight into the spirit and methods of empirical science. The reader, therefore, will do well to note the precise logical character of Kant's procedure in framing and supporting these hypotheses, which constitute his special contribution to biology. In the first place, he assumes following Buffon with no evidence at all, that two species incapable of producing fertile offspring when mated, thereby testify that they can have had no common ancestors. He thus, with a single dogmatic phrase, "there can be only one cause of this" infertility, begs the entire question of the transformation of species, which had been already raised in his time by writers of the first eminence, whose work was well known to him. Further, in order to reconcile his doctrine of the impossibility of any real modification of nature's "original model" for each species with his doctrine of the descent of widely divergent races and varieties from a single species, he adopted the hypothesis of the latent pre-existence of "germs" anticipatory of the subsequent changes of milieu which the species was to undergo, and destined to take command of the reproductive process when the proper occasions arrive, while the other germs obligingly retire into inactivity.[21]

---

[21] Cf. "On the Use of Teleological Principles": "Wherever the ancestors of a race accidentally came and persisted, there was developed the germ latent in their organization with special reference to that neighborhood (*Erdgegend*) and capable of adapting them to that climate."

This, which remained to the end of his days one of Kant's most cherished notions, had most of the faults of which a scientific hypothesis is capable; and it had not even the ambiguous merit of serving the purpose for which it was designed. It was intended as a support to the anti-evolutionistic dogma which Kant had made his own: "every natural kind remains true to its original nature"; yet it was admittedly consistent with an immense and indefinable degree of divergence, on the part of the descendants of a given pair, from the characters of their ancestors. As Kant himself observed, it assigned many of the species of the systematists to a common descent. But if the "reproductive faculty" of the primeval wolf was—as Kant grants that it may have been—capacious enough to contain special "germs" for the subsequent production, not only of wolves, but also of jackals, pug-dogs, greyhounds, dachshunds, hyenas, and bull-dogs, there appeared to be no adequate reason for assigning any particular limit to the original capacity, and the consequent eventual versatility, of that faculty in any organism whatever. It was entirely open to Kant, without abandoning his theory of anticipatory germs, to regard the wolf in turn as the development of a germ implanted in still earlier ancestors, which the wolf and his diverse present progeny share in common with a group of organisms still more various; and so on *ad indefinitum*. Since the immutability of "nature's original model" was to be sufficiently salved by the simple device of supposing that model to have virtually contained within itself, and in course of time, under changing external conditions, to have extruded from itself, a vast assortment of other extremely dissimilar models, there was nothing in the most thoroughgoing theory of the transformation of species which could be inconsistent with an immutability of so elusive and so elastic a character. Kant's rejection of evolutionism was thus not justified even by those singular embryological speculations into which his desire to reject that theory seduced him.

3. *The Review of Herder's "Ideen."* In 1785 Kant published a review of Herder's *Ideen zu einer Philosophie der Geschichte der Menschheit*. Herder, as I have elsewhere shown, was not a believer in the transformation of species; but he may perhaps be without exaggeration described as a near-evolutionist. He set forth in the *Ideen* the theory of a gradual production of organisms in an ascending series in which little by little the form and powers of man were approximated. Through all this "graded scale of beings"

was conspicuous that "unity of type" which the work of Daubenton and Buffon in comparative anatomy had brought to light. The successive emergence of ever higher forms Herder ascribed to some innate potency in "nature" tending to progress and to the constant increase and diversification of life. Just how he conceived this to operate in the actual formation of organisms it seems impossible to make out; one is obliged to doubt whether he ever framed any definite ideas on the subject. But on the unity, yet inexhaustible diversity, of nature's productive power, and on the strange way in which, as he supposed, all animals and plants, and perhaps even snow-flakes and other inorganic things, are fashioned after a single archetype of form, Herder had much to say that was eloquent and impressive, if not very clear. In reviewing the book, therefore, Kant was naturally led to touch upon the subject of organic evolution. The passage runs as follows:

> As for the graded scale (*Stufenleiter*) of organisms, one cannot so severely reproach the author because it will not consent to extend far enough to match those conceptions of his which reach far beyond the limits of this world. For the use of it even in relation to the kingdom of nature here on earth likewise leads to nothing. The slightness of the degrees of difference between species is, since the number of species is so great, a necessary consequence of their number. But a *relationship* between them—such that one species should originate from another and all from one original species, or that all should spring from the teeming womb of a universal Mother—this would lead to ideas so monstrous that the reason shrinks from before them with a shudder. Such ideas cannot with justice be imputed to our author.

It is surely one of the humorous incidents in the historiography of science that more than one grave historian should have found, in the writings of this very period when Kant repudiated evolutionism with the tremulous emotion of a child frightened by a hobgoblin, the idea of evolution playing "the same role as in contemporary science."

4. *Conjectural Beginning of Human History*.[22] In this essay of 1786 Kant was not dealing with the general issue concerning the origin of species; he was, as the title indicates, seeking an answer

---

[22] *Muthmasslicher Anfang der Menschengeschichte*; in *Kant's Werke*, Royal Prussian Academy edition, VIII, pp. 107-123.

only to the question how one species, our own, took its start on its remarkable historical career. But in this was implicit the biological problem whether this species—whatever be true of the others—can be supposed to be descended from other species. The fact that Kant thought this particular problem could be solved in isolation from the general problem is perhaps itself significant; it suggests a predisposition on his part to assume that *Homo sapiens* has powers and attributes which distinguish him so widely from other members of the animal kingdom that they cannot be explained as " evolved " from those of the latter. At all events, Kant's solution of the problem of the origin of man, in this essay, though somewhat involved, is entirely unequivocal. And it is original only in the sort of methodological preamble which he characteristically prefixes to it.

He calls his solution, with seeming diffidence or modesty, a " conjecture," but he did not in fact present it as merely such. For the essay begins by remarking that the question is simply one of historical fact, and that where historical facts are concerned a conjecture unsupported by any sort of recorded testimony—by any " actual report (*wirkliche Nachricht*)" cannot be seriously considered as a source of knowledge (*ein ernsthaftes Geschäft*). A conjecture is " a flight of the imagination," a *Luftreise*. It is permissible to frame a hypothesis on the beginning of human history, and no doubt some such conjectures are less fanciful and unlikely than others; and Kant admits that he had himself formed one which had seemed to him to have considerable probability. But he does not expect others to accept it unless historical evidence can be adduced to prove it true.

But—Kant continues—we have such evidence: nothing less than " the original sacred record (*heilige Urkunde*)," contained in the Book of Genesis. And this, he asserts, is completely in accord with his " conjecture." He begs his readers to examine the course which he " has taken on the wings of the imagination, though not without guidance furnished by reason and observation " (i. e., the account which he is about to give of the beginning of human history), and to see for themselves whether it does not follow " precisely the same line as had been previously delineated by that original historical record."

He therefore proceeds to outline the hypothesis which he finds corroborated by the Scriptural narrative. " If one is not willing to run wildly into conjectures, one must make the beginning [of man's history] one which the human reason is incapable of deducing from any antecedent natural causes (*Naturwarsachen*); therefore, as the *existence* of man, and of man as already of adult size, since he must do without the care of a mother; and as the existence of a human *pair* so that he may propagate his species; and of only a *single* pair, in order that war may not at once break out, as it would if men not akin lived near one another." [28] Kant's German here is somewhat clumsy, even for him, but what he is saying is plain enough: the "first parents" of our species had themselves *no* parents of any species; they just suddenly began to "exist" full-grown, and no "natural" causes can be conceived which would explain their abrupt appearance on this planet. That this account of the "beginning of human history" was, so far as it went, in accord with the biblical account is obvious. It is equally obvious that it was not in accord with the theory, which had been suggested or advocated by a number of Kant's precursors or contemporaries, that the human species is descended from pre-existent animal species. The passage just quoted was an unqualified rejection of that theory; and Kant's rejection of it was—at least nominally—not based upon any arguments drawn from assumed biological data, but solely upon the fact that the theory was contradicted by the "historical evidence"—the *wirkliche Nachricht*—to be found in the second chapter of the "First Book of Moses."

As Moses was not supposed to have been an eyewitness of the interesting events related in that chapter, Kant's use of it as "historical evidence" logically presupposed that it—and presumably the canonical writings as a whole—were written under divine inspiration. But—though he does not explicitly say so—he was clearly unwilling to assume this. For he excluded the first chapter of Genesis from the "original sacred record" in which he sought support for his conjecture; that record consisted, he says, of chapters 2-6. He had apparently recognized that the two biblical versions of the story of creation were inconsistent. And even in the chapters on which he relied for verification of his conjecture, his use of them was highly selective, and the texts which he cited were usually interpreted in an analogical rather than in their natural and literal sense. He has no anthropomorphic god "walking in the garden in the cool of the day" and no serpent "more subtle than any beast

---

[28] Ibid., p. 110, italics Kant's.

of the field," possessing the gift of speech and using it to tempt Eve to eat of the forbidden fruit. Such naïve literalism in biblical exegesis was no longer possible for an "enlightened" philosopher of the late eighteenth century [24]—and most evidently impossible for the author of the *Kritik der reinen Vernunft* (of which the amended second edition was published in the same year as the *Mutmasslicher Anfang der Menschengeschichte*).

But on the fundamental and unescapable question of the mode of origination of the human species, the question with which Kant, in this portion of the essay, was concerned, the second chapter of Genesis—if it was to be used at all as "historical evidence"—must be construed literally. And since he here assumed that such evidence was indispensable if any objectively verifiable answer to the question was attainable, Kant was necessarily led to the anti-evolutionistic position which we have already seen him expressing. But he was manifestly not led to it unwillingly. The "original sacred record" merely confirmed his own "conjecture," of which he usually speaks as if he had formed it before and independently of any reading of the Book of Genesis—which, of course, he had not. But his unconvincing antedating of the conjecture suggests that he was "citing Scripture for his purpose"—was finding in it a sanction for an opinion which he would in any case have held to be probable and preferable to any other, on what he believed to be "philosophical" grounds of "reason and observation"—though he does not explain what those grounds were.

Since the purpose of the present study is simply to collate and analyze Kant's opinions and arguments concerning the theory of organic evolution, this summary of the only portion of the essay on "The Beginning of Human History" which is relevant to that subject should perhaps end at this point. But the summary may be somewhat misleading to readers unacquainted with the essay as a whole; for only a small portion of it has to do with the genetic problem of the "origin" of the human species. A more nearly descriptive—though obviously an inconveniently lengthy—title for this writing of Kant's would have been "Conjectures Concerning the Early Stages of the Development of Man's Mental and Moral Faculties." And Kant might have set forth his ideas on this theme without first proving (to his own satisfaction) that the original

---

[24] Cf. Kant's *Was ist Aufklärung*, 1784 in *Werke*, VIII, pp. 33 ff.

human pair had no ancestors but was supernaturally created with certain specifically human "predispositions" and potentialities. However, Kant thought it needful to explain once more (or rather, to prove that no one could explain) man's genesis, and then passed on to an essentially different topic—to expound his view as to man's psychological character and his behavior immediately after his creation and the changes which these underwent, the new mental powers and emotional susceptibilities which developed in him, during the earlier centuries of the history of the race. These later developments were, of course, a sort of "evolution" in the psychological sense, but not in the biological sense. They did not change man into another species; they were merely the unfolding of the original latent potencies embodied in him at his creation.

Kant still, however, bases his account of the early history of man and human society upon a combination of "conjectures" and of "historical evidence" derived from the narrative of the Book of Genesis down to the Confusion of Tongues. He divides the historical account into three stages or epochs, but of the first of these he gives the reader no definite description, fearing that his conjecture concerning this will seem to them too extravagant to be believed; but he obliquely intimates that at the outset man's mental powers and moral qualities were hardly developed at all. Kant's really definite description begins with the second epoch. In this man could walk upright and possessed power to form some general concepts and had terms to express these. For biblical evidence as to this last Kant quotes the passage in the second chapter of Genesis in which Adam gives names to all the animals in Eden. Men lived in families, had affection for their offspring and imparted to them such knowledge as they had. But throughout this period man's actions were controlled solely by "instinct," which enabled him to know, e. g., which plants were poisonous and which were safely edible, etc. But he had no "reason," i. e., no ability to infer from particular experiences to more general and more remote effects of courses of action. They took no thought for the morrow. While inexperienced man obeyed the voice of nature (i. e., instinct), he got on very well by means of it, but he did not yet possess *reason* (*Vernunft*). This, however, soon began to develop in him, and with it came not only a radical alteration but also a vast multiplication of his appetites and desires—even with respect to food. For "it is a property of the reason that it can, with the aid of the imagination, invent desires which not only do not arise from any natural impulses but are opposed to them. This in the beginning received the name of concupiscence [or cupidity, *Lüsternheit*]; it produced little by little a whole swarm of useless propensities which are "entirely contrary to nature." At the same time, through his newly acquired reason, man learned to "compare himself with his fellows," or companions (*Genossen*), that is, with those who hitherto had been his companions, the other animals in Eden, and this comparison gave him a lofty feeling of superiority (*Üppigkeit*) to all other creatures. Thus Kant passes from the theme of the genesis of man to that of his fall from innocence. The latter, as the reader will note, was caused by the supervention of "reason" in man and its gradual domination of his "instincts." This was the counterpart in Kant's conjectural history, of the eating of "the fruit of the tree of the knowledge of good and evil," and one might say that Kant's name for the biblical Satan was *die Vernunft*, were if not for the fact that the Fall for which "the reason" was responsible was in Kant's view also an Ascent; it had both bad and good consequences, and in final analysis the latter were more important, and were—or were destined to be—dominant over the evil consequences.

5. *The Essay "On the Use of Teleological Principles in Philosophy."* To the title of this article, published in 1788, the contents do not altogether closely correspond.[25] Part of it is, indeed, a prelude to the examination of the conception of purposiveness in nature given two years later in the *Critique of Judgment*; but a greater part consists in a defence of the theories of his two papers on the idea of "race" against certain critics. For the purposes of the present inquiry those theories have already been sufficiently expounded. But it is worth while noting that, in the case of one of his critics, Forster, Kant supposed himself to be confronted with a definite evolutionary theory, on which he felt obliged to pass judgement. The articles of what Kant understood to be Forster's "system" were these:

> The earth in travail, giving birth to animals and plants from her pregnant womb, fertilized by the sea-slime; a consequent

---

[25] There is a reference to the species question in a fragment in the *Lose Blätter* (I, 137 f.), assigned by Reicke to 1787. This is probably merely a draft for part of the essay here considered. The fragment is in the usual vein; Kant speaks in it, for example, of "the inconceivable constancy of species, in the midst of so many causes affecting them and modifying their development."

multiplicity of local originations of organic species, Africa having its own separate species of men (the negroes), Asia another, and so on; as a deduction from these assumptions, the relationship of all organic species in an imperceptibly graded series, from man to the whale, and so backwards (conjecturally even to the lichens and mosses)—and a relationship not of similarity merely, but of actual derivation from a common stock.[26]

On this Kant remarks as follows:

These ideas will not, indeed, cause the investigator of nature to shrink back from before them with a shudder, as from before a monstrosity [27] (for there are many who have played with them for a time, though only to give them up as unprofitable). But the investigator *will* be frightened away from them upon a serious scrutiny, by a fear lest he be lured by them from the fertile fields of natural science to wander in the wilderness of metaphysics. And for my part I confess to a not unmanly terror in the presence of anything which sets the reason loose from its first and fundamental principles and permits it to rove in the boundless realms of imagination.

Kant's alarm, it is evident, was aroused by all three of the hypotheses which he ascribed to Forster. But he particularly disapproved of any attempt to inquire into the origin, the laws of genesis, of organisms in general, or of the original "stock" from which any species is descended. Such inquiries "lie beyond the province of any possible physical science." For science is competent to discover only relations of efficient causation; but organisms, being material systems "in which every part is at once cause and effect of every other part," admit only of "a teleological, not at all of a physico-mechanical, mode of explanation."

6. *The Critique of Judgment*. The principal source of the belief that Kant was an evolutionist in biology is a celebrated passage in

[26] Kant's language clearly seems to ascribe these ideas to Forster, but quite without justification from anything in Forster's article. So far from fathering this system, Forster mentions it as an example of an over-ambitious hypothesis, beyond the reach of verification by man, and therefore beyond the limits of true science (*Teutsche Merkur*, 1786, pp. 57-86, 150-166). And in his *Kleine Schriften*, III, p. 335, Forster emphatically asserts the immutability of "the principal features of the primitive form (*Urbild*) of every species."

[27] Kant refers to a passage of Forster's in which these expressions are jestingly used. But, as it happens, they were originally Kant's own expressions, occurring in the review of Herder's *Ideen* already cited.

the *Critique of Judgment* (1790), Sec. 80. This passage is, unfortunately, usually quoted with its most important part—an appended footnote—omitted. That Kant's true position may clearly appear (in so far as a position which is involved in a scheme of elaborate self-contradictions can ever be clear), it is necessary to cite the text here nearly in full:

> It is praiseworthy to go through the great creation of organized natures with the aid of comparative anatomy, in order to see whether there may not be in it something resembling a system, even in the principle of generation of such beings. For otherwise . . . we are obliged to give up in discouragement all pretension to *natural insight* in this field. The agreement of so many species of animals in a certain common plan which appears to underlie not only their skeletal structure but also the arrangement of their other parts—so that, upon the basis of an original outline of wonderful simplicity a great variety of species could be produced merely by the shortening of one member and the lengthening of another, the diminution of this part and the elaboration of that—all this gives our minds a ray, though a feeble ray, of hope that something may here really be done with the principle of the mechanism of nature—apart from which there can be no natural science as such. This similarity of forms—so great that, amidst all their diversity, they seem to have been produced according to a common original type—gives force to the surmise of an actual relationship between them, by virtue of their generation by one primal mother (*Urmutter*)—through the gradual approximation of one animal species to another, from that in which the principle of purposiveness seems best established, i. e., man, down to the polyp, and from this even to the mosses and lichens, and, finally, down to the lowest stage of nature known to us, namely, to crude matter; from which matter and its forces, according to mechanical laws, . . . the entire system of nature (which in organized beings is to us so incomprehensible that we feel constrained to think another principle for it) seems to descend.[28]

> Here it remains open to the *archeologist* of nature to derive from the surviving traces of her earliest revolutions, according to any natural mechanism known to him or conjectured by him, the whole of that great family of creatures (for so we should

---

[28] The entire hypothesis mentioned down to this point, it will presently appear, Kant really rejects as not only untrue but absurd. For it is a hypothesis implying "equivocal generation" and the reducibility of organic processes to mechanical laws.

have (*müsste*) to think of it, if the above-mentioned relationship is to have any ground). He can suppose the womb of Mother Earth . . . to have given birth at first to creatures of less purposive form; these in turn to have brought forth others (*diese wiederum andere [Geschöpfe] gebären lassen*) better adapted to the places where they originated and to their relations with one another; until finally Nature's womb, grown torpid and ossified, produced only species that underwent no further modification; so that the number of species from that time forward remained just what it was at the moment when Nature's potency in the production of forms reached its end. Only, he must still in the end ascribe to this universal mother an organization purposively predisposed for the production of all these creatures. Otherwise the purposiveness of form characteristic of the products of the animal and vegetable kingdoms would be inconceivable.

Now this passage, though it painstakingly avoids all positive affirmation, doubtless sounds as if Kant intended by it, if not to indicate his own conversion to transformism, at least to issue to others a dispensation to embrace that doctrine. But the following note, attached to the end of the second paragraph, puts a different face upon the matter:

> We may call a hypothesis of this kind a daring adventure (*ein gewagtes Abenteuer*) of the reason; there are doubtless few investigators of Nature, even of the most acute minds, to whom the hypothesis has not at times presented itself. For *absurd* it is not—in the sense in which *generatio aequivoca*, is absurd. It would after all be a case of *generatio univoca*, in the most general sense of the word, since the hypothesis supposes that every organism is derived from another organism, though the one may differ from the other in species; as if, for example, certain water-animals transformed themselves little by little into marsh-animals and these in turn, after some generations, into land-animals. *A priori*, in the judgment of reason alone, there is nothing self-contradictory in this. Only, experience shows no example of such a thing. According to experience, all generation is not only *generatio univoca* (in contrast with generation of the organic out of the inorganic), but also *generatio homonyma*, in which the parent produces progeny having the same organization as itself. *Generatio heteronyma* [i. e., transformation of species], so far as our knowledge of Nature through experience reaches, is nowhere found.

This is certainly not the language of a believer, still less that of an advocate. True, Kant's position has significantly changed since two years previous. He has at last fairly discriminated the question concerning transformation from that concerning equivocal generation, and has learned that the admission of a common descent of different organic species is not necessarily inconsistent either with his hypothesis of "purposive predispositions" or with those doctrines of the completely teleological character of organisms, and of their independence of all merely external causes of modification, which that hypothesis was designed to safeguard. He no longer condemns transformism on *a priori* grounds as a philosophical monstrosity. Its truth or falsity becomes a question to be settled by empirical evidence. But he also appears to say as plainly as possible that all the known empirical evidence is against the theory. No contemporary of Kant's, reading this passage in the *Critique of Judgment* as a whole, was likely to find in it encouragement to risk that "bold adventure of the reason" of which it speaks. Moreover, in the next section (Sec. 81) Kant, in discussing various embryological hypotheses, unmistakably gives his own endorsement to the opinion that "the Supreme Cause of the world ... would, in the original products of his wisdom, have supplied merely the predispositions by which an organic being produces another of like kind *and the species perpetually maintains itself.*" Throughout the remarks upon embryology contained in this section Kant seems to take the constancy of specific forms for granted.²⁹

The chief topic of this second or biological part of the *Critique of Judgment* is, of course, that question which had been present to Kant's mind ever since his adoption of a theory of the evolution of the inorganic world "according to mechanical laws." Could organisms also be mechanistically "explained," or only teleologically? It would require too much space to set forth and discuss adequately Kant's extremely diverse utterances on this question in his last important treatise. But when all those utterances are considered together, they do not seem to indicate any essential departure from

---

²⁹ Brock in commenting upon Sect. 80 of the *Critique of Judgment* observes that Kant takes cognizance directly only of the hypothesis of saltatory mutation, and is silent concerning the possibility of transformation through the summation of slight individual variations. This remark seems to me scarcely justified by Kant's language. By *generatio heteronyma* he means the change of one species "little by little" (*nach und nach*) into another; though he evidently had only vague ideas of the rate at which, and the mode in which, this change might be supposed by the partisans of transformism to take place. (Cf. Brock in *Biol. Centralblatt*, Bd. 8, p. 644.)

the position which we have found him all along maintaining. It is true that he now insists with the utmost emphasis that without the conception of mechanism there is no such thing as science. "It is infinitely important for reason, in its explanation of Nature's processes of production, . . . not to pass beyond the mechanism of Nature" (Sec. 78). He even declares that "apart from causality according to mechanical laws organisms would not be products of Nature at all" (Sec. 81). But he also continues with equal emphasis to insist that "absolutely no human reason (in fact, no finite reason like ours in quality, however much it may surpass it in degree) can hope to understand the production of even a blade of grass by mere mechanical causes. . . . It is absolutely impossible for us to derive from Nature itself grounds of explanation for purposive combinations," such as living beings are (Sec. 78). In short, we *must* regard organisms as part of the cosmic mechanism; and we *cannot* so regard them. How these two assertions are to be harmonized is a thing "which our reason does not comprehend. It lies in the supersensible substrate of Nature, of which we can determine nothing positive, except that it is the being-in-itself of which we merely know the appearance" (Sec. 81). Kant, in short, had by this time acquired the deplorable habit of affirming both sides of a contradiction and leaving it to "the supersensible" to reconcile them. Passages from the last *Critique* may therefore be cited which seem to conflict with his earlier assertions of a sheer gap between the inorganic—the realm of mechanism—and the organic—the realm of teleology. But equally copious repetitions of those assertions may also be found. And upon the definite question of the possibility of "equivocal generation," Kant, as the footnote already cited shows, remained true to his often-repeated opinion; the very notion of such a thing was to him an absurdity.

7. *The Anthropology of 1798.* In his seventy-fourth year Kant returned to the subject of anthropology. His *Anthropologie in pragmatischer Hinsicht* does not, indeed, deal chiefly with the questions to which his earlier anthropological writings are devoted; the greater part of it is a rather miscellaneous but not uninteresting combination of his "critical" psychology and ethics with the purely temperamental convictions, tastes, and prejudices of a septuagenarian bachelor professor, on matters of every-day life and social intercourse. Thus we find laid down, quite as a maxim of applied science, the practical observation that "eating alone (*solipsismus*

convictorii) is not healthy for philosophers," though relatively harmless for mathematicians and historians. If any *philosophirende Gelehrten* are addicted to eating in solitude, they will surely desist when they learn that they are practising *solipsismus convictorii*. But there is one brief passage in the book (in a footnote) which sometimes has been quoted as expressing Kant's belief in the possibility of the transformation of species. The human infant, he observes, comes into the world with a cry. This is characteristic of no other animal; and since it must, so long as man remained in the "wild" state, have been dangerous to both mother and child, by inviting attack from predatory carnivorous beasts,

> we must suppose that in the primitive wild state of nature, with respect to this class of animals, . . . this outcry of the new-born was unknown and that there subsequently supervened a second epoch in which the parents had attained the degree of civilization necessary for the household life. This thought leads us far; for example, it suggests the question whether this epoch may not, on the occasion of some great revolution of nature, be followed by a third, an epoch in which the orang-outang or the chimpanzee would perfect the organs which serve for walking, touching, speaking, into the articulated structure of a human being, with a central organ for the use of the Understanding, and would gradually develop itself through social culture.

This passage obviously does not say, or suggest, that the human race as it already exists is descended from the orang-outang or chimpanzee; but it does undeniably suggest the possibility and, indeed, the probability, that one or the other or both of these anthropoids may hereafter acquire the organs and functions which are now peculiar to man. But a distinguished neo-Kantian philosopher, the late Professor William Wallace, has asked whether Kant may not have "cautiously put the future instead of the past, and hinted at what probably has been rather than what may one day be." [30] If this interpretation were adopted, we should have to conclude that Kant in his old age finally accepted the doctrine (though he was too timid to express it openly) that what we call the human race is descended from a species of apes which assumed the human form and developed the organs of articulate speech and the mental powers and sensibilities now found only in *Homo sapiens*.

But Wallace's suggested interpretation of the passage cited is,

---

[30] William Wallace, *Kant*, 1882, p. 115.

unfortunately, wholly erroneous. Kant was, as any educated eighteenth-century reader would at once have recognized, not cryptically depicting the future of the anthropoids when he really meant their past; he was briefly summarizing, and intimating his acceptance of, a famous theory which one of the most celebrated naturalists of the period, Charles Bonnet, had propounded chiefly in his *Palingénésie philosophique, ou Idées sur l'état passé et sur l'état futur des êtres vivans* (1770). Bonnet was the most boldly speculative naturalist of his generation; he also, like Kant, but more insistently, maintained that his hypotheses were fully in accord with the teachings of the Scriptures. He was, moreover, accustomed to argue directly from the moral attributes of God listed by the theologians—Benevolence, Justice, etc.—to specific conclusions about animals, especially about their *état futur*. When describing the "past state" of any species, he was obliged, being learned in the natural science of the period, to take account of certain evidence, or supposed evidence, derived from the observation of existing species or from such paleontological knowledge (still very limited) as was then available—though even this he felt it necessary to reconcile with the Book of Genesis. But when he turned to the description of the "future state" of a species—man or the chimpanzee or the orangoutang—he could, and did, infer simply from the divine attribute of Benevolence that there is a better time coming for these (and all other) species.

This optimistic prediction, however, can be understood only in connection with another theory of Bonnet's, namely, that the whole of "Nature" passed through a series of "epochs," all of them of long duration but each of them brought to an end by a "revolution," a vast cosmic cataclysm in which all "gross bodies," (including, of course, the bodies of organisms) are destroyed by fire. (This theory hardly seems deducible from the Benelovence of the Creator; it was, in fact, based not on theological grounds, but on supposed geological and astronomical evidence). But in all of these epochs, there exists, apparently in the reproductive organs of every animal, a *petit corps organique*, an infinitesimally small, submicroscopic body which contained, or was associated with, the "soul" of the individual, and was the "germe" from which a new body would spring in the next epoch. It was a minuscule "preformation" of that body, for Bonnet still accepted the *emboîtement*, or nest-of-boxes, genetic theory of Leibniz; but the *germe* was not conceived by Bonnet, or by Leibniz, as a precise predelineation of the body which it was eventually to become; it was (like Kant's *Anlagen* of "races") an aggregate of diverse "potencies" corresponding to the future organs and structures which were to be "evolved," i. e. unfolded, from it on some appropriate future occasion. The same "little indestructible body" however, retained a sort of physical memory of all the "modifications" that have taken place in the gross bodies with which it has been united, and this is the "physical" basis of the "*Personality*" of the animal. Thus, not only human beings but all individuals of every species are—in an extremely dubious sense, but in a sense which seemed to Bonnet very important —immortal. To deny this would be to deny the "Goodness of the Author of Nature." A benevolent God would never permit any one of his creatures to perish utterly. Every single one of them will, therefore, after the present physical world is burned up, experience a "palingenesis," a rebirth, in the epoch that is to follow. Nor is this the only or the most wonderful of the inferences which Bonnet draws from the same theological premise. After the next "revolution of the globe," every individual will be promoted to the next higher rank in the Scale of Being. The Orang-outang which is at present leading a rude simian life in the forests of Borneo (or confined in a Europan zoological garden) will then have all the attributes, physical and psychological, of man.

Bonnet does not profess to know, and does not believe that the finite mind of man will ever know, just how many of these cosmic epochs there have been and will be, though he thinks it possible that the number of those that have existed in the past—each preceded by a catastrophic "revolution" and followed by a fresh creation of the inorganic world in a new form and by a renascence of organic bodies from their indestructible "germs"—may have been very great. But—as the whole scheme of his philosophy requires—he is sure about at least three "epochs" or "worlds," which may perhaps constitute the entire history of "Nature." The first was the world immediately preceding the one in which we live. This may have been, and probably was, very different from the present world in its physical structure and in the kinds of living beings inhabiting it; but about this we cannot with certainty know anything, except that the indestructible "germs" of existing organisms were already present in it. "*Nous n'avons point Historien de ce premier Monde.*" It was destroyed, in the manner already indicated, whereupon occurred the "first revolution," that is, the creation of a second world, the one now existing; and about this, Bonnet thinks, we have a good deal of "historical" as well as other knowledge, derived primarily from the first chapter of the First Book of Moses (which, unlike Kant, Bonnet evidently preferred to the second chapter), supplemented by scientific observations and inferences from "physical geography" (i. e. geology). But this "world" will in its turn be destroyed by fire (conclusive proof of this prediction Bonnet finds in the Second Epistle of Peter),[31] and after this "second revolution" a third and better "epoch" will follow, and—the germs being, as always, unaffected by any such catastrophe—every *individual* organism, from the lichens to man, will be reborn in a less "imperfect" condition, with respect both to its body and its "soul." But all will hold the same rank relatively to the others. Thus—to conclude with Bonnet's own words:

> the same gradation which we see today between the different orders of organized beings will, without doubt, be found in the future state of our globe, but it will show different proportions, and will be determined by the degree of *perfectibility* of each species. Man, then transported to a dwelling place more suitable to the eminence of his faculties, will leave to the ape or the elephant that first place which he occupies among the animals of our planet. In this universal restoration of the animals, therefore, it will be possible to find Leibnitzes and Newtons among the apes or the elephants, and Renaults and Vaubans among the beavers. The more inferior species, such as the oysters, polyps, etc., will be in comparison with the more elevated species in this new hierarchy what birds and quadrupeds are in comparison with man in the present hierarchy. Perhaps, again, there will be a continual and more or less slow progress of all species towards a still higher perfection, so that all the degrees of the Scale will be constantly changeable, in a fixed and constant ratio; I mean to

---

[31] *II Peter:* Chap. 3, vs. 7. "But the heavens that now are, and the earth, have been stored up for fire, being reserved against the Day of Judgment and destruction of ungodly men." Ibid., vs. 12-13: "the heavens being on fire shall be dissolved, and the elements shall melt with fervent heat, but . . . we look for new heavens and a new earth." The reader will note Bonnet's amazingly free mixture of biblical citations and inferences of his own from the attributes of Deity. The author of the Epistle which he quotes was talking in terms of orthodox eschatology, and his "new heavens and new earth" are reserved for redeemed souls; Bonnet, however, is not writing about future rewards or punishments of men, but about the future condition of all now existing organisms, without moral distinctions.

say that the *mutability* of each degree will always have its reason in the degree immediately preceding it.[32]

Only to a superficial reading can Kant's reference to this passage of Bonnet's exhibit him in the guise of an evolutionist in biology. For, in the first place, there is no indication that he conceives even these extensive modifications of form and function as transforming the animals of which he speaks into new "natural" species, in his own sense of that term. In the second place, the passage does not suggest that the *existing* human species is descended from the apes. For in the "second epoch" mentioned, we already find our human ancestors living the household life; and the "third epoch," characterized by such striking improvements in the orang-outang and the chimpanzee, is subsequent to the second, and, in fact, still in the future.[33] Finally, even to this hopeful anticipation of a "good time coming" for the apes at some future "revolution of nature," Kant does not really subscribe; he merely expresses some passing wonder whether something of the sort "might not" occur. As a matter of fact, his publication of so vague and inept a passage as this after Maupertuis, Buffon, Diderot, Erasmus Darwin, and Goethe [34] had written, shows that in his declining years he had not lost that constitutional aversion from the proper hypothesis of organic evolution which we have found to be characteristic of him from the beginning of his career. Also from the beginning, it is true, we have seen in him, as we see here, a constant vague inclination towards evolutionistic modes of thought. But through all that half-century, which constituted the period of the true beginnings of

---

[32] Bonnet, *Palingénésie*, I, 1770, pp. 203-204.

[33] Only by disregarding the natural construction of Kant's language can the sentence about the "third epoch" be interpreted as referring to past time. William Wallace (from whose skilful rendering of the passage I have borrowed some phrases) asks: "Has Kant cautiously put the future instead of the past, and hinted at what probably has been rather than what may one day be?" (*Kant*, p. 115). But why should Kant in 1798 have felt obliged to hint so obliquely at an idea familiar to his contemporaries for half a century, which Buffon had hinted at a good deal more plainly, and several celebrated writers had adopted? The desire to avoid theological opprobrium could hardly have been a motive for taking so evasive and misleading a way of imparting his real view. For theological opprobrium was as likely to attach to certain opinions which he frankly accepted—and probably to the hypothesis of the *future* transformation of apes into rational beings—as to the hypothesis of their past transformation.

[34] Goethe's first unequivocally evolutionary utterance seems to be found in his *Vorträge über die . . . allgemeine Einleitung in die vergleichende Anatomie*, 1796. Cf. Wasielewski, *Goethe und die Descendenzlehre*, p. 27.

biological evolutionism Kant, our analysis has shown, never once professed belief in the transformist hypothesis; nor did he ever show an ability to apprehend clearly either the precise meaning or the force of the considerations which could even then be adduced in favor of that doctrine.

EIGHT

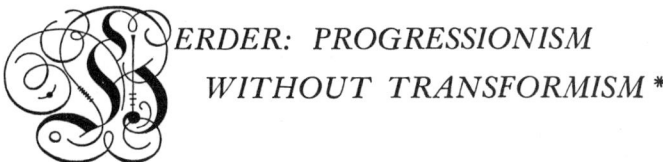ERDER: PROGRESSIONISM WITHOUT TRANSFORMISM *

ARTHUR O. LOVEJOY

If certain of the French *philosophes* have received less credit than is their due for their evolutionist opinions, Herder has often been praised for an early profession of faith in the doctrine of the transformation of species, though in fact he repeatedly and quite explicitly repudiated it. A writer of this century, Bärenbach,[1] has written a book to show that Herder was a precursor of Darwin, and declares that in his *Ideen zur Philosophie der Geschichte der Menschheit*, " he laid down the fundamental laws of the modern development theory, and of the Darwinian theory in particular, and gave clear expression to the law of the evolution of organisms and the theories of natural selection and the struggle for existence." Professor Osborn's account of Herder's relation to the theory apparently follows Bärenbach, and as a result is highly misleading. Herder, says Osborn, probably was helped to his evolutionism by " coming under the influence of Kant's earlier views." But " Herder was less cautious than his master, and appears almost as a literal prophet of the modern natural philosophy. In a general way he upholds the doctrine of the transformation of the lower and higher forms of life, of a continuous transformation from lower to higher types

---

* Reprinted with considerable revision from articles in *Popular Science Monthly*, 1904, pp. 327-336, and in *Journal of Philosophy*, 1922, Vol. XIX (14), pp. 381-385.

[1] *Herder als Vorgänger Darwin's*; cf. the same writer's monograph on Herder in *Der neue Plutarch*, VI. My citation is from the latter work. The author was a Hungarian, and a philosopher of Kantian affiliation.

and of the law of perfectibility." "In his *Ideen*, published in Tübingen in 1806, . . . we see that Herder clearly formulated the doctrine of unity of type which prevailed among all the evolutionists of the period immediately following."

These few sentences contain an undue proportion of errors, and the whole exposition of Herder's views from which they are taken is substantially wrong. It is worth while, therefore, to attempt a more accurate account of Herder's attitude towards evolutionism than is to be found in current writings on the subject. In a matter of this kind, even accuracy about dates is not wholly to be disdained; and it should be observed that the *Ideen* were published, not at Tübingen in 1806, but at Riga and Leipzig in 1784-5. Again, Herder, although once a pupil, was hardly a disciple of Kant; the author of the *Metakritik* would assuredly have been surprised to hear Kant called his "master"; and it is sufficiently clear, from Herder's own language, that the influence which led him to employ such expressions as have caused some to consider him an evolutionist, was that of Buffon and other naturalists of the century, not that of Kant. During Herder's student days (1762-4) in Königsberg, indeed, it is improbable that Kant's influence could have encouraged a belief in organic, as distinguished from cosmic, evolution. Herder, in the *Ideen*, argues at length against what he calls "the unproved and totally self-contradictory paradox" that animal species can depart from their divinely appointed character and that man is directly related to the ape.

Yet Herder's book is certainly full of *aperçus* that come near to the evolution theory; and it doubtless helped to produce a state of mind favorable to the acceptance of the theory. Some passages in the *Ideen*, read by themselves, might easily seem to justify the classification of Herder with the thoroughgoing evolutionists. Where he stands in the matter may perhaps best be shown by setting down in catalogue-fashion the several contentions that he advanced in regard to the history of the animal kingdom and the relation of the lower species to man.

1. Herder clearly recognized that there had been a sequence of temporally successive forms appearing upon the globe, beginning with simpler forms and proceeding to those most highly organized; "from stones to crystals, from crystals to metals, from these to plants, from plants to animals and from animals to man, we see the

form of organization ascend; and with it the powers and propensities of the creature become more various, until finally they all, so far as possible, unite in the form of man" (Bk. V, chap. 1). Each new type is dependent for its existence upon the prior existence of simpler types—dependent usually, indeed, in a very plain sense, since the newcomer commonly has the earlier-born creatures for its necessary food. "It is manifestly contrary to Nature that she should bring all creatures into existence at the same time. The structure of the earth and the inner constitution of the creatures themselves make this impossible. Elephants and worms, lions and infusoria, do not appear in equal numbers, nor could they be created, in consistency with their natures, at one time or in equal proportions. Millions of shellfish must needs have perished before our bare rock of earth could be made a fruitful soil for a finer type of life; a world of plants is destroyed each year in order that higher beings may be nourished thereby. Even if one wholly disregards the final causes of the creation, yet even in the very raw material of Nature there lies the necessity that one being should come out of many, that in the revolving cycle of creation countless multitudes should be destroyed so that through this destruction a nobler but less numerous race might come into being" (Bk. X, chap. 2). "Man, therefore, if he was to possess the earth and be the lord of creation, must find his kingdom and his dwelling-place made ready; necessarily therefore, he must have appeared later and in smaller numbers than those over whom he was to rule" (ibid.). Most of this Herder might have got from Buffon; and there is obviously nothing in these passages which necessarily implies the mutability of species, nothing which is inconsistent with the doctrine of special but gradual creation. Nor is there even in such a passage as this: "From air and water, from heights and depths, I see the animals coming nearer to man, and step by step approximating his form. The bird flies in the air; every deviation of its structure from that of the quadruped is explicable from its element. The fish swims in the water: its feet and hands are transformed into tail and fins," etc. (Bk. II, chap. 3). If Herder had not elsewhere seemed to deny such a theory, we might at first sight be disposed to construe this passage as an assertion of the literal transformation of species. But when the words are closely scrutinized it is evident that they require no such interpretation. They say no more than that animals came into being in a progressive order in which the human type was steadily approximated, and in which each form was adapted to its environment.

2. In this connection, Herder liked to dwell upon the homologies of form and structure observable in all vertebrates, and indeed, as he thought, in all creatures, even those that are outwardly most dissimilar. There is a certain *Hauptform* or *Hauptplasma* in which the whole animal kingdom agrees. "It is undeniable that, amid all the differences of the living beings on the earth, a certain uniformity of structure and, as it were, a standard form, appears to prevail, which yet is transformed into the richest diversity. The similarity of the skeletal structure of land animals is obvious; . . . the inner structure makes the thing especially evident, and many outwardly uncouth forms are in the essentials of their internal anatomy exceedingly like man. The amphibia deviate farther from this standard; birds, fishes, insects, aquatic animals—the last of which merge in the vegetal or inorganic world—deviate still farther. Beyond this our eyes cannot penetrate; but these transitions render it not improbable that in marine forms, plants, and even in the so-called inanimate things the same basis of organization may rule, though infinitely more rude and confused. In the eye of the Eternal Being, who sees all things in one connected whole, it may be that the form of the ice-particle, as it is generated, and that of the snowflake that is formed upon it, may have an analogous resemblance to the formation of the embryo in the womb. We can therefore accept it as a general law that, the nearer they approach man, the more do all creatures resemble him in their essential form; and that Nature, amid the infinite variety which she loves, seems to have fashioned all the living things upon our earth after a single original model (*Hauptplasma*) of organization" (Bk. II, chap. 4).[2] Herder had also learned from the comparative anatomists that it is a corollary to this similarity of structure that organs which function usefully in certain species appear also in other species where they have little or no apparent use or function; in other words, he knew of the existence of vestigial and rudimentary organs. "What Nature had given to one animal as a merely accessory feature (*Nebenwerk*) she has developed into an essential feature in another; she brings it into

---

[2] This passage is quoted in part in *From the Greeks to Darwin*—being the only citation from Herder there given; but the translation is singularly inaccurate, and in one place makes Herder appear to say the opposite of his real meaning. First ed. 1894, pp. 103-4; second ed. 1929, pp. 153-4.

plain view, enlarges it, and makes the other organs—though still in perfect harmony—subservient to it. Elsewhere again these subordinate parts predominate; and all organized beings appear as so many *disjecti membra poetae*. He who would study them must study one in another; where an organ appears neglected or concealed, let him turn to some other creature in which Nature has perfected and plainly displayed it " (loc. cit.).

3. Herder had further learned from Buffon that, within the limits of the specific type, a species may vary widely under differing climatic influences. "Those species that inhabit nearly all parts of the globe, are differently formed in almost every climate," etc. (Bk. II, chap. 3).

4. The author of the *Ideen* also recognized, and frequently dilated upon, that fact in nature which later suggested the specifically Darwinian form of the theory of descent—the fact, namely, that Nature turns out more aspirants for life than she can provide with the means of living, and that there results from this a universal struggle for existence between species and between individuals. Herder had, in fact, been profoundly impressed by the way in which the life-processes of nature seem to be the expression merely of a blind, striving *Wille zum Leben* (in the language of a later school), careless of the single life, tending only to the production of the greatest possible number of living beings, each of them competing with all the others. The discovery of this impressive and sinister aspect of nature was, certainly, the main source, alike of the most important scientific hypothesis of the nineteenth century, and of certain of the most significant and characteristic developments of nineteenth-century philosophy—especially of philosophical pessimism. The emphasis which Herder lays upon this class of facts is therefore interesting and noteworthy: " Where and when each being *could* arise, there it arose; energies (*Kräfte*) pressed in through every gate of entrance and formed themselves to life " (Bk. X, chap. 2). " Nature employs infinitely many germs; . . . she must needs therefore reckon upon some loss, since all things crowd upon one another (*alles zusammengedrängt ist*), and nothing finds room completely to develop itself " (Bk. II, chap. 2). " The whole creation is at war, and the most conflicting powers lie close to one another. . . . Each being strives with each, since each itself is hard-pressed for life; it must save its own skin, and guard its own existence. Why does

Nature act thus? Why does she thus crowd her creatures one upon another? Because she aimed to produce the greatest number and the greatest variety of living things in the least possible space; so that one subdues another, and only through the equilibrium of opposing powers is peace brought about in the creation. Every species cares for itself as if it were the only one in existence; but by its side stands another which keeps it within bounds; and it was only in this adjustment of warring species that creative Nature found the means of preserving the whole" (Bk. II, chap. 3). In all this Nature (for Herder almost invariably personifies) takes no account of the individual, but rather sacrifices him ruthlessly to her "one great end, which is—not the little end of the sentient creature alone, but—the propagation and continuance of the species." And in this connection Herder anticipates Schopenhauer in picturing the pleasures of the love of the sexes, and the romantic illusions connected with that love in man, as merely a subtle trick whereby "Nature" cajoles the individual to sacrifice himself to her larger aim. Schopenhauer's famous chapter on the "Metaphysics of the Love of the Sexes" is little more than an amplification of a passage in the second book of the *Ideen*. Always, Herder perceives, when the reproduction and increase of life is at stake, Nature turns Machiavellian, and plays upon the egoism of the individual for her own very different ends. "It is," he writes, "particularly humiliating to man that in the sweet impulses which he terms love, and to which he attributes so much spontaneity, he obeys the laws of Nature almost as blindly as a plant. . . . Two creatures sigh for each other, and know not for what they sigh; they languish to become one, which dividing Nature has denied; they swim on a sea of deception. Sweetly deceived creatures, enjoy your time; yet know that ye accomplish not your own little dreams, but, pleasantly compelled, the great aim of Nature. . . . As soon as she has secured the species, she suffers the individual gradually to decay. Hardly is the season of love over before the stag loses his proud antlers, the bird its song and much of its beauty, the plants their fairest colors. The butterfly sheds its wings and expires, while alone and unweakened it might have lived through half the year. This is the course of Nature in the development of beings out of one another; the stream flows on, though one wave is lost in the wave that succeeds it" (Bk. II, chap. 2).

But though Herder thus clearly remarked these characteristics of Nature's dealings, he did not deduce from them either the biological or the philosophical consequences which have since become familiar. It did not occur to him to find in the facts of the over-production of organisms and the struggle for existence an explanation—such as Maupertuis had already proposed—of that progressive production and selection of more highly organized and better adapted beings, the reality of which he so fully recognized. Nor was he led, by his apprehension of a universal tendency to the maximum production of living things, to erect the notion of "unconscious will" into the central conception of a metaphysical system.

5. In his accounts of the beginnings of human society, and of the moral instincts without which society could not exist, Herder observes that these beginnings were made possible by a prior peculiarity in the physiological constitution which distinguishes the human species—namely, by the greater length of the period of helpless infancy. This, of course, is an idea upon which many evolutionistic moralists and sociologists have latterly delighted to dwell. Herder sets forth how the prolongation of infancy was the condition and the chief cause of man's moral nature—how it provided the true training school in which the featherless biped was fitted for the social state: "The first society arose in the paternal habitation, bound together by the ties of blood, of mutual reliance, and of love. Thus to destroy the savagery of men and to habituate them to domestic intercourse, it was necessary that infancy in our species should continue for some years. Nature held them together by tender bonds, so that they might not separate and forget one another, like the beasts that soon reach maturity. The father becomes the teacher of his son, as the mother has been his nurse, and thus a new tie of humanity is formed. Herein lay the ground of the necessity of human society, without which it would have been impossible for a human being to grow up, and for the species to multiply. Man is thus born for society; this the affection of his parents tells him, this the years of his longer infancy show" (Bk. IV, chap. 6).

This conception, however, was by no means a new idea of Herder's own. It is therefore a little curious to find the idea put forward by evolutionist writers of the end of the nineteenth century as a fresh and striking discovery. Even so learned a man as the late John Fiske supposed it to be a great novelty, and even conceived that he himself was the original discoverer of it. In the preface to his book *The Destiny of Man* Fiske wrote: "The detection of the part played by the lengthening of infancy in the genesis of the human race is my own especial contribution to the doctrine of evolution, so that I feel somewhat uncertain as to how far that subject will be understood." But the thing had been detected, not merely by Herder, but also by Pope, some fifty years before him. Pope wrote, in the Third Epistle of the *Essay on Man* (1733):

> Thus beast and bird their common charge attend,
> The mothers nurse it, and the sires defend;
> The young dismissed, to wander earth or air,
> There stops the instinct, and there ends the care.
> A longer care man's helpless kind demands,
> That longer care contracts more lasting bands. . . .
> Still as one brood, and as another rose,
> These nat'ral love maintained, habitual those:
> The last, scarce ripened into perfect man,
> Saw helpless him from whom their life began,
> While pleasure, gratitude and hope combined
> Still spread the interest, and preserved the kind.

Pope got the suggestion of this from one of Bolingbroke's *Fragments or Minutes of Essays*, which appear to have been written for the purpose of thus providing the poet with material. But Pope here sees the point better than his philosophical mentor. Bolingbroke was attempting, on the one hand, to show that " man is connected by his nature . . . with the whole tribe of animals, and so closely with some of them, that the distance between his intellectual faculties and theirs, which constitutes as really, though not so sensibly, as figure the difference of species, appears, in many instances, small, and would probably appear still less, if we had the means of knowing their motives, as we have of observing their actions." [3] On the other hand, he is replying to those theologians who loved to dilate upon the miseries of the " natural state of mankind." He writes accordingly:

> I say then, that if men come helpless into the world like other animals; if they require even longer than other animals to be nursed and educated by the tender instinct of their parents, and if they are able much later to provide for themselves, it is because they have more to learn and more to do; it is because they are prepared for a more improved state and for greater happiness. . . .

---
[3] " Fragment L," *Works*, 1809, ed.; VIII, p. 231.

> The condition wherein we are born and bred, the very condition so much complained of, prepares us for this coincidence (of social and self-love). . . . As our parents loved themselves in us, so we love ourselves in our children, and in those to whom we are most nearly related by blood. Thus far instinct improves self-love. Reason improves it further.[4]

Bolingbroke here emphasizes the value of a long infancy "in giving time for educative influences to work upon the plastic brain," but he suggests less plainly than Pope a relation between this physiological peculiarity of the human species and the origin of society.

There is in this, however, no evidence of originality on Pope's part; for the more significant aspect of the matter had been pointed out nearly half a century earlier by Locke, in Sections 79-80 of the *Second Treatise of Government*:

> The end of conjunction between male and female being not barely procreation, but the continuation of the species, this conjunction betwixt male and female ought to last, even after procreation, so long as is necessary to the nourishment and support of the young ones, who are to be sustained by those that got them until they are able to shift and provide for themselves. . . . And herein, I think, lies the chief, if not the only reason, why the male and female in mankind are tied to a longer conjunction than other creatures, *viz.*, because the female is capable of conceiving, and, *de facto*, is commonly with child again, and brings forth too a new birth, long before the former is out of a dependency for support upon his parents' help. . . . [Thus] the father is under an obligation to continue in conjugal society with the same woman longer than other creatures, whose young, being able to subsist of themselves before the time of procreation returns again, the conjugal bond dissolves itself. . . . Wherein one can not but admire the wisdom of the great Creator who . . . hath made it necessary that society of man and wife should be more lasting than that of male and female among other creatures, that so their industry might be encouraged, and their interest better united, to make provision and lay up goods for their common issue.

How far beyond Locke the same idea can be traced I do not know, but it is certain that the theory of the value of longer infancy which Fiske presented in 1874 as something "entirely

---

[4] "Fragment L," ibid., p. 240. Pope versifies the last four sentences cited in the *Essay on Man*, Ep. III, lines 149, 124, 133-4.

new in all its features "[5] had been clearly set forth as early as 1689, in one of the most familiar classics of English political philosophy.

But unfortunately this pleasing theory had been subjected to some rather damaging criticism by Rousseau in 1755, in a note appended to his *Discourse on the Origin of Inequality*. Referring to the passage of Locke cited above, Rousseau remarks, in substance, that, if it is a question of explaining the origin of the family, the thing primarily to be accounted for is the beginning of the permanent cohabitation of male and female during the nine months between copulation and the birth of the child. If the parents did not live together—i. e., if the habit of family life had not already been formed—during this period, why should the primitive human male have come to the aid of the female "*after* the *accouchement*"? "Why should he aid her to rear an infant which he does not even know to be his, and the birth of which he has neither purposed nor foreseen?" Locke, in seeking in the length of human infancy an explanation of the beginnings of the permanent family, had forgotten, Rousseau intimates, another characteristic of the human species—the long period of gestation. When this is borne in mind— as Rousseau's criticism implies—it becomes evident that the proposed explanation presupposes the thing to be explained. The helplessness of the human infant certainly would not have united the parents unless they had already, for a considerable period, been united; and if their union had endured for so long, it is not obviously necessary to invoke additional explanations to account for its having endured longer—especially as the period of helplessness, Rousseau suggests, was probably much briefer in the case of primitive man. At all events, the first and great transition—that from casual matings to relatively lasting cohabitation of the sexes—is left unexplained by the theory in question. Thus—for these and other reasons—Rousseau concludes that *le raisonnement de Locke tombe en ruine*.

He would doubtless have pronounced a similar judgment on later examples of the same argument. For the weakness of the argument should have been still clearer by the time it was revived in the late 19th century. It was then well known to naturalists that the family is not peculiar to man; and that, apparently, "in the higher apes monogamy is the rule, the male and female roaming at large in a family party."[6] The gorilla, for example, "lives in a society con-

---

[5] *Outlines of Cosmic Philosophy*, I, p. viii.
[6] Pycroft, *The Courtship of Animals*, 1914, p. 25.

sisting of male and female, and their young of various ages and the family group inhabits the recesses of the forest. . . . The male animal spends the night crouching at the foot of the tree, and thus protects the female and their young, which are in the nest above, from the nocturnal attacks of leopards." [7] Fiske ignored the zoological knowledge of his time in declaring that " while mammals lower than man are gregarious," it is only in man that there " have become established those peculiar relationships which constitute what we know as the family." [8] The fact is simply that some species are of a monogamous or monandrous habit, and that the ancestors of man were probably of such a species. *Why* an animal has this characteristic we do not know. The theory of natural selection would, indeed, suggest that *if* the young of a species remain helpless for a long period, that species is more likely to survive if the male remains with the female and aids her to defend the young. But this would only mean that a primarily unfavorable variation—helpless infancy—was accompanied by another variation which in some degree offset its disadvantages. The latter variation cannot have been *caused* by its effect—the better protection of the young; and it therefore is not explained by that effect. And it must, as Rousseau's observation suggests, have manifested itself primarily as an instinct to continue with the same mate or (in the case of the male, in some species) mates, *before* the birth of offspring. Fiske fell into an extraordinary inversion of causal relations, and at the same time missed the fact that really needed to be accounted for, when he wrote that " one effect of lengthened infancy of stupendous importance " was that, among our " half-human forefathers," as " helpless babyhood came more and more to depend on parental care, the fleeting sexual relationships established among mammals were gradually exchanged for permanent relations."

As an explanation of the origin of the family, then, Fiske's theory was neither new nor true. Nor did it show the " value," in the sense of the indispensability, of prolonged infancy, even as a means to man's greater intellectual attainments. It was not evident that the

---

[7] Hartmann, *Anthropoid Apes* (1885), p. 231. Garner, however, describes both gorilla and chimpanzee as monandrous but not usually monogamous. " The chimpanzee," he asserts, " keeps his children with him until they are old enough to go away and rear families of their own." (*Apes and Monkeys*, 1900, pp. 99, 232.)

[8] *The Meaning of Infancy*, 1909, p. 29.

continued plasticity requisite for the learning-process need be inseparable from physical helplessness; and in fact, it is not.

6. In all these cases the receptive and pregnant mind of Herder had grasped and elaborated separate ideas which (with the exception of the last) have since become familiar as arguments for or deductions from the doctrine of organic evolution. But did he himself accept that doctrine? Did he believe in the mutability of species and in the literal descent of man from lower forms of life? If I am able to interpret his utterances correctly, he did not. There are several passages in the *Ideen* in which Herder discusses the relation of man to the animal kingdom; and, in order that the reader may have the means of deciding for himself what Herder's position was, I will cite them at some length. "There are those," he says (Bk. III, chap. 6),

> who have, I will not say degraded man to the rank of a beast, but have denied to him the character of his race, and would make him out to be a degenerate animal (*ausgeartete Thier*) which in striving after a higher perfection has wholly lost the distinctive qualities (*Eigenheit*) of its species. This, however, is manifestly contrary to the truth and to the evidence of natural history; man obviously has characteristics that no animal possesses, and performs actions of which both the good and the evil belong to him alone. . . . Since every animal remains true upon the whole to the character of its species, and since we alone have free will instead of necessity for our ruling power, then this difference must be investigated as a fact—for fact it undeniably is. The other questions—how man came by this distinctive characteristic; whether it was his from the beginning, or is adventitious and acquired: these are questions of a purely historical sort. Now, setting aside all metaphysics, let us confine ourselves to physiology and experience.

Herder then points out the anatomical peculiarities of man, particularly those which, as Daubenton had shown, are connected with his greatest peculiarity, the upright attitude. And in view of these considerations Herder concludes thus:

> Would the human animal, if he had been for ages in an inferior state—and if he had been formed as a quadruped in his mother's womb, with wholly different proportions—would he have left that state of his own accord and have raised himself to an erect

posture? Out of the faculties of a beast, which would ever be drawing him backward, could he have made himself a man, and, even before he became a man, have invented human speech? If man had ever been a four-footed animal, if he had been such for thousands of years, assuredly he would remain such still; and nothing but a miracle of new creation could have made him what he now is. Why, then, should we embrace unproved, nay, totally self-contradictory, paradoxes, when the structure of man, the history of his species and, as it seems to me, the whole analogy of the organization of our earth, lead us to another conclusion? No creature that we know has ever departed from its original organization and adapted itself to another contrary to it; for it can operate only through the powers that inhere in its organization, and Nature is abundantly able to hold each living being fast in that state to which she has assigned it. In man everything is adapted to the form he now bears; from this everything in his history is explicable; without it nothing is capable of explanation. . . . Why should we humble in the dust the crown of our high calling, and shut our eyes to that central point in which all the radii of Nature's circle seem to converge?

In a later passage (Bk. VII, chap. 1) Herder directly discusses, only to reject, the theories of those (he probably has Monboddo especially in mind) who assert a kinship, or an identity of species, between the apes and man. " I could wish," he writes,

> that the affinity of man to the ape had never been urged so far as to cause people to overlook, in seeking a graded scale (*Leiter*) of being, the actual steps and intervals without which no scale can exist. What, for example, can the rickety orang-outang explain in the figure of the Kamchatkan, the pigmy chimpanzee in the size of the Greenlander, the pongo in the Patagonian? for all these forms would have arisen from the nature of man if there had been no such thing as an ape upon the earth. . . . In point of fact, the apes and man were never one and the same species (*Gattung*). For each race Nature has done enough, to each she has given its own proper heritage. The apes she has divided into as many species and varieties as possible, and extended these as far as she could. But thou, O man, reverence thyself! Neither the pongo nor the gibbon is thy brother; the American and the Negro are. These then thou shouldst not oppress nor kill nor rob, for they are men like thee; but with the ape thou canst not enter into fraternity.

Herder expresses himself in a similar vein in a preface which he wrote for a German translation (1784) of the work in which Lord Monboddo set forth, among other things, his theory of the close relationship of man and monkey.[9] Herder praises cordially Monboddo's taste for ancient art, and his large and philosophical way of dealing with the problem of the origin and development of human language; but he warns the reader against Monboddo's views on the orang-outang. The opinion that "*Affe und Mensch e i n Geschlecht sei,*" Herder marks as "an error which even the facts of anatomy contradict."

In another chapter of the *Ideen* Herder speaks of the transitions (*Uebergange und Ueberleitungen*) and metamorphoses (*Verwandlungen*) through which Nature leads the successive orders of animals, in a fashion which seems at first sight plainly to imply the derivation of higher from lower species by ordinary descent; yet in the same paragraph he pauses to insist upon the fixity of specific types:

> It may appear that such transitions are incompatible with the definiteness of form to which every species remains true, and in which not the smallest bone undergoes alteration. But the reason for this invariability is apparent; since every creature can receive its organization only from other creatures of its own species. Our orderly Mother Nature has thus plainly predetermined the way by which any organic power should come into actual existence (*Wirksamkeit*); and thus nothing can escape from its once-determined form (Bk. V, chap. 3).

It is hard to see how any reader of Herder's generation could have understood these utterances, when taken all together, in any other sense than as an assertion of the essential immutability of species and a denial of man's descent from simian or any other animal ancestors. How Herder would have explained the gradual appearance of progressively higher forms is, undeniably, somewhat incomprehensible. But the truth is that his whole treatment of the subject is poetical, vague, and not very careful of consistency, rather than explicit, definite, and scientific. The theory of descent was, at the time he wrote, almost a commonplace of current biological discussion; but his attitude towards it was certainly hostile. He was most of all eager to show that the characteristics and poten-

---

[9] See Lovejoy, "Monboddo and Rousseau," *Essays in the History of Ideas*, 1948, pp. 38-61, especially pp. 45-54.

cies of so noble a creature as man cannot have been derived by inheritance from any merely animal ancestors. His positive doctrine was—to use the terms customary in the first half of the nineteenth century (see below, 382)—a " progressionist " form of " the theory of special creation." Every species was brought into existence by a separate act of the Creator. But the embarrassing implications of the later special-creation theory Herder seemingly avoided by referring to the Universal Cause, not as " God," but as " Nature." That he was not a forerunner of Darwin is evident from the fact that Herder's only explanation of the progressive march of the history of organisms is teleological; it was the " purposes of Mother Nature " that caused everything to go through the stages of development which we know. But Darwin's most fundamental conviction was that teleological explanations are to be excluded from natural history.

NINE

FOSSILS AND THE IDEA OF A PROCESS OF TIME IN NATURAL HISTORY

FRANCIS C. HABER

Of all the phenomena presented by natural history, none attracted the attention of naturalists more than the prodigious quantity of marine shells to be found in the depths of the earth and in the highest mountains, wrote Baron d'Holbach in his article "Fossiles" in the *Encyclopédie* (1757).[1] Similar comments were also made by others in the literature of natural history around the middle of the eighteenth century. Holbach noted further that there was a multitude of works on fossils, but that in explaining them, naturalists, even some who were otherwise enlightened, continued to be partisans of plastic forces or of the Noachian Deluge, despite observational evidence which controverted both views.

In speculating about the fossil enigma, the position of religious officials, Protestant and Catholic alike, was a strong deterrent to moving outside the accepted bounds of orthodoxy, but most naturalists seem to have been as austerely orthodox as the guardians of dogma.[2] With Mosaic history standing directly across their path,

---

[1] *Encyclopédie; ou, dictionnaire raisonné des sciences, des arts, et des métiers,* VII (1757), 209-11. Articles by Holbach were indicated by the mark (—).

[2] Lynn Thorndike has made this point in "*L'Encyclopédie* and the History of Science," *Isis*, VI (1923), 361-86. He also notes how the chronology of the Old Testament was often accepted as the opening chapter of human history in articles of the *Encyclopédie*. The question of deception in order to get the work past the censors needs to be considered in this connection, however. See Joseph Edmund Barker, *Diderot's Treatment of the Christian Religion in the Encyclopédie* (New York: King's Crown Press, 1941).

the investigators of fossils, rather than push on, scurried off in strange directions, or returned to earlier positions. Virtually every suggestion which had ever been made to explain the origin of fossils found some advocate in the early eighteenth century. Careful, but pious, naturalists tried to avoid the premise of a slow deposition in the process of time, which would conflict with Mosaic history, but they could not find any other natural means of accounting for fossils and fossil strata which would fit Mosaic history.

One course of compromise was to avoid the Deluge altogether and place the origin of fossils at the time of Creation. This had the merit of lending to the fossil strata the appearance of a natural origin without the necessity of explaining their deposition as a process of time. Elias Camerarius in 1712 suggested that God had supplied the varied forms of fossils at Creation to furnish the interior of the earth with foliage corresponding to that on the exterior. Theodore Arnold in 1733 ventured the opinion that at Creation infinitesimal particles were brought together to form the outlines of all the creatures and objects destined to occupy the earth—a sort of trial run. There were many such attempts at this kind of compromise, but one of the most learned was that of Father Bertrand in 1765. He displayed an intimate knowledge of the history of thought regarding fossils in his *Recueil de divers traités sur l'histoire naturelle de la terre et des fossiles* (1766), but he rejected all previous explanations because they did not conform to either the observations of natural data or to Scriptural history. Aristotelian alternations of land and sea would not have left uniform strata, he reasoned, and they would have required an enormous course of centuries in any case. The Flood, on the other hand, was not long enough, for the strata were not the work of a cataclysm. Plastic forces could not imitate nature down to such details as the worn teeth of fish, while shells were an integral part of the stone in which they are found, so both must have been formed together. But there simply was not enough time in the history of the earth for the natural production, orderly deposition, and solidification into stone of the masses of shells. Father Bertrand suggested weakly that the Creator had arranged the fossils in their places for some impenetrable purpose in the Great Design of Nature. Indeed, his whole work proved to be an elaborate exposition of natural theology, that hybrid of science and theology about which much has been written, but which requires a few remarks in connection with fossils.

As it was forged under Cartesianism, the view of the universe in natural theology was embroidered with analogies drawn from mechanics, particularly timepieces, and Christian anthropomorphism. In the rising temper of late seventeenth-century cosmic optimism, the attributes of God were redefined. The angry, intervening Jahveh gave place to a clever Master Craftsman and benign Master who had so contrived the world-machine that everything had its proper place and function. All was so ingeniously harmonized that the machine would run on indefinitely without any need of adjustment once it had left the hands of the Maker. The ultimate purpose, or final cause, of the Great Contrivance was to provide for the convenience and sustenance of man. This was made explicit by the writers in the natural theology movement, but implicit in its exposition was the attribution to God of a pride of workmanship. Certainly the pious writers would not accuse God of being vain, but as they extolled the Spectacle of Nature and the fine workmanship everywhere displayed in it, the attitude can be detected that they thought God had done some showing off to dazzle man with His great powers, thereby insuring man's respect, subservience, and admiration.

Homage to the truths of revelation become increasingly hollow in natural theology as its advocates turned to evidences of design in nature as a proof of deity. Natural theology thus became an apology for the new mechanical philosophy and helped in making the transition from the theologically dominated spirit of the mid-seventeenth century to the rational-scientific philosophy of the eighteenth century, absorbing Newtonianism and rejecting Cartesianism in its progress. At the same time, as the utilitarianism of secular life asserted itself, natural theology gave a strong religious sanction to the pursuit of that scientific inquiry which had no apparent practical benefits. This dual role can be discerned in the exceedingly popular *Physico-Theology* of William Derham:

> Many of our useful Labours, and some of our best modern Books shall be condemned with only this Note of Reproach, That they are about trivial Matters, when in Truth they are ingenious and noble Discoveries of the Works of GOD. And how often will many own the World in General to be a Manifestation of the Infinite Creator, but look upon the several Parts thereof as only Toys and Trifles, scarce deserving their Regard? But in

the foregoing (I may call it) transient View I have given of this lower, and most slighted Part of the Creation, I have, I hope, abundantly made out, that all the Works of the LORD, from the most regarded, admired, and praised, to the meanest and most slighted, are Great and Glorious Works, incomparably contrived, and as admirably made, fitted up, and placed in the World. So far then are any of the Works of the LORD, (even those esteemed the meanest) from deserving to be disregarded, or contemned by us, that on the contrary they deserve ... to be *sought out, enquired after*, and *curiously*, and *diligently pryed into* by us; as I have shewed the Word in the Text implies.[3]

As the Works were pryed into, the enormity and complexity of the Creation rapidly emerged before eighteenth-century man, and he was staggered at the infinite pains the Master Craftsman had taken in outfitting his abode. It is small wonder that hymns of praise were sung about the Benevolence of the Creator—" upon a transient View of the Animal World in General only, we have such a Throng of Glories, such an enravishing Scene of Things as may excite us to Admire, Praise, and Adore the Infinitely Wise, Powerful, and Kind CREATOR. . . ."[4] In surveying the largess, however, it would have been most ungracious of man to carp about flaws in it. There were no "rude bungling Pieces" in Nature. "And so far are we from being able to espy any Defect or Fault in them, that the better we know them, the more we admire them; and the farther we see into them, the more exquisite we find them to be."[5] Turning to the Principal Fabrick, the Terraqueous Globe, we find Derham extending this attitude to the earth's strata:

And so for all the other Parts of our Terraqueous Globe, that are presumed to be found Fault with by some, as if carelessly order'd, and made without any Design or End; particularly the Distribution of the dry Land and Waters; the laying of the several Strata, or Beds of Earth, Stone, and other Layers before spoken of; ... I have before shewn, that an Infinitely Wise Providence, an Almighty Hand was concerned even in them; that they all have their admirable Ends and Uses, and are highly instrumental and

---

[3] William Derham, *Physico-Theology, or a Demonstration of the Being and Attributes of God, from His Works of Creation*, 15th ed. (Dublin, 1754), p. 431. See also p. 37.
[4] Ibid., p. 265.
[5] Ibid., p. 38.

beneficial to the Being, or Well-being of this our Globe, or to the Creatures residing thereon.[6]

Nature became a sanctuary in natural theology—"The World thenceforth becomes a Temple and life itself one continued act of adoration."[7] Innumerable theologians took up the study of natural history, and under the influence of natural theology, they regarded the Creation as a complete mechanism. Their work, marked by a passive wonderment, was concerned with gathering examples of design in the contrivances of nature, contrivances which were doing exactly what they had been designed to do. The admirable ends and uses of fossils were somewhat elusive, but in the uncritical spirit of the orthodox naturalists, it was easy to assume that they were Reliques of the Deluge or Medals of Creation, as fossils were called, serving a moral function by reminding man of his early transgressions and punishment.

Orthodox naturalists became all the more anxious to discover in fossils a scientific corroboration of the testimony of Moses as the *philosophes* attacked the supernaturalism in the Old Testament. Almost instinctively, they adopted the doctrinaire philosophy of empiricism which had become rampant as the ideas of Newton and Locke saturated the fabric of eighteenth-century thought. They condemned system-making, never doubting for a moment that Mosaic cosmogony in its garb of natural theology was among the self-evident truths of science, and they contented themselves with the naming and cataloguing of the artifacts in the temple of nature.

An influential expression of the tendency to use empiricism to close off inquiry into fundamental questions was the work of the Abbé Noël Antoine Pluche, the leading French popularizer of natural theology. His *Spectacle de la nature*, first published in 1732 and often republished, was perhaps the most widely read book of its kind in France during the eighteenth century.[8] Pluche undertook the refutation of the entire range of speculation on the origin of

[6] Ibid., p. 82.
[7] William Paley, *Natural Theology: or, Evidences of the Existence and Attributes of the Deity, Collected from the Appearances of Nature*, 1st ed., 1802 (Boston, 1837). This statement appears as a legend under a frontispiece showing a pious man standing on an eminence with the Bible in hand, gazing with adoring eyes at the spectacle of nature.
[8] According to D. Mornet, *Les sciences de la nature en France, au XVIIIe siècle: un chapitre de l'histoire des idées* (Paris, 1911), p. 8.

the world which conflicted with Mosaic cosmogony in his *Histoire du ciel* (1743-1753), forcefully emphasizing the sufficiency of the empirical method in science, utility as the goal of science, and the vanity of trying to unlock all of nature's secrets. He singled out the fossil enigma as a particularly mischievous source of false systems, and, with an air of finality, he assigned the origin of fossil deposits to the Flood, about four thousand years ago. He closed the discussion with the observation:

> The natural conclusion of the comparison we have made of the thoughts, either of the ancients or the moderns, on the origin and end of all things, with what Moses teaches us is that NOT ONLY IN RELIGION, BUT ALSO IN PHYSICS, WE MUST RESTRICT OURSELVES TO THE CERTAINTY OF EXPERIENCE AND THE MODERATION OF REVELATION.[9]

Experience, the great touchstone of the physical sciences, needed the analogy of historical process before it could become fruitful in the reconstruction of a world which lay completely outside experience, but it was precisely this historical process which the timeless world-machine view of the universe excluded. As a result, many *philosophes* and deists were in the same camp with the proponents of natural theology on the fossil enigma, despite their mission to rid the world outlook of anthropomorphism, supernaturalism, and final causes. Voltaire, for instance, did all in his power to combat the idea of development in nature. In 1746, he sent to the Academy of Bologna a *Dissertation sur les changemens arrivés dans notre globe, et sur les pétrifactions qu'on prétend en être encore les témoignages* in which he accused the natural philosophers of wanting great changes in the scene of the world much as the people craved spectacles.

It seemed to Voltaire a complete inversion of reason and experience to elaborate a great system of earth changes in order to explain a few shells like those of marine animals living in the seas of the Indies which naturalists claimed they had found in the mountains of Europe. Instead of trying to get the seas up into the mountains, he thought it was more reasonable to question the identity of the shells. As for ammonites, known for ages as serpent stones, it was obvious to him that they were coiled snakes which had been petrified, or stones which had formed in such a shape. The small shells,

---

[9] Noël Antoine Pluche, *Histoire du ciel considéré selon les idées des poëtes, des philosophes, et de Moïse*, vol. II (The Hague, 1740), pp. 476-7.

apparently from the seas near Syria, which had been found in the mountains of France and Italy, were, he remarked, probably dropped by pilgrims and crusaders who had the custom of wearing shells in their hats. Where there were masses of shells, he suggested that mountain lakes had dried up. But rearranging the earth to account for a few shells was nonsense. The mountains were not, as Burnet had said, a ruins. "This chain of rocks is an essential piece in the machine of the world."[10] Nor could marble and metals be dissolved by the Deluge as Woodward claimed, while a shift in the axis of the earth, pretended by so many, was disproved by the astronomical observations on precession.

There is then, no system which can give the least support to this idea, so widely prevalent, that our globe has changed its face, that the ocean has occupied the earth for a very long time, and that men have formerly lived where porpoises and whales are today. Nothing which vegetates or which is animated changes; all the species have remained invariably the same; it would be very strange that the grain of the millet had eternally conserved its nature, and the entire globe varied its.[11]

After Benoît de Maillet's *Telliamed, ou entretiens d'un philosophe indien sur la diminution de la mer avec un missionaire français* was published in 1748 and Buffon's *Theory of the Earth* appeared in 1749, Voltaire was provoked into writing further pieces in defence of the position he had taken in the *Dissertation*.[12] Both Maillet and Buffon had projected geological theories of great transformations of the earth's surface to explain the fossil remains. Their theories

---

[10] *Oeuvres complètes de Voltaire*, Kehl ed., xxxi (1785), 384. This edition mistakenly gives the date of the *Dissertation* as 1749. See J.-M. Quérard, *La France litteraire* (Paris, 1839), x, 297.

[11] Ibid., pp. 385-6.

[12] Charles Lyell, in the historical sketch of the progress of geology in *The Principles of Geology*, 9th ed. (New York, 1857), pp. 54-5, accused Voltaire of bad faith and inconsistency in the treatment of fossils in order to ridicule those religionists who wanted to make fossils a proof of Mosaic history. There was more consistency and sincerity in Voltaire's articles on fossils than Lyell admitted. As Voltaire's editor remarked, Voltaire never changed his mind on the ideas he first put forward in the *Dissertation*, in spite of the attacks of naturalists (ibid., p. 18). Voltaire's attacks were directed towards both the diluvialists and those who were using fossils as a basis of systems "devoid of probability, contradicted by facts, or contrary to the laws of mechanics." Strangely, Lyell did not include Maillet in his historical sketch, although he was mentioned in the text (p. 572).

threatened Voltaire's whole concept of a completed and permanent world, but in addition, Buffon had cleverly poked fun at the anonymous *Dissertation*, not knowing until later that Voltaire was its author. Voltaire was not appeased by Buffon's hurried expressions of respect, especially since they were not accompanied with any concessions on the point at issue.

In his *Singularités de la nature*, Voltaire scorned the naturalists' method of giving a name to an object, and then assuming that the object corresponded fully to the name. Some gallant, for instance, had given the name *Conchae Veneris* to a shell whose shape resembled a female organ, but this hardly proved that the shell was the remains of a lady, he slyly argued. Similarly, giving the name of Ammonite or Nautilus to a shell did not mean that it really was the remains of such a species, and it was therefore no proof that tropical seas had covered Europe. He also ridiculed the ideas of Maillet and Buffon that mountains were formed by the flux and re-flux of the seas over thousands of centuries.

What then is the true system? the one of the great Being who has made all, and who has given to each element, to each species, to each genus its form, its place, and its eternal functions. The Great Being who has formed the gold and iron, the trees, the plants, man, and the ant, has made the ocean and the mountains. Men have not been fish, as *Maillet* says; all is probably what it is by immutable laws. I cannot repeat too often that we are not gods who can create a universe with a word.[13]

Voltaire continued his assault on the system-makers in a long article, " Des coquilles, et des systèmes bâtis sur des coquilles," in the *Dictionnaire philosophique* and in the " Dissertation du physicien du Saint-Fleur," in *Les Colimaçons*, belaboring Buffon and Maillet for their opinion that calcareous stone was comprised of the remains of marine shells. He also poked fun at the system of Bernard Palissy, who maintained the same thing, and impugned his motives. The character of Palissy was revealed in the title of his book, *Le moyen de devenir riche*, Voltaire sneered, and, like the rest of the system-makers, he was a charlatan.[14] The high priest of the Enlightenment

---

[13] *Oeuvres complètes de Voltaire*, XXXI, 418 (in chapter XI, " De la formation des montagnes ").

[14] Ibid., p. 486. The article on Coquilles, *Dictionnaire philosophique*, is in vol. XXXIX, pp. 140-56. Palissy's work was *Recepte véritable, par laquelle tous les hommes de la France pourrant apprendre à multiplier et augmenter leurs*

was annoyed to see the wonderfully simple, precise system of the Newtonian world-machine, which he had done so much to establish among the French, challenged on the basis of some old shells.

Telliamed (de Maillet spelled backwards) belatedly marked the summation of seventeenth-century progress in geology and the point of rupture with the attempts to bypass the conclusion that fossils were a product of nature's history. Maillet (1659-1738) was a Cartesian (his work was published posthumously). He was well-read in ancient literature and the leading works on fossils, had traveled widely in the Mediterranean regions, and had apparently done some original investigation into the processes of sedimentation. His literary style was successfully modeled after Fontenelle's *Conversations on the Plurality of Worlds*. Speaking through his Indian philosopher, Telliamed, in six days of conversation with a Christian, Maillet unfolded his system. It was remarkably free of final causes, anthropomorphism, and supernaturalism, but filled with Cartesian fantasy and with credulity.

Maillet assumed that the universe was eternal and that there was a constant transmigration of matter throughout the various whirlpools of the Cartesian system. Some suns lost their fire and picked up bodies of water, while others were inflamed and cast off ashes. Our own earth, he thought, had taken its origin from the ashes of other celestial bodies gathering at the center of a whirlpool. After a terrestrial globe had formed, its surface was covered by bodies of waters picked up from the outer regions. The waters circulated over the earth for countless ages, drifting the fine ash-like material until they finally gave shape to the primordial mountains. The waters were, however, steadily evaporating back into the outer regions, and eventually the mountains were exposed. Life then appeared in shallow coastal water and plants began to grow on land. Some sea animals trapped in the marshes, after the repeated failures of many, mastered flight or walking, to give rise to birds and land animals. Maillet pointed to the basic similarity in the anatomy of some of the sea, land, and air species which made it likely that such a development had occurred through a specialization of parts. As the waters continued to lower, the debris of land erosion and the remains of organisms were deposited on the bed of the sea, forming in time a succession of strata.

thresors (1563). The title of a 1636 edition of this work commenced, "*Le Moyen de devenir riche.*"

The system of Maillet had been instigated by his observing fossil shells in the hills of Egypt, and the core of *Telliamed* was concerned with geological proofs of the diminution of the sea, which was his answer to the fossil enigma. Around this core he fabricated much speculation in order to make the diminution reasonable. Maillet worked in catastrophes, such as the taking on of additional bodies of water from outer regions and the shifting of the earth's axis, but his geology was essentially uniformitarian. At one point, dealing with the time required by the process of slow changes, he made the statement (later echoed by Hutton and Lyell), " But, continued Telliamed, not to enter upon a question, which you look upon to be necessarily connected with your religion, . . . let us be here content not to fix a beginning to that which perhaps never had one. Let us not measure the past duration of the world, by that of our own years." [15]

Maillet recognized in the succession of strata the progression of life, suggesting that the first appearance of a species could be approximately dated from his estimate that the sea had diminished at the rate of three feet four inches every thousand years. Unfortunately for his reputation, after several pages of provocative general discussion in the " Sixth Day " chapter on the origin of man and animals through evolution, he appended as proofs many far-fetched stories of the simple transformism of men and land animals out of sea animals. His mermaids and mermen only proved his gullibility and brought ridicule upon his whole system.

The reaction to *Telliamed* was violent. The orthodox were scandalized by it and gave a stock retort, like that of Dezallier d'Argenville, author of books on fossils and subsequently a contributor to the *Encyclopédie*: " What a folly in this author to substitute Telliamed for Moses, to bring man out of the depths of the sea, and, for fear that we should descend from Adam, to give us marine monsters for ancestors! Only a kind of godlessness could invent such dreams." [16] However, *Telliamed* was a fascinating book,

---

[15] *Telliamed; or, the World Explain'd . . . —A Very Curious Work* (Baltimore, 1797), p. 194. An English edition was printed in London, 1750, and the best French edition was published at the Hague in 1755 by the Abbé le Mascrier, who had the original manuscript in his possession, and was disappointed with the first edition which he had permitted Jean Antoine Guer to prepare.

[16] *L'histoire naturelle éclaircie dans une de ses parties principales, la conchyliologie*, . . . (Paris, 1757), p. 74. See also Hugh Miller, *The Foot-Prints*

and it was well publicized by its opponents. It brought into sharp focus the issue of the evolution of the earth and of organic life through the slow operation of natural forces, in spite of its cosmological trappings and the appendage of fantastic reports of transformism. It was still being refuted in the middle of the nineteenth century by anti-evolutionists. Henceforth, the orthodox regarded fossils with suspicion, for they had now seen how fossils could be used to promulgate views in direct opposition to their belief in the finished Creation of Mosaic cosmogony and natural theology. The suspicion was converted into alarm by Buffon and Holbach.

Although the geological processes described by Buffon in his *Theory of the Earth* bore a resemblance to those given in *Telliamed*, especially the formation of mountains beneath the sea by the flux and reflux of tides and currents, Buffon was a disciple of Newton. His theory on the formation of our planetary system, by a comet striking the sun obliquely and knocking off a mass of matter which separated out into satellites according to the law of gravitation, was so sound that it remains today as a respectable hypothesis. Buffon's geology was not so well conceived, but it was superior to preceding works. He attempted to solve the fossil enigma by natural processes, and as a result he postulated that strata had been deposited slowly in the course of time. The previous attempts of Burnet, Whiston, Woodward, Bourguet, Scheuchzer, Steno, Ray, and Leibniz to explain the presence of shells on continental lands were reviewed, and he attributed their failure to a reliance on the authority of Scripture, adding, "The notion that shells were transported and left upon the land by the Deluge is the general opinion, or rather superstition, of naturalists." [17] The universal Deluge was an established fact, he continued, but it was the direct operation of the Deity, not the effect of a physical cause, and should not be blended with bad philosophy.

Buffon's attempt to elevate Scripture out of natural science did not impress the Faculty of Theology at the Sorbonne, which took action and in 1751 furnished Buffon with a list of reprehensible statements in his cosmogony. Buffon published a retraction in the

of the Creator, 3rd London ed. (Boston, 1856), pp. 243-4, for a 19th-century attack on Maillet.

[17] *Natural History*, ed. William Smellie (London, 1791), I, 130-1.

fourth volume of his *Natural History*, saying that he believed very firmly in the text of Scripture and all that is reported in it about creation, both as to order of time and matters of fact.[18] In spite of his abjuration, Buffon continued on his way to a new theory of the earth. He had only known Leibniz' work through brief summaries,[19] but with the publication of the *Protogaea* in full in 1749, he was stirred to a reconsideration of igneous forces, while the researches of Jean Jacques Dortous de Mairan seemed to prove the existence of a central residual heat in the earth.[20] On the assumption that the earth had evolved from an incandescent, molten mass, in conformity with Newton's law of gravitation, and following a suggestion in the *Principia* itself,[21] Buffon turned to the question of the *duration* of the primitive heat in the earth. He engaged upon a series of ingenious experiments in his laboratory to determine the rate of cooling of balls made of various materials. Projecting his computations from the laboratory models to the bodies in the planetary system, Buffon calculated the duration of each. The results were published in his *Introduction à l'histoire des minéraux* in 1774

[18] See *Oeuvres complètes de Buffon*, ed. Pierre Flourens (Paris, 1853-4), XII, 350-5, or Jean Piveteau, *Oeuvres philosophiques de Buffon* (Paris: Presses Universitaires de France, 1954), 106-9.

[19] In addition to the short statement in *Acta Eruditorum* (Leipzig, Jan., 1693), pp. 40-2, about Leibniz's *Protogaea*, some of Leibniz's views on fossils were reported by Fontenelle in the *Histoire de l'académie royale des sciences*, 1706 (Paris, 1708), pp. 9-10. (The latter appears in English translation in John Martyn's *Philosophical History and Memoirs of the Royal Academy of Sciences*, II [London, 1742], 356-8.) Leibniz also expressed his views on fossils in the transactions of the *Akademie der Wissenschaften zu Berlin* in 1710. He served as first president of this society, which was founded in 1700 on the plan of Leibniz, as the *Societas Regia Scientiarum*. Bertrand, in *Recueil de divers traités sur l'histoire naturelle de la terre et des fossiles* (Avignon, 1766), pp. 41-2, mentions some of the works from which Leibniz may have gained his ideas on fossils; among them is listed Agostino Scilla's *La vana speculazione disinganata dal senso* (Naples, 1670), a work which strongly influenced Maillet.

[20] Mairan published articles on the subject in the *Mémoires* of the French Academy in 1702 (pp. 161-80) and 1719 (pp. 104-35). The idea of central heat was developed in his *Dissertation sur la Glace* (Paris, 1749) and again in the *Mémoires* of 1765, "Nouvelles recherches sur la cause générale du chaud en été & du froid en hiver, en tant qu'elle se lie à la chaleur interne & permanente de la terre," pp. 143-266.

[21] Book III, Prop. XLI, Prob. XXI, in which Newton speculates on the time it would take for a ball of iron the size of the earth to cool. Buffon stated that he was inspired by this statement to undertake experiments on the rate of cooling of iron. *Oeuvres*, IX, 86.

and were incorporated in modified form in his classic of French prose, *Des époques de la nature*, in 1778.

Buffon's *Époques* was the first history of the earth to give an estimate of actual elapsed periods of time to explain the fossil strata. His dimension of time was derived from the principle of refrigeration, but it was closely coordinated with the fossil evidence of geological changes. Out of deference to Scripture, Buffon divided his history into seven epochs. In the first, the planetary system took form out of the mass struck from the sun. The earth, a rotating molten mass assumed the shape of an oblate spheroid, lost its incandescence as its surface cooled and hardened, and was solidified to the center at the end of about 3,000 years. During the second epoch, all the volatile substances were driven out of the hot earth to form the atmosphere, chains of mountains arose from wrinkles as consolidation progressed, and the principal mineral veins were sublimated. By the end of thirty or thirty-five thousand years, the earth had cooled sufficiently for the vapors of the atmosphere to condense on the crust. In the third epoch, the seas were formed from these condensing vapors. Marine animals appeared in the seas and flourished in great numbers, and because the waters were very warm for ten or fifteen thousand years, the first species included large and different kinds than those existing now.

At the end of 50,000 years, sedimentary strata of clay, shale, coal, and fossils had been formed, while the action of water on the exposed land had given it a new face. A period of volcanic action and the retreat of the waters from the lands characterized the fourth period, ending 60,000 years after the earth's formation. The continued cooling of the earth brought about the extinction of the first huge marine species, but in the northern regions the land had become cool enough for the existence of land animals. In this still tropical climate, they assumed giant sizes and were dominated by elephants, rhinoceroses, and hippopotamuses (an idea suggested from the discovery of mammoths in Siberia). This fifth epoch ended around the year 65,000 after the earth's formation, and in the next epoch, the tropical fauna and flora moved towards the equatorial zone with the cooling climate. The continents separated about the year 70,000, and man entered on the scene, introducing the historic period of five or six thousand years. Buffon speculated that the refrigeration would continue, and when the temperature of the earth was reduced to one twenty-fifth of what it now is, all life would perish from the earth. This fateful event would occur in the 168,000 year after the formation of the earth.

Buffon had anticipated a storm of protest at his breach of Biblical chronology. "How do you reconcile, it will be said, this high antiquity that you give to matter with the sacred traditions which give to the world only six or eight thousand years?"[22] He did all that he could to smooth the way with the orthodox and still press forward with his system. The Days of Creation, he explained, were periods of indeterminate length, six spaces of time, and only in the last epoch in which man was created do the genealogies represent a true chronology. He made several digressions in the *Époques* to reassure his readers that his system was not in contradiction with Moses, and at the same time to impress upon them the necessity for enlarging their view on time. In one such digression, he wrote:

> I must reply to a kind of objection which has already been made to me on the very long duration of time. Why throw us, it has been asked of me, into a space as vague as a duration of 168,000 years? Because, according to the view of your plan, the earth has aged 75,000 years, and organic nature must still subsist for 93,000 years: is it easy, is it even possible to form an idea of all or of the parts of so long a course of centuries? I have no other reply than the exposition of the monuments and the consideration of the works of nature: I will give the details of this and the dates of the epochs which follow from it, and it will be seen that, far from having increased the duration of time unnecessarily, I have probably shortened it far too much.
>
> And why does the mind seem to get lost in the space of duration rather than in that of extension, or in the consideration of measures, weights and numbers? Why are 100,000 years more difficult to conceive and to count than 100,000 pounds of money?[23]

It was only necessary to envisage the number of centuries needed to produce all the shelled animals whose remains were in the earth, then the number of years for the deposition of the remains, then the time for their petrifaction, and it would be seen how great a duration was needed in building up these fossil strata alone. The "facts deposited in the archives of nature" were indisputable, Buffon

---

[22] *Oeuvres*, IX, 473.

[23] Ibid., IX, 493. In the manuscript copy Buffon had written 400,000 or 500,000 instead of 168,000 years. See *Des époques de la nature*, ed. Lucien Picard (Paris [1894]), p. 64.

maintained, and the order of time indicated by them was no less real than the epochs of civil history, although they were not marked by fixed points, nor limited by centuries which we can count with exactitude. The natural periods could be compared with one another, and perhaps one day, he hoped, contemporary dates and intermediate epochs would be ascertained.

Buffon's appeal for time in the process of the formation of the strata was forceful, but he emphasized again and again that he had used as little time as possible in his system. In the *manuscript* copy of the *Époques* he wrote, "when I counted only 74,000 or 75,000 years for the time passed since the formation of the planets, I gave notice that I constrained myself in order to oppose received ideas as little as possible." To explain the phenomena satisfactorily, he continued, it would be necessary to assign to the first periods of cooling alone, not some thousands of years, but a million, if not more.[24]

Although it was not appreciated for many years, by taking a path of compromise with Mosaic history, Buffon furnished a means of reinterpreting Genesis so as to make room in it for ideas of time and change. Once the orthodox were forced by the geological record to abandon their concept of a finished world of six thousand years' duration, they found the *Époques* to be a useful source of face-saving expedients. But to Buffon's contemporaries the work was an extremely dangerous and seductive threat to organized religion, the more so because of its brilliant style, the great reputation of its author in natural history, and the novelty of the subject matter.

In the *Époques* Buffon appears to have retreated from some of his earlier evolutionism, and he even allowed for the special creation of species. One problem he faced as a result of his principle of refrigeration was that, by assuming nature degenerated in its vigor with the dissipation of heat, the largest forms of a species appeared first. This introduced a series of devolutions in the succession of species which, along with doubts accruing from Buffon's thinking about evolution from a biological point of view, left the *Époques* somewhat ambiguous about the actual evolution of species. Nevertheless, the work as a whole promulgated an evolutionary view of nature, which his critics recognized would, if adopted, mark the end of the timeless world-machine and substitute a natural process of changes for external deity in the work of creation.

---

[24] *Époques*, ed. Picard, op. cit. See also *Oeuvres*, IX, 443.

Nature being contemporaneous with matter, space, and time, her history is that of all substances, all places, all ages; and although it appears at first view that her great works never alter or change, and that in her productions, even the most fragile and transitory, she always shows herself to be constantly the same, since her primary models regularly reappear before our eyes in new representations; however, in observing her closer, it will be seen that her course is not absolutely uniform; it will be recognized that she admits sensible variations, that she receives successive alterations, that she even lends herself to new combinations, to mutations of matter and of form; that, finally, much as she seems to be fixed as a whole, she is variable in each of her parts; and if we encompass her in all her extent, we can no longer doubt that she is today very different from what she was at the beginning and from what she became in the succession of time: it is these various changes that we call her epochs. Nature has existed in different states; the surface of the earth has successively taken different forms; the heavens themselves have varied, and all the things in the physical universe, like those in the moral world, are in a continual movement of successive variations.[25]

The *Époques* was an advanced step in the idea of a genetic cosmic evolution first put forward by Descartes, and then elaborated by Leibniz. Buffon also continued their tradition of reconciling science with religion, but the *Époques* was confounded with the more radical theories of Holbach and Diderot. The views on nature of both Holbach and Diderot had been early influenced by the first volumes of Buffon's *Natural History*, which inaugurated a new approach to the study of nature. Buffon had decried the calculating and counting method of the physical sciences in biology, called for the examination of biological data on a comprehensive plan, and advocated the idea of growth and connection within nature. This latter idea had also been expounded by Leibniz through his concepts of the monad, the chain of being, and the doctrine of continuity—all advances by degrees in nature, and nothing by leaps. Leibniz had also speculated on the evolution of species in his *Protogaea* after observing extinct ammonite fossils. These ideas of Leibniz were further developed by Maupertuis, and the concepts of both Maupertuis and Buffon were absorbed by Diderot in his thinking on the continuity and growth of nature.[26] In this whole development of a

[25] *Oeuvres*, ix, 456.
[26] See Ernst Cassirer, *The Philosophy of the Enlightenment*, tr. F. C. A.

dynamic cosmology among the *philosophes* there was infinitely more involved than fossils, but the bearing fossils had on ideas of past changes in the earth was not without its importance.

In his 1757 article in the *Encyclopédie* on "Fossiles," Holbach had expressed his sympathy with those ancients who believed that the sea had formerly stayed on our continents for a long course of ages. Any other system, he felt, was subject to insuperable difficulties in explaining the fossil strata of the earth. He developed his own views at greater length in two sections of the article " Terre " (1765) on the strata of the earth and the revolutions of the earth. The alteration of the earth's axis had been responsible for revolutions of the earth's surface in ages long past, he conjectured, but there was also a continuing revolution being effected through the actions of winds, waters, volcanoes, and the constant changes in nature.

> We see all these causes, often combined, perpetually acting on our globe; it is not surprising then, that the *earth* shows us, almost at every step, a vast mass of debris and ruins. Nature is busy destroying in one part in order to produce new bodies in another. . . . let us conclude then, that the earth has been and still is exposed to continual revolutions, which contribute without cease, either suddenly, or little by little, to change its face.[27]

This thought on the constant changes in nature was expanded in Holbach's *Système de la nature* (1770), the " Bible of Atheism," to which Diderot may have added the notes and some pages of text.[28] Morley has said of the work, " No book has ever produced a more widespread shock. Everybody insisted on reading it, and almost everybody was terrified." [29] There were no final causes in the system, no external deity working on dead matter, no fixed species, no favored species—only endless trials, failures, and survivals in an eternal chain of cause and effect in the busy workshop of nature.

Matter was eternal, in Holbach's system, but all of its forms were the products of time and change. Nature was always casting matter

Koelln and J. P. Pettegrove (Boston: Beacon Press, 1955), Chapter II, esp. pp. 73-92.

[27] *Encyclopédie*, XVI, 166.

[28] See "Extrait de la correspondance de Grimm," pp. xi-xvi, Paris edition of 1821. The title page also states that the new edition has notes and corrections by Diderot, and this is discussed in John Morley's *Diderot and the Encyclopaedists* (London, 1878), II, 173.

[29] Morley, op. cit., II, 174.

into new combinations, which could only survive if they met general laws and the particular circumstances and relationships of the environment. What we call monsters are only productions inadequately organized or adapted for survival. All the individuals of a species are unique and vary from one another, however slightly, and the species have all undergone, and would continue to undergo, change in an infinity of successive developments as changing external conditions destroyed the suitability of some forms and made the survival of new variations possible. It is not clear, however, to what extent Holbach thought there was a genetic relationship between the species themselves.[30]

Holbach thought that the earth was a product of time, not an eternal form, and that all of the organized beings on it were particular productions suitable to it alone. The earth had passed through many changes, however, and these were recorded in the earth itself. Vast continents had been engulfed by the seas, whose sojourn on the very place we now inhabit is attested by shells, the remains of fish, and other marine productions. The elements had long disputed the empire of our globe. The revolutions and overthrows of nature had left a vast mass of debris and ruins which can be seen everywhere.[31] Confronted with this series of changing conditions, man must have changed his form many times, Holbach conjectured, or he would have perished ages ago.

The appearance of novity in man's civilization seemed to contradict the development through the long ages assumed by Holbach. Maillet had met this objection by suggesting that calamities, civil and natural, had periodically wiped out the gains made by man in the arts and sciences. Maupertuis had expressed a similar idea, referring to the fossil masses as an indication of great catastrophism which might have broken the continuity of civilization.[32] Holbach likewise supposed that the revolutions which convulsed nature in the past had periodically destroyed peoples, leaving only a few survivors who had to struggle against new conditions so desperately that only a lingering memory of previous progress in the arts and sciences was retained. Commenting on the terror which these revolutions must have struck in the heart of man, Holbach went on to

[30] *Système de la nature* (Paris, 1821), I, 97 ff.
[31] Ibid., pp. 446-7.
[32] *Essai de cosmologie*, in *Oeuvres de M. de Maupertuis*, new ed. (Lyon, 1756), I, 71-2, and *Système de la nature*, ibid., II, 153-4.

trace the origin of nature worship, mythology, superstition, and finally religion, from primitive man's desire to protect himself against the ravages of nature by enlisting the aid of supernatural powers.[33]

The affinity of Holbach's system with Diderot's ideas on nature was so obvious that it was at first thought that Diderot had written the *Système de la nature*, and Diderot judiciously left Paris in case a warrant was issued against him. In his brilliant *Pensées sur l'interprétation de la nature* (1754), Diderot had suggested the "embryo" theory of nature. What we take for the history of nature, he wrote, is only the very incomplete history of an instant. An individual is born, grows, exists, declines and perishes. Perhaps, he observed, it is the same with entire species. Although faith teaches us that animals left the hands of the Creator as we now see them, would not the philosopher, left to his own thoughts, suspect,

> that from all eternity animality has had its particular elements scattered throughout the mass of matter; that it finally came about that these elements united; that the embryo formed from these elements passed through an infinitude of organizations and developments; that it has had, in succession, movement, sensation, ideas, thought, reflection, conscience, sentiments, passions, signs, gestures, sounds, articulation, language, laws, sciences, and arts; that millions of years elapsed between each of these developments; that it will, perhaps, undergo still more development and extensions which are unknown to us; . . . that it could disappear forever from nature, or else, continue to exist under a form and with faculties completely different from those which characterize it at this instant of duration?[34]

Holbach's system brought to a climax the mechanical materialism towards which the *philosophes* had been working. It was pure atheism organized to combat the religious sentiments of the age, and it can hardly be considered a work in natural history. The whole movement which spawned the embryo theory of nature was complex and many-sided, owing as much to Empedocles as to natural science, but there can be no doubt that the awareness of the fossil

---

[33] Ibid., p. 468 ff.
[34] *Oeuvres complètes de Denis Diderot* (Paris, 1821), II, 217-8. Cf. Henry Fairfield Osborn, *From the Greeks to Darwin* . . . , 2nd ed. (New York, London: Charles Scribner's Sons, 1929), pp. 171-2; Lester G. Crocker, *The Embattled Philosopher; A Biography of Denis Diderot* (Michigan State College Press, 1954), pp. 139-40; and Cassirer, op. cit., p. 91.

masses and the problem they posed for cosmogony made a contribution towards this facet of Enlightenment philosophy, if only to confirm, with apparent proofs of catastrophes throughout a long course of ages, ancient ideas on the process of nature.

The embryo theory of nature and its concept of time and change, could have immediately enriched geological thought, but it apparently had the opposite effect in the eighteenth century. The *philosophes* were not original investigators in geology, while most of those who were such showed a strong inclination to be orthodox. The association of fossils with systems of atheism raised an opposition to the dynamic view of nature among the geologists, and from the publication of *Telliamed*, there was a growing concern displayed in the writings on fossils to prove that all geological evidence supported the system of Moses. It was recognized, too, that if the age of the earth could be fixed in accordance with Biblical chronology, the atheistic systems of cosmogony would be cut at the root. In a period of six thousand years, only an external deity could have accomplished the vast work apparent in the earth's crust, a natural process in time was eliminated, and the operation of miracles in nature was vouchsafed. By the time Buffon's *Époques* appeared, only eight years after Holbach's *Système*, the duration of the earth had become a critical theo-scientific issue, and the acceptance of a natural process in time in the geological sciences could only be won by the relentless demands of natural evidence, which bit by bit was to force concessions from the unwilling empiricists under the spell of literal Mosaic cosmogony.

The philosophy of empiricism, whatever its shortcomings as a tool of received ideas, did stimulate field work. Fossils remained in the mineral classification, the third kingdom of nature, throughout the eighteenth century, and traveling naturalists added greatly to the store of knowledge about the identification and distribution of fossils. Linnaeus served as a model in field work, and he exerted an influence on the classification of fossils, although he was less successful here than in botany and biology. His travels around the Baltic Sea lands had familiarized him with many fossil deposits, and he remarked of the stone in one area, that it contained more fossils than porridge had grains. From his observations, he concluded that the fossil strata could not have been the work of one universal deluge, but were the work of time. In the last edition of his *Systema Naturae*, he wrote:

That all matter was primordially in a state of fluidity, and that the earth arose from the bosom of the waters, we have the testimony of Moses, Thales, and Seneca. And it is manifest, that the sea enveloping the chaotic nucleus, produced by slow and gradual means the continent, which by continually exhaling its dews into clouds, is regularly moistened by aetherial, rectified, deciduous showers. Genuine remains of the general deluge, as far as I have investigated, I have not found; much less the adamitic earth: but I have every where seen earths formed by the dereliction or deposition of waters, and in these the remains of a long and gradual lapse of ages.[35]

In line with his views on the succession of strata, Linnaeus put forth ideas which were later developed by Tobern Bergman and Abraham Gottlob Werner in connection with "formations," but his attempt to classify the materials of the earth by chemical and physical properties reflected the predominance of mineralogical interests in the geological sciences. Mining the natural resources of the earth had, of course, grown with the expansion of manufacturing, and the discovery, identification, and utilization of the substances of the earth was of far more importance to mining engineers than cosmogony. Their primary concern was with the constitution of the earth's crust as it is, not as it was, and their pragmatic approach coincided neatly with the doctrinaire philosophy of empiricism. The existence of remains of organisms in strata was only an empirical fact in the analytical approach of mineralogy, but strata themselves were an important factor in mining, because they grouped like kinds of material. Stratigraphy led to cosmogony.

After his appointment to the Freiburg Mining Academy in 1775, Werner became the central figure in geology, raising it to a recognized science, and instilling in a generation of students excellent observational techniques. But behind Werner's impressive pragmatic science was an uncritical acceptance of the premise that water was the *prime* agency of geological action. He taught that all the substances of the earth were once dissolved in the universal, primordial waters, and that from time to time solid matter was precipitated or crystallized out of the waters to form the strata of the

---

[35] Charles Linné, *A General System of Nature* . . . , tr. from the edition by Gmelin, Fabricius, Willdenow, etc. by William Turton (London, 1806), VII, 3. For the geological views of Linnaeus, see A. G. Nathorst, "Carl von Linné as a Geologist," *Annual Report*, 1908, of the Smithsonian Institution, pp. 711-43.

earth. He did envisage a succession in the deposition of materials, and he, like Linnaeus, became aware of a time-process in the formation of the strata. In his *Allgemeine Betrachtungen über die festen Erdkörper*, delivered as a popular lecture in 1817, Werner remarked, " Our earth is a child of time and has been built up gradually." [36] However, in 1817, such a view had been fairly well established in geological thought, and Werner's most brilliant students, D'Aubuisson de Voisins, Leopold von Buch, and Alexander von Humboldt, had done much in establishing it, after they abandoned the Neptunist teachings of their master. During the early, and most influential, period of Werner's teaching, the lapse of time was susceptible of a great latitude of interpretation. There was no necessary time-cycle in precipitation, as there is in the growth span of an organism, and in a system of such precipitations, the whole process could have taken place very quickly, even within the forty days of the Deluge. The leading principle of Neptunism, that water has been the prime agency of geological action, was also compatible with Mosaic cosmology, and had probably been derived from it, so it is not surprising that most of the Wernerians were also thoroughly convinced Diluvialists during the latter eighteenth century. That the Neptunists had a strong theological bias was fully displayed by their violent attacks on the Plutonists (or Vulcanists), who claimed that heat was also an important geological agency.

Evidences of volcanic action in the structure of the earth had often been mentioned in seventeenth and early eighteenth-century speculations on natural history and cosmogony, but after the middle of the eighteenth century a more systematic and objective examination of volcanic materials in the earth was undertaken by naturalists such as Giovanni Arduino, Jean Étienne Guettard, Nicolas Desmarest, Lazaro Spallanzani, Rudolph Eric Raspe, Girard Soulavie, Peter Simon Pallas, Guy S. Tancrede de Dolomieu, and Faujas de Saint-Fond. Even the Neptunists would agree that there had been occasional volcanic action in the past, but these Plutonists went further and gave an igneous origin to basalts, pointed out the extensive layers of lava in the strata of certain regions, particularly the

---

[36] Cited in Frank Dawson Adams, *The Birth and Development of the Geological Sciences* (Baltimore: Williams and Wilkins, 1938), p. 221. See also Kirtley F. Mather and Shirley L. Mason, *A Source Book in Geology* (New York and London: McGraw-Hill Book Co., 1939), on Werner, von Buch, and others.

Auvergne, and discovered the intrusion of once-molten material between sedimentary strata. When Desmarest, Soulavie, and Saint-Fond maintained that long periods of time must have occurred between the depositions of the layers of lava in the strata of Central France, Plutonism was brought into direct conflict with Biblical chronology and the Diluvial doctrine of aqueous origins of the strata. In addition, Buffon had based his *Époques*, so closely associated with the atheism of the *philosophes* in the orthodox mind, directly on plutonic principles, and the Wernerians took upon themselves the task of refuting all evidence which would give heat a prominent role in geology. When the Neptunists were confronted with the works of James Hutton, they launched a virtual religious crusade against Plutonism.

In 1788, Hutton's "Theory of the Earth" appeared in the *Transactions of the Royal Society of Edinburgh*. It was expanded and published in two volumes as the *Theory of the Earth with Proofs and Illustrations* in 1795, but the early article is of special interest because it reveals more of the non-empirical elements in Hutton's thought. Whereas Buffon had written his *Époques* from the dynamic view of nature, Hutton's thought was a development of natural theology and the timeless world-machine view. With shades of Derham, Hutton wrote:

> When we trace the parts of which this terrestrial system is composed, and when we view the general connection of those several parts, the whole presents a machine of a peculiar construction by which it is adapted to a certain end. We perceive a fabric, erected in wisdom, to obtain a purpose worthy of the power that is apparent in the production of it.[37]

The final cause of this machine was man's welfare, and Hutton even pointed to the distribution of the minerals as a proof of it—the more useful being the most common and accessible to man. However, he envisaged the movement of the parts in the earth machine on a scale which had been previously applied only to the operations of the heavens by the natural theology writers. At the same time, he clearly discerned the incompatibility of natural theology with Mosaic cosmogony in geology, the one extolling the

[37] "Theory of the Earth; or an Investigation of the Laws observable in the Composition, Dissolution, and Restoration of Land upon the Globe," *Royal Society of Edinburgh, Transactions*, 1 (1788), 209.

regularity of God's laws in nature, the other subjecting the earth to supernatural fiats.

Hutton was impressed with the tremendous role that fossils played in making up the bulk of the earth's strata. He accepted the conclusion that calcareous strata were composed of the remains of sea animals, and that these remains were the relics, not of one catastrophe, but of a quiet deposition over vast periods of time. These strata, then, had resulted from the destruction of a succession of former worlds, and Hutton made the process of decay his first principle in the operation of the earth-machine. On the bottom of the sea, the detritus was consolidated by the pressure of the water above and, he conjectured, by the heat of the earth coming from below the sea bed. This same heat also, in some manner, elevated the consolidated strata, giving rise to new worlds, and thus repairing the decay of the old ones. The renovation process was Hutton's second principle. Hutton postulated a world always in existence as his third principle, although the present world is actually undergoing both decay and repair continuously.

In the destruction of a world, Hutton made clear, "we are not to suppose, that there is any violent exertion of power, such as is required to produce a great event in little time; in nature, we find no deficiency in respect to time, nor any limitation with regard to power." And, he continued,

> Nature does not destroy a continent from having wearied of a subject which had given pleasure, or changed her purpose, whether for better or worse; neither does she erect a continent of land among the clouds, to shew her power, or to amaze the vulgar man: Nature has contrived the productions of vegetable bodies, and the sustenance of animal life, to depend upon the gradual but sure destruction of a continent; that is to say, these two operations go hand in hand.[38]

The process of destruction could be observed everywhere in operation, but the process of renovation was out of reach for investigation, and Hutton pressed it on grounds of necessity. "If no such reproductive power, or reforming operation, after due inquiry, is to be found in the constitution of this world, we should have reason to conclude, that the system of this earth has either been intentionally made imperfect, or has not been the work of infinite power and

[38] Ibid., p. 294.

wisdom." [39] In presenting his system, Hutton hoped, "We shall thus also be led to acknowledge an order, not unworthy of Divine Wisdom, in a subject which, in another view, has appeared as the work of chance, or as absolute disorder and confusion." [40] The *Theory of the Earth* was a work on natural theology, as well as geology, but as such it was not received with any show of welcome, though Hutton should have been hailed as the Newton of Geology. His uniformitarianism, his recognition of the role of heat in the consolidation of strata, his wealth of empirically sound observations, have since established Hutton's reputation, but his conclusion that in the system of the world, "we find no vestige of a beginning,— no prospect of an end," [41] was repugnant to the theological temper of geologists in his own age.

The Neptunists attacked all those points in Hutton's geology which made the endless ages possible. That the age of the earth was made the fundamental issue has been noted by Adams, and also by Gillispie, who states:

> ... In the Vulcanist-Neptunist debate, the antiquity of the earth was the issue that transformed a discussion among scientists into a dispute between zealots, even though the ostensible difference between the two schools centered around the primacy of heat as opposed to water in the formation of the crust of the earth.[42]

Of the many Neptunists who distinguished themselves by zealous onslaughts against Hutton, Richard Kirwan was typical. After denying the evidence for a great antiquity of the earth presented by Hutton, the empirical Kirwan struck the telling blow, " I have been led into this detail by observing how fatal the suspicion of the high antiquity of the globe has been to the credit of the Mosaic history, and consequently to religion and morality; a suspicion grounded on no other foundation than those weaknesses I have here exposed." [43]

Hutton and his popularizer, John Playfair, protested that the system did not deny a beginning and an end to the earth, but in

---

[39] Ibid., p. 216.
[40] Ibid., pp. 210-11.
[41] Ibid., p. 304.
[42] Charles Coulston Gillispie, *Genesis and Geology, A Study in the Relations of Scientific Thought, Natural Theology, and Social Opinion in Great Britain, 1790-1850* (Cambridge: Harvard University Press, 1951), p. 43.
[43] "On the Primitive State of the Globe and its Subsequent Catastrophe," *Royal Irish Academy, Transactions*, VI (1797), 307.

the understanding of their contemporaries, this made little difference, for the concept of a time process which Hutton espoused was fatal to the supernaturalism of Mosaic cosmogony. Valuable as this time process was to dynamic geology, however, it was non-historical. As Playfair said, comparing the system of Hutton with that of Buffon, in it "no latent seed of evil threatens final destruction to the whole; and where the movements are so perfect, . . . they can never terminate themselves."[44] In the mechanical operations of Hutton's system, the identity of fossils was lost in the massive operations of endless time, so that the work of the Neptunists in stratigraphy proved to be as essential as the uniformitarianism of Hutton for the discovery of a chronology of the earth—but the actual key to the chronology lay in the correlation of "index" fossil species with particular strata.

The first significant recognition that fossil species could be used as a label for strata appears to have been made by the Abbé Sauvages de la Croix, Professor of Medicine and later Professor of Botany at the University of Montpellier, in a "Mémoire contenant des observations de lithologie, pour servir à l'histoire naturelle du Languedoc, & a la théorie de la terre," read before the French Academy in 1749 and 1750, and published in the *Mémoires* of 1746 (1751). In Holbach's article on fossils in the *Encyclopédie*, mention is made of the plan of Guillaume-François Rouelle to study fossils by groups in a district so that similar groups anywhere could be ascertained by means of a few characteristic species. Although Rouelle, known best as a chemist, apparently did not undertake the project, he did exert considerable influence as a teacher. Lavoisier studied chemistry under him, and also did a creditable work on geology, probably at the instigation of Rouelle.[45] The relationship of fossil species to particular strata was also brought out by the Abbé Giraud-Soulavie in his seven-volume *Histoire naturelle de la France méridionale* (1780-84), and on the basis of this relationship, he ascertained five successive ages in the calcareous strata of Vivarais. But William

---

[44] *Illustrations of the Huttonian Theory of the Earth* (Edinburgh, 1802), p. 483, n. xxiv.

[45] For a contemporary appreciation of Roulle's influence in geology, see Nicolas Desmarest's article (if a work which grew to four volumes may be called that) in the *Encyclopédie méthodique* on "Géographie physique." Lavoisier's work, on littoral and pelagic beds ("following the example of M. Rouelle") appeared in the *Mémoires de l'Académie des sciences* for 1789 (1793), pp. 351-71.

Smith, because of the comprehensiveness of his work, is generally credited with establishing the identification of strata by particular fossil species. Although his work was not published in a completed form until 1815, Smith had made available to a few friends an outline of his ideas as early as 1799. Many of his conclusions were, as a result, presented to the public by his friend, the Reverend Joseph Townsend, in a work entitled, *The Character of Moses Established for Veracity as an Historian, Recording Events from the Creation to the Deluge* (Bath, 1813, 1815). Neither Smith nor Townsend grasped the idea that time was involved in laying down the successive strata, and both thought that they had contributed support to Mosaic cosmogony. Townsend charged Hutton with holding and spreading infidel opinions by giving the earth an antiquity of millions of years and by dismissing supernatural agents in explaining the disorder of the solid parts of the earth.

At the same time that Smith was outlining the strata of Britain, a similar task was being carried out in the Paris Basin, though with a different emphasis. Smith was a practical engineer who had hit upon a useful system for tracing coal measures and ascertaining the direction of strata for the draining of marshland. Zoology was the paramount interest of the investigators of the Paris Basin, Lamarck and Cuvier, who, along with Brongniart, founded the modern science of paleontology. (In Germany, Baron von Schlotheim was also laying the base of scientific paleobotany, early recognizing the importance of fossils for stratigraphy.) They studied the individual fossils as well as the correlation of particular species with the strata, and from the order of succession of the animal forms, they were able to discover a chronological order in the strata. Appraisals of the life histories of the fossil species also made evident to them the time process in the formation of the strata, and consequently, they called for a revolution in the conception of the earth's age.

Lamarck projected a theory of the earth to explain the origin of fossil strata in the *Hydrogéologie* (1802), a theory which was a variant of the *philosophe* concept of natural process, as given by Diderot and Holbach, expanded to meet the cosmogonical problems presented by his work on invertebrate fossils. Undisguised contempt for Mosaic cosmogony was shown by Lamarck in this treatise, and he invoked unlimited time for the processes of nature. The generation and destruction of organized matter were principal forces in

Lamarck's conception of natural process, or, as applied to geology, the degradation of organisms into dead matter, and the integration of dead matter by living organisms. In contrast to the pre-established harmony between the forces of destruction and renovation in Hutton's system, Lamarck thought that the life principle expanded along lines of indefinite progression. The fossil record convinced him that there was an evolution of life through an infinity of variations and gradations. He regarded all the great transformations of the earth's surface as a result of almost invisible changes multiplied by time. "For Nature time is nothing, and is never a difficulty; she always has it at her disposal, and it is for her a power without limits with which she does the greatest things like the smallest." [46]

Throughout the *Hydrogéologie*, Lamarck stressed the importance of seeing the power of time in geological processes, as in the following passages:

> Oh! how great is the antiquity of the terrestrial globe! and how little the ideas of those who attribute to the globe an existence of six thousand and a few hundred years duration from its origin to the present!
>
> The natural philosopher and the geologist see things much differently in this respect; because, if they consider ever so little, first, the nature of fossils spread in such great numbers in all parts of the exposed globe, either at heights, or at considerable depths; second, the number and disposition of the beds, as well as the nature and order of the materials composing the external crust of the globe, studied in a great part of its thickness and in the mass of the mountains, how many occasions they have to be convinced that the antiquity of this same globe is so great that it is absolutely outside the power of man to appreciate it in any manner! [47]
>
> How much this antiquity of the terrestrial globe will grow in the eyes of man, when he has formed a just idea of the origin of living bodies, as well as the causes of the development and gradual perfecting of the organization of these bodies, and especially when he has conceived that, time and circumstances having

---

[46] *Hydrogéologie, ou recherches sur l'influence générale des eaux sur la surface du globe terrestre, sur les causes de l'existence du bassin des mers, de son déplacement, de son transport successif sur les différents points de ce globe, enfin sur les changemens que les corps organisées vivants exercent sur la nature et l'état de cette surface* (Paris, 1802), p. 67.

[47] Ibid., p. 88.

been necessary to bring into existence all the living species such as we see them, he is himself the latest result and the present *maximum* of this perfecting, whose end, if there is one, cannot be known! [48]

In a memoir published in 1805, Lamarck sounded a note of despair over the reception of his theories, apparently feeling that his conception of a vast antiquity of the earth, so essential to his whole system of earth and species changes, was a major obstacle:

> These considerations, I know, having never been presented elsewhere than in my *Hydrogéologie*, and not having obtained the serious examination that I believe they deserve, can only appear extraordinary even to the most enlightened persons.
>   Indeed, man, who judges the greatness of duration only relative to himself and not to nature, will undoubtedly never really find the slow mutations which I have just presented and consequently he will believe it necessary to reject without examination my opinion on these great subjects.[49]

At about the same time that French readers were becoming aware of Lamarck's views on the time-process in nature, English readers were reacting to the somewhat similar ideas of Erasmus Darwin, as given in *Zoonomia; or, the Laws of Organic Life* (1794-96). If Hutton had dramatized the role of Cleanthes in Hume's *Dialogues Concerning Natural Religion*, Darwin took the part of Philo, " The world plainly resembles more an animal or vegetable, than it does a watch or a knitting-loom." [50] Darwin was well-read on the speculation over fossils, was familiar with the works of Buffon and Hutton, and accepted the idea of a central heat in the earth. His idea of an embryo-state universe was shaped upon the *philosophe* model. Would it be too bold to imagine, he wrote, that in the great length of time since the earth began to exist, perhaps millions of ages before the commencement of the history of mankind, all warm-blooded animals arose from one living filament which the Great First Cause endowed with animality, with the power of acquiring new parts and of

---

[48] Ibid., pp. 89-90.

[49] " Considérations sur quelques faits applicables a la théorie du globe, observé par M. Peron dans sons voyage aux Terres australes, et sur quelques questions géologiques qui naissent de la connoissance de ces faits," *Annales du Muséum d'Histoire Naturelle*, VI (1805), 50.

[50] David Hume, *Dialogues concerning Natural Religion*, 2nd ed. (London, 1779), p. 131.

delivering down these improvements by generation to its posterity, world without end? [51]

An exaggerated conception of the extent of fossil remains in the earth was present in Darwin's speculation on the powers of generation. The following statement indicates the influence of fossils, and the philosophy of Hume, on his cosmogony:

> The late Mr. David Hume, in his posthumous works, places the powers of generation much above those of our boasted reason; and adds, that reason can only make a machine, as a clock or a ship, but the power of generation makes the maker of the machine; and probably from having observed, that the greatest part of the earth has been formed out of organic recrements; as the immense beds of limestone, chalk, marble, from the shells of fish; and the extensive provinces of clay, sandstone, ironstone, coals, from decomposed vegetables; all which have been first produced by generation, or by the secretions of organic life; he concludes that the world itself might have been generated, rather than created; that is, it might have been gradually produced from very small beginnings, increasing by the activity of its inherent principles, rather than by a sudden evolution of the whole by the Almighty fiat.[52]

The *Zoonomia* had reached a third edition by 1801, and in it the theory of a generated world was sharpened, but this view was vigorously countered the next year by the publication of William Paley's *Natural Theology*, the grand synthesis of the argument for the existence of God from the evidences of design in nature. The alternative to design, Paley warned, was chance, and he underscored the threat of atheism in the philosophy of chance, adding,

> There is another answer, which has the same effect as the resolving of things into chance; which answer would persuade us to believe, that the eye, the animal to which it belongs, every other animal, every plant, indeed every organized body which we see, are only so many out of the possible varieties and combinations of being, which the lapse of infinite ages has brought into existence; that the present world is the relic of that variety; millions of other bodily forms and other species having perished, being by the defect of

---

[51] *Zoonomia* . . . , 3rd ed. (London, 1801), II, 240. This statement does not appear in the first edition.
[52] Ibid., p. 247.

their constitutions incapable of preservation, or of continuance by generation.[53]

The systematic investigation of the earth by geologists was by this time bringing to light such an abundance of evidence of a succession of changes in the earth's surface structure that the orthodox were hard pressed to explain them with Mosaic history, but they were undaunted. The conservative reaction from the excess of the French Revolution undoubtedly checked many naturalists from following the path of evidence outside the pale of orthodoxy. On the other hand, it was no longer possible for a sincere geologist, however orthodox, to avoid seeing that there were facts in his science which stood in conflict with the strictly literal interpretation of Genesis, and a new phase of trying to harmonize the Holy Scriptures with geology was opened.

Among the empirical Diluvialists, the work of Jean André Deluc, recognized as a leading geologist in his own day, presaged the rise of nineteenth-century Mosaic geology. The entire course of Deluc's career, some fifty years, was devoted to the accumulation of facts which would confirm from nature the veracity of Moses. He opposed the conclusions of Hutton and others that our continents are of indefinite antiquity, and by *facts* sought to prove that "our continents are of such small antiquity, that the memory of the revolution which gave them birth must still be preserved among men; and thus we are led to seek in the book of Genesis the record of the history of the human race from its origin." [54] However, he could not find natural facts which would obliterate the conclusion that long lapses of time had entered into the formation of the earth, so he solved the problem by drawing a cut-off line at the Deluge for the empire of natural knowledge. Before the Deluge, Deluc maintained, the earth was in a period of gestation, and the stages of this pregnancy were told in the first verses of Genesis. Each of the days may have involved long periods of time, as the fossil memorials suggested, but since the earth was still in the shaping hands of the

---

[53] Paley, *Natural Theology* . . . , Chapter v, par. 4. The previous paragraph contains remarks on fossils shells. Comment on the argument against gradations in the process of time was made in *Edinburgh Review*, 1 (Jan. 1803), 301.

[54] *An Elementary Treatise on Geology* (London, 1809), p. 82. Gillispie, op. cit., p. 64, remarks, "More than anything else, however, the Vulcanist antiquity of the earth bothered Deluc, frightened him really. It was his King Charles's head."

Creator, and subject to the supernatural forces He was likely to employ, this period was out of reach for scientific deduction. However, the present earth was given birth at the Deluge about two thousand years ago, the continents were fixed as permanent features, supernatural forces ceased, and the world was left to run on natural laws. Knowledge of the postdiluvial world could be derived from scientific investigation, but man could not hope to know more of the antediluvian world than was revealed to Moses.

Mosaic geology was distinguished from previous attempts at harmonizing Scripture with cosmogonies framed to take into account the fossil masses by the fact that its advocates in scientific circles had at their command an extensive body of observational data, and they were in possession of methodological techniques and standards which needed only to be applied energetically for geological progress to result. Of those eighteenth-century advocates of a generated universe, the only one who brought a sound knowledge of the fossil species themselves to the support of his evolutionary theories was Lamarck. His influence was totally eclipsed by his countryman and rival, Georges Cuvier, who placed Mosaic geology on what appeared to be a firm and conclusive scientific basis.

Taking for his field the little-known fossil quadrupeds, and applying to them the principles of comparative anatomy, Cuvier astonished the world by reconstructing prehistoric forms of animal life whose existence had never been suspected. He could not fail to see that the age of the earth was more than six thousand years, and in the " Introduction " to his *Recherches sur les ossemens fossiles* (1812), later published separately as a *Discours sur les révolutions de la surface du globe,* Cuvier reiterated Lamarck's plea for more time in geology:

> Genius and science have burst the limits of space; and a few observations, explained by just reasoning, have unveiled the mechanism of the universe. Would it not also be glorious for man to burst the limits of time, and, by a few observations, to ascertain the history of this world, and the series of events which preceded the birth of the human race? [55]

It was a noble proclamation, but Cuvier could not, or would not, rise fully to the occasion. He had no feeling of sympathy for the

---

[55] *Essay on the Theory of the Earth,* tr. Robert Jameson (Edinburgh, 1822), pp. 4-5.

views of either Hutton or Lamarck, who had burst the limits of time, and after he had cracked open Biblical chronology enough to let in the epochs of fossil formations in which he was interested, he undertook the strengthening of Mosaic history, following paths marked out by both Buffon and Deluc.

A substantial part of Cuvier's *Discours* was taken up with his attempt to prove the reality of the Deluge and the short history of man on the earth. He reviewed the traditions of a deluge among the various nations, and asserted that such a confluence of traditions in support of the Deluge could not be mere chance. Chronologies and ancient astronomical systems with pretensions of a greater antiquity than Biblical chronology were examined by Cuvier with all the tedious pedantry of the best scriptural apologists, and he concluded that the testimony of Moses excelled them all. In his view, the Deluge connected antediluvian natural history with civil (post-diluvian) history, but between the two he posed a watershed, after the manner of Deluc, between the operations of supernatural and natural forces.

It so happened that the formations of the Paris Basin which Cuvier studied had some alternate strata of marine fossils, fresh-water fossils, and material without a fossil content. These alternations, he concluded, were the result of sudden revolutions in the state of the sea. " These repeated irruptions and retreats of the sea have neither been slow nor gradual; most of the catastrophes which have occasioned them have been sudden," he wrote.[56] As to explaining these ancient revolutions by now existing causes, he said, " unfortunately this is not the case in physical history; the thread of operation is here broken, the march of nature is changed, and none of the agents that she now employs were sufficient for the production of her ancient works." [57] Cuvier thought that we could seek only in vain among the natural forces still operating to find an explanation of the revolutions memorialized in the earth, and that was why so many conflicting theories of the earth had plagued geology. Supernatural causes alone were sufficient to effect the revolutions.

The deposits made between revolutions were regarded by Cuvier as the product of natural forces and time, but by breaking the " thread of operations " by means of catastrophes, Cuvier defended the role of supernaturalism in the antediluvian world, while at the

[56] Ibid., p. 15.
[57] Ibid., p. 24.

same time, he erected a defence against evolution. When, as the greatest authority on comparative anatomy, he " proved " that every species had a limited ability to vary, he further vouchsafed the divine origin of species. And his vigorous espousal of the Deluge as an actual geological event mollified the orthodox, so that it seemed safe now to reinterpret Mosaic history. Thus Cuvier provided a safety valve between the irrefutable proofs of an ancient earth and literalist Sacred history, between the push of geology and the drag of theology.

Theologians and geologists who were anxious to avoid a conflict welcomed Cuvier's interpretation of the past. There was a quick reinterpretation of the meaning of the first verses of Genesis, either through extending the time of the Days of Creation or by interpolating indefinite ages at the "beginning" before the Days commenced. Among English naturalists, James Parkinson wrote the first two volumes of his excellent *Organic Remains of a Former World* (1804, 1808) with the usual deference to Biblical chronology, but before publishing the third volume in 1811, he had read some of the *mémoires* of Lamarck and Cuvier, and in this last volume, he admitted that the ancient remains must " have been the work of a vast length of time." [58] Parkinson then expressed pleasant surprise at how well this apparent contradiction to Biblical chronology was explained by expanding the time of the Days of the Creation. Thomas Chalmers, a Scottish divine, reviewed Jameson's English translation of Cuvier's *Discours* and accepted the great antiquity of the globe, upon which geologists were agreed, but he doubted if the text of Genesis warranted taking liberties with the time involved in a Day. However, he hastened to point out, Genesis did not say when or how long the *beginning* was: " Moses may be supposed to give us not a history of the first formation of things, but of the formation of the present system." [59] The leadership of Chalmers among the fundamentalists went a long way towards mitigating the opposition of theologians to an expanded time span for the earth.

[58] *Organic Remains*, III, 449. J. Challinor, in " The Beginnings of Scientific Palaeontology in Britain," *Annals of Science*, VI (1948), 46-53, rates Parkinson as one of the founders of scientific paleontology.

[59] *Works of Thomas Chalmers* (Glasgow, 1836-42), XII, 370. The review appeared in the *Christian Instructor* for 1814. In a lecture in 1804, Chalmers had gained some notoriety for asserting, in reference to geology, " The writings of Moses do not fix the antiquity of the globe. . . ." See William Hanna, *Memoirs of the Life and Writings of Thomas Chalmers* (1849-52), I, 81.

There was, nevertheless, a flood of literature from the orthodox who thought that to extend the antiquity of the earth was going to open the floodgates of atheism. Concessions followed rapidly, however, when the Reverend William Buckland, who became the Dean of English Geology, announced,

> ... the grand fact of *an universal deluge* at no very remote period is proved on grounds so decisive and incontrovertible, that, had we never heard of such an event from Scripture, or any other authority, Geology of itself must have called in the assistance of some catastrophe, to explain the phenomena of diluvian action which are universally presented to us, and which are unintelligible without recourse to a deluge exerting its ravages at a period not more ancient than that announced in the book of Genesis.[60]

Shortly afterwards he presented in his *Reliquiae Diluvianae* (1823) what seemed to be overwhelming proofs of the physical occurrence of the universal Deluge.

Diluvialism had never enjoyed a more unassailable position in cosmogony than during these years under the leadership of Cuvier in France and Buckland in England. By pivoting their base firmly on the Deluge as a glorious vindication of the veracity of Moses, geologists carried out a broad flanking movement which swept away serious resistance to the demand for time in the antediluvian period of history. On the whole, there was a feeling, in this decade of the 1820's, that what had threatened to become an open warfare between theology and science had passed over into a harmonious blending of truth from both quarters.

The spell of the Cuvier-Buckland compromise was broken in 1830 by the publication of Sir Charles Lyell's first volume of the *Principles of Geology*. Lyell " was taught by Buckland the catastrophical or paroxysmal theory," [61] but after considerable reading and observation had come around to the uniformitarian philosophy. Some recent work, notably that of Constant Prévost and George Poulett Scrope, had already laid the groundwork for a new expression of the philosophy that steady and continuous operation of natural forces in the course of time could effect all the changes visible in the earth's structure. But the merit of Lyell's *Principles* was in its compre-

---

[60] *Vindiciae Geologicae; or, the Connexion of Geology with Religion Explained* [Oxford, 1820], p. 24. Cited in Gillispie, op. cit., p. 107.

[61] *Life, Letters, and Journals of Sir Charles Lyell, Bart.*, (London, 1881) II, 6-7. Letter of Lyell to Whewell, March 7, 1837.

hensive and persuasive exposition of the doctrine. To pave the way towards its acceptance, Lyell prefaced the work with an exceptionally fine, though far from definitive, history of theories of geology. Among those prejudices which had retarded the progress of geology most, he gave first place to "prepossessions in regard to the duration of past time," because undervaluing the quantity of past time created an apparent coincidence of events which were in fact widely separated in time, and made it difficult for men to perceive the aggregate effects of causes which have operated throughout millions of years. Seeing disconnected geological monuments in close contact with each other, without the intermediate events, made the passage from one state of things to another appear so violent that the idea of revolutions naturally suggested itself, but the prejudice about time had led geologists into the fallacy of accepting the gaps in the fossil record as gaps in the processes of nature itself. When Lyell interpolated time and process into the gaps, all supernatural cataclysms, including the Deluge, and therewith the "broken thread of operations" which had been a proof of the sudden appearance of new species, were rendered superfluous in geology.

Praise and condemnation met the *Principles of Geology*; praise for Lyell's presentation of data, and condemnation for the conclusions he drew from them. A new effort was called for in order to bring geology and revelation back into harmony, and in the next two decades a voluminous literature aimed at trying to assimilate the chronological revolution forced upon the orthodox by Lyell's uniformitarianism. Buckland capitulated to his student. Speaking of the doubts experienced by many learned and pious men, he wrote:

> These doubts and difficulties result from the disclosures made by geology, respecting the lapse of very long periods of time, before the creation of man. Minds which have been long accustomed to date the origin of the universe, as well as that of the human race, from an era of about six thousand years ago, receive reluctantly any information, which, if true, demands some new modifications of their present ideas of cosmogony; and, as in this respect Geology has shared the fate of other infant sciences, in being for a while considered hostile to revealed religion; so like them, when fully understood, it will be found a potent and consistent auxiliary to it, exalting our conviction of the Power, and Wisdom, and Goodness of the Creator.[62]

---

[62] Buckland, *Geology and Mineralogy Considered with Reference to Natural Theology* (London, 1836), I, 8-9.

In 1830, and for a number of years afterwards, Lyell believed in the recency of man and the fixity of species. The implication of evolutionism in his system of natural process apparently did not occur to him, and to some extent this may have been because, as a geologist, he was dealing with dead materials and because his uniformitarianism had the same kinship as Hutton's with the contrivance-world of natural theology. Time without beginning in Lyell's system was more a term in a mechanical equation than a factor in the processes of growth and development. However, the outline of the tree of life was beginning to appear from the work of paleontologists, and estimates of the duration of life were becoming more concrete.

Practical geologists by and large considered only the tangible remains of species and continued to think that the breaks in the continuity of the fossil record represented gaps in the thread of life, but as the extent of prehistoric life was uncovered, time captured the imagination of writers. They looked out over the enormous vista of creation rising before their eyes, like steps, disappearing over the horizon in their Temple of Nature, and in the first excitement of discovering this new dimension, an extravagance was lavished in praise of the power and majesty of the Creator. But the enlarging vista required a continuous series of revisions in the harmonizing efforts of Mosaic geologists.

With each step in the progress of paleontology some conflict had arisen, and especially troublesome had been the order of creation, but by the 1840's there was an increasing acceptance of the "Miltonic hypothesis." Under this theory, it was believed that after the Lord had taken Moses up to the mountain top to survey Creation, Moses, from his limited knowledge of science and its terminology, could only describe in rude terms the vision presented to him. Hugh Miller and John Pye Smith in England and J. H. Kurtz in Germany were influential in developing the Miltonic hypothesis in Mosaic geology, an interpretation which virtually removed Genesis from the field of practical geology. This movement was undoubtedly important in preparing the way for the "Higher Criticism," and in bringing to a close this geological phase in the struggle between literal theology and science, soon forgotten in the great crisis over evolution itself for which it had prepared the way.

While the Cuvier-Buckland compromise was the reigning geological system, Charles Darwin, in 1828, entered Cambridge to become a clergyman. There, he enjoyed the friendship of Professor Henslow, a man so orthodox he would have " grieved if a single word of the Thirty-nine Articles were altered;" but Darwin did not show an interest in geology until after he had graduated. He then took a geological trip with Adam Sedgwick, who had written in 1825:

> The sacred records tell us—that a few thousand years ago "the fountains of the great deep were broken up "—and that the earth's surface was submerged by the water of a general deluge; and the investigations of geology tend to prove that the accumulations of alluvial matter have not been going on many thousand years; and that they were preceded by a great catastrophe which has left traces of its operation in the *diluvial detritus* which is spread out over all the strata of the world.[63]

In 1831, Darwin still intended to become a clergyman; "nor," he wrote, "was this intention and my father's wish ever formally given up, but died a natural death when, on leaving Cambridge, I joined the *Beagle* as naturalist." [64] His views on religion were indeed sound at this early period, as he indicated with amusement in his autobiography. "Whilst on board the *Beagle* I was quite orthodox, and I remember being heartily laughed at by several of the officers (though themselves orthodox) for quoting the Bible as an unanswerable authority on some point of morality." [65] On the voyage, his "chief favorite" reading was Milton's *Paradise Lost*, and on his excursions away from the ship, when he could take only a single volume, Darwin usually chose Milton, whose powerful support of the Mosaic account of creation could hardly have disturbed his orthodoxy. At Cambridge, he had memorized Paley's *Evidences* and enjoyed his *Natural Theology*. So far, there was hardly a ripple of unconventionality in the young naturalist.

Darwin's point of departure from orthodoxy on this voyage was, of course, his reading of the first volume of Lyell's *Principles of Geology*, urged upon him by Professor Henslow with the warning, " on no account to accept the views therein advocated." [66] The

---

[63] " On Diluvial Formations," *Annals of Philosophy*, n. s. x (1825), 34-5. Also cited in Gillispie, op. cit., p. 113.
[64] *Life and Letters of Charles Darwin, including an Autobiographical Chapter*, ed. Francis Darwin (New York & London, 1925), I, 39.
[65] Ibid., I, 277. [66] Ibid., I, 60.

efficacy of the *Principles* in throwing light on the geology of South America quickly converted Darwin to uniformitarianism.

Later, in 1844, Darwin remarked to Leonard Horner, "I always feel as if my books came half out of Lyell's brain, and that I never acknowledge this sufficiently; . . . for I have always thought that the great merit of the *Principles* was that it altered the whole tone of one's mind. . . ."[67] Over and above the methods of reasoning and the factual content in the *Principles*, the alteration of Darwin's whole tone of mind was unquestionably his gaining a sense of the time process in nature. With a few misgivings, Darwin easily bypassed the controversy over the antiquity of the earth and catastrophism, going from Milton to Lyell in one easy step. Darwin's isolation on the *Beagle*, far from the angry presses, probably made the assimilation of uniformitarianism much easier. However, had Henslow not prevailed upon Darwin to take along the *Principles*, at the end of the voyage he might have written an endorsement to the views of the expedition's commander, Captain Fitzroy, who concluded from the marine fossils found in the lands of Patagonia that "if Patagonia was covered [by the sea] to a great depth, all the world was covered to a great depth; and from these shells alone my mind is convinced (independent of Scripture), that this earth has undergone an universal deluge."[68]

The fact that Lyell did not adopt the theory of evolution of species on his own initiative shows that more was required for an evolution theory than his *Principles of Geology*. It is also true that Darwin was led into speculations on evolution to a large extent by his biological observations on the modification of closely related species in various regions of South America and the Pacific Islands. Again, in seeking a natural principle to explain the modifications, he turned to the study of methods of artificial selection used by gardeners and breeders before the idea of natural selection was suggested to him by reading Malthus. Yet, in assessing the influential factors leading up to the *Origin of Species*, the alteration of the whole tone of Darwin's mind on the matter of time and natural process deserves credit as a major step towards his theory of evolution, an influence acknowledged by Darwin himself:

---

[67] *More Letters of Charles Darwin* . . . , ed. F. Darwin and A. C. Seward (New York, 1903), I, 117.

[68] *Narrative of the Surveying Voyages of His Majesty's Ships Adventure and Beagle between the years 1826 and 1836* . . . (London, 1839), II, 666.

Independently of our not finding fossil remains of such infinitely numerous connecting links, it may be objected that time cannot have sufficed for so great an amount of organic change, all changes having been effected slowly. It is hardly possible for me to recall to the reader who is not a practical geologist, the facts leading the mind feebly to comprehend the lapse of time. He who can read Sir Charles Lyell's grand work on the Principles of Geology, which the future historian will recognise as having produced a revolution in natural science, and yet does not admit how vast have been the past periods of time, may at once close this volume.[69]

---

[69] *Origin of Species*, 6th ed. (Modern Library ed.), p. 236.

THE NINETEENTH CENTURY

TEN

LAMARCK AND DARWIN IN THE HISTORY OF SCIENCE

CHARLES COULSTON GILLISPIE

I

For myself, as for many historians of science, the first reading of Professor Lovejoy's *The Great Chain of Being* came as a revelation of what the philosophical history of ideas about nature might accomplish. So too, in a more restricted sense, did his articles on early evolutionists introduce us to the history, or perhaps the prehistory, of the concept of evolution. I count it a great honor, therefore, to contribute a paper to a volume which restores those essays to print. And I am sure it will be taken as a testimonial to the stimulus afforded by Professor Lovejoy's scholarship if I say that *The Great Chain of Being* appears to leave a gap, which his evolutionary papers do not fill, between the "temporalizing" of the principle of plenitude as the program of nature, and the actual establishment of the concept of evolution on a scientific footing in the work of Charles Darwin. It is this gap across which I shall try to toss a line, and I make bold to hope that Professor Lovejoy may recognize his own guidance in the direction it takes, for it may be that my paper will bring the question back to the starting point of *The Great Chain of Being*. My purpose is to define the sense in which Darwin did, and the sense in which he did not, have predecessors. And since the foremost of the pre-Darwinian evolutionists was certainly Lamarck, this may best be accomplished by comparing the places which they occupy in the whole structure of the history of science.

Most students will agree that Darwin's work had two aspects, empirical and theoretical. First, it definitively established the mutability of organic species in their descent out of the past. Secondly, it explained these variations by the concept of natural selection. Religious fundamentalists might deny the fact of evolution. But this reaction was intellectually trivial, and where philosophical offense was taken, it was rather the view of the world implicit in the theory of natural selection which wounded humane sensibilities more deeply, and which was repudiated as inadmissible or meaningless or both, for the two complaints come down to the same thing and turn on the eternal question of what a scientific explanation really says.

There is, perhaps, a certain inconsequence which besets controversy of this sort. Having rejected Darwin's evolutionary principles, most of his opponents thought it worthwhile to impugn his originality. Among moralists Samuel Butler, and among scientists (though for different reasons) the French, put it about that Lamarck had had everything essential to an evolutionary biology. And in the face of the enterprise on which the contributors to the present volume are engaged, it would be difficult indeed to claim the fact of the evolutionary variation of species as a Darwinian discovery. It is true that Darwin disposed of a greater fund of species than had Lamarck. Moreover, the seating of the chronology of earth history in paleontological indices gave biologists by way of return the succession of species in geological time. It was for lack of this information that Lamarck had had to establish his order in the scale of increasing morphological complexity.

It is to be doubted, however, whether the uniformitarian philosophy of Charles Lyell was as essential to Darwin's success as is usually said, or as I once said myself.[1] For Lamarck had been his own Lyell. His *Hydrogéologie* prepared the ground for his theory of evolution with a uniformitarian earth history as uncompromising as Lyell's, if not so well founded.[2] More generally, it might be argued—indeed, I do argue—that in the relative cogency with which the two theories organize actual biological information, Lamarck's presentation in the great *Histoire naturelle des animaux sans vertèbres* is the more interesting and elegant.[3] It is analytical and informs a systematic taxonomy, whereas Darwin simply amassed detail and pursued his argument through the accumulated observa-

[1] *Genesis and Geology* (Harvard University Press, Cambridge, 1951).
[2] Paris, 1802.
[3] Paris, 1815-1822.

tions in a naturalist's commonplace-book. To be single-minded and relentless is not necessarily to be systematic, and the merit of Darwin's approach must be sought elsewhere.

Nevertheless, despite the greater formal elegance of Lamarck's ultimate presentation, his theory failed to compel assent. It scarcely even won attention. Those most competent to judge, Lamarck's own scientific colleagues, treated his ventures into theory as the embarrassing aberrations of a gifted observer, to be passed over in silence. "I know full well," he once observed bitterly, "that very few will be interested in what I am going to propose, and that among those who do read this essay, the greater part will pretend to find in it only systems, only vague opinions, in no way founded in exact knowledge. They will say that: but they will not write it." [4] Cuvier and Lamarck were able to work together in actual taxonomy. But they could never agree on the structure of nature.

No such humiliating judgment of irrelevance awaited Darwin's theory. (Even though my argument is that Darwin's original contribution was the theory and not the evidence, I shall in the interests of economy perpetuate the injustice which makes Wallace's role in the history of science little more than an object lesson in the agonizing generosity of creative minds.) Huxley's description of his own reaction is well known. Once stated, the force of the concept leapt out at him like the pattern from the pieces of the puzzle of adaptation, so that all he could say to himself was, "How extremely stupid not to have thought of that." [5] The right answer, it presented itself in that combination of unexpectedness and irresistibility which has often been the hallmark of a truly new concept in scientific history.

One sometimes reads, however, that the force of Darwin's ideas derived from their mechanistic character in an age which identified the scientific with the mechanistic. I cannot think this quite correct. The one thing Darwin did not, and could not, specify was the mechanism of variation or heredity. All he could do was postulate its naturalistic mode. Ultimately, of course, his hypothesis was vindicated by discoveries in genetics of a materialistic character. But that is quite another matter—the *bête-machine* belongs to the 18th century (or to the 20th), but not to the 19th. It was through no metaphor or analogy that Darwin prevailed. He prevailed

[4] *Recherches sur l'organisation des corps vivants* (Paris, 1802), p. 69.
[5] Leonard Huxley, *Life and Letters of T. H. Huxley* (2 v., Appleton, New York, 1901), I, p. 183.

because his work turned the study of the whole of living nature into an objective science. In the unlikely guise of a Victorian sermon on self-help in nature, on profit and loss, on progress through competition, there was clothed nothing less than a new natural philosophy, as new in its domain as Galileo's in physics. Darwin, indeed, abolished the distinction which had divided biology from physics at least since Newton, and which rested on the supposition (or defense) that the biologist must characteristically study the nature and the wisdom of the whole rather than the structure of the parts.

II

Lamarck, too, conveyed a philosophy of nature in his theory of the development of life, but it stands in the same relationship to Darwin's as does Hegel's historical dialectic to that of Marx. It is no compliment to Lamarck's own conception of his lifework, therefore, to make him out an unappreciated forerunner of Darwin. I recently asked a friend who is a biologist specializing in evolution what he and his colleagues understood by Lamarckism, and the first thing he said was the inheritance of acquired characteristics, and after that a lingering temptation in biology generally to indulge in an "Aristotelian vitalism."[6] Now vitalism and the mode of acquisition and transmission of variations were, indeed, the points on which scientific discussion of evolution turned in the later decades of the 19th century and before the establishment of modern genetics. And it is most natural that biologists should have looked upon Lamarck in the perspective of their science, which takes the shape of evolution from Darwin. But in doing so they have first missed and then misrepresented the point of Lamarck's work, which was neither Aristotelian nor vitalistic, and which instead was meant to establish, not simply the subordinate fact of transmutation, but a view of the world. For Lamarck's theory of evolution was the last attempt to make a science out of the instinct, as old as Heraclitos and deeply hostile to Aristotelian formalization, that the world is

[6] I wish to acknowledge the kindness of the colleague in question, Dr. Colin S. Pittendrigh, who explained to me very patiently various aspects of the outlook of modern evolutionists. Section V of this essay has specially profited from this discussion—though I should be distressed if any mistakes which remain were attributed to anyone but me.

flux and process, and that science is to study, not the configurations of matter, nor the categories of form, but the manifestations of that activity which is ontologically fundamental as bodies in motion and species of being are not. This is no longer a familiar view. It is not even recognizable. And it may be helpful, therefore, to reconstitute it by moving from what is familiar in Lamarck to what is less so, and by this means to trace the formation of his theory of evolution backward from its definitive formulation in taxonomy to its origin in an insufficiently understood pattern of 18th-century resistance to the implications of Newtonian physical science. Moreover, since Lamarck was himself sympathetic to the *idéologues*, and a disciple of Condillac in relating mind to nature through the associationist psychology, a genetic analysis of his theory of evolution will conform to his own conception of scientific explanation.[7]

For Lamarck began as a theorist and only ended as a taxonomist. The great *Histoire naturelle des animaux sans vertèbres* appeared between 1815 and 1822. It is presented as exemplifying the evolutionary theory, which was already fully developed in the *Philosophie zoologique*, published in 1809. But Lamarck had first adumbrated the notion of transmutation of living species in his course at the *Muséum d'histoire naturelle* in 1800. The date is significant. Since we know that in 1797 he still believed in the immutability of species, this interval of three years has always been taken as the critical and creative period in Lamarck's life, when he revolutionized his concepts and founded evolutionary thought.[8] The circumstance is curious for another reason Although Lamarck was fifty-seven years old in 1800, he was only beginning his career as a zoologist. Since 1793 when, known to science as a botanist, he was appointed to the new chair of zoology at the reorganized *Muséum*, his thoughts had been absorbed by writings on chemistry, geology, and meteorology. These interests still figure in the *Philosophie zoologique*, and are usually ignored by scientific readers. That book has three divisions. Part I treats of natural history, Part II of physiology, and Part III of psychology. But under the inspira-

---

[7] I have already published the substance of this section as "The Formation of Lamarck's Evolutionary Theory," *Archives internationales d'histoire des sciences*, IX (1956), pp. 323-338, where the reader will find somewhat more circumstantial and bibliographical detail. The conclusions were revised somewhat to publish in *Actes du VIIIe Congrès International d'Histoire des Sciences* (Florence, 1956), pp. 544-548.

[8] Marcel Landrieu, *Lamarck* (Société zoologique de France, Paris, 1909), pp. 297-302. This work contains a complete bibliography of Lamarck.

tion of Cabanis, the latter two divisions handle a single theme, the physical basis of life and consciousness.

Concise statement was never Lamarck's own way. It is, nevertheless, possible to abstract from Part I of the *Philosophie zoologique* a summary of the evolutionary theory in its final form. In living nature, according to this zoological philosophy, inheres a plastic force—indeed, living nature *is* a plastic force—forever producing all varieties of animals from the most rudimentary to the most advanced by the progressive differentiation and perfection of their organization. If this action of organic nature were omnipotent, the sequence would be altogether regular, a perfect continuum of organic forms from protozoa to man. But the innate tendency to complication is not the only factor at work. Over against it, constraining it into certain channels of necessity which we mistakenly take for natural species, works the influence of the physical environment. The dead hand of inorganic nature causes discontinuities in what the organic drive toward perfection would alone achieve. These appear as gaps between the forms of life. Changes in the environment lead to changes in needs; changes in needs produce changes in behavior; changes in behavior become new habits which may lead to alterations in particular organs and ultimately in general organization. But the environment cannot be said to act directly on life. On the contrary, in Lamarck only life can act, for life and activity are ultimately one. Rather, the environment is a shifting set of circumstances and opportunities to which the organism responds creatively, not precisely as an expression of its will (although Lamarck's admirers interpreted him in that fashion) but as an expression of its whole nature as a living thing. And it was rather as a consequence than as a statement of his view of nature that Lamarck laid down two corollaries which he described as laws: that of the development or decay of organs through use or disuse, and that of the inheritance of the characteristics acquired by organisms in reacting to the environment.[9]

According to Lamarck himself, *Philosophie zoologique* was the elaboration of an earlier work, *Recherches sur l'organisation des corps vivants*, of 1802. Here, too, the main evolutionary principles may all be found, though it is even less possible than in the later work to take them for the point of the argument. The body of

[9] The best summary is in E. Guyénot, *Les sciences de la vie en France aux 17e et 18e siècles* (A. Michel, Paris, 1941), pp. 418-439.

the treatise is devoted to physiology and psychology, and the theory of evolution appears as a preface. Moreover, the emphasis is different. The position is rather that species do not exist than that they are mutable. What interested Lamarck at this stage in the development of his opinions was the whole tableau of the animal series. We are to see it, not as the chain or ladder, but as the escalator of being. For nature is constantly creating life at the bottom. And life fluids are ever at work differentiating organs and complicating and perfecting structures. And there is a perpetual circulation of organic matter up the moving staircase of existence, and of its lifeless residue spilling as chemical husks back down the other side, the inorganic side. Here, too, Lamarck states laws. But they are different laws from those of 1809. They generalize the facts, not about evolution in time, but about the whole zoological scale of being, which exists both in time and at any time. The first law states that there is indeed a regular series in nature, and the second that it resides, not in species, but in organic "masses" which he defined as the life stuff distributed among the different systems of organization.[10]

Moving back again, two years this time to the inaugural lecture of 1800, in which Lamarck first spoke of transmutation, one meets with yet another emphasis. He there advanced the evolutionary thesis, so he tells us, as a pedagogical device, by which to lead his students' minds back down the path which nature herself had followed in (so to say) producing them. Consistently enough, he presents the animal series as a study in degradation, not development. The theses of 1802 do appear, as do most of the evolutionary principles of 1809, but they occur only as very summary propositions in support of the main contention. This is that natural history must begin with the fundamental distinction between living and non-living bodies, between organic and physical nature.[11]

Now, not only was this the argument of Lamarck's debut as a zoologist, it was also the argument of his final assault upon the new chemistry, *Mémoires de physique et d'histoire naturelle* of 1797, in which he referred in passing to the unchanging character of species.[12] The *Mémoires* resumed the attack which as a young man

---

[10] *Recherches sur l'organisation des corps vivants*, pp. 39-41.
[11] *Système des animaux sans vertèbres* (Paris, 1800).
[12] *Recherches sur les causes des principaux faits physiques* (2 v., 1794), II, 214, to which compare *Mémoires de physique et d'histoire naturelle* (1797), pp. 270-271.

he had launched twenty years before in the *Recherches sur les causes des principaux faits physiques*. The central dynamical proposition is that which he later developed into the escalator of being: that all inorganic composites are residues of life processes, perpetually repairing the decay and disintegration which are all that physical nature holds in the way of process, and as perpetually doomed by mortality in their drive to bring living order to a world of chaotic physical necessity. Returning for a moment to the *Philosophie zoologique*, this is perfectly consistent with the theory of evolution, in which irregularities in the animal scale mark the casualties in the conflict between organism and brute environment. This relationship between organic nature as order and physical nature as disorder, a situation of both opposition and dependence, is fundamental to Lamarck's thought, which in this respect is almost dialectical.

Nor is the inconsistency on species other than trivial. In a short essay of 1802, Lamarck himself tells us how he came to alter his view on what he then saw as a detail.[18] All he did between 1797 and 1800 was to assimilate the question of animal species—or rather their non-existence—to that of species in general. For in Lamarck the word has not lost its broader connotations. It still carries the sense of all the forms into which nature casts her manifold productions in all three kingdoms (or rather in both divisions). He had long been impressed with the perpetual crumbling decay of the surface of the earth. He had long shared Daubenton's opinion that there are no permanent species among minerals. The only entities in organic nature are the "*molécules intégrantes*" and the masses which form in the play of circumstance and universal attraction.

This makes a striking parallel to the view that Lamarck came to hold of the living world. In both organic and inorganic nature, there is nothing but process linking the individual—the particular animal, the particular molecule—and the system of organization—mammalian quadruped, granitic structure—into which it is temporarily cast. This explains Lamarck's pleasure in the concept of masses as links in the double chain of systems along which materials move, from mollusc to man, from limestone to granite. It is natural enough to think of the principle of granite as mass. What Lamarck did was to think of the principle of mammal in the same fashion.

[18] *Recherches sur l'organisation des corps vivants*, pp. 141-156.

For the notion was still very widespread in the 18th century that minerals are molded by some plastic force, that they are bred in the womb of the earth. Lamarck did not express this old instinct. But he cannot have been unaware of it, and it is implicit in his chemistry, where he refused to believe that a molecule can be "as ancient as the world." And the interest of this chemistry is that in asserting the indefinite variability of chemical composition, it contained the germ of what, when it was transferred to natural history, was to become Lamarck's evolutionary theory.

In Paris, Lamarck complained, the chemists teach that the integral molecule of every compound is invariant, and consequently that it is as old as nature. It follows that species are constant among minerals. As for himself, he was convinced that the integral molecule of every compound can change in its nature, that is to say in the number and proportion of the principles which constitute it. To deny this is to deny the phenomena of chemistry—the fermentations, the dissolutions, the combustions—which leave the molecules in some different condition.[14]

Furthermore, if attention be turned from evolutionary natural history to the other aspects of the *Philosophie zoologique*, the physiology and psychology, these too emerge as derivative from an archaic chemistry of qualities, both in manner and substance. The cardinal principle of this chemistry was that only life can synthesize. Conversely, in the life process the physiology of growth consists in retention during youth of what is needed from the materials which the organism passes through its system. Aging and death follow on the progressive hardening of the pliant organs by their lifelong digestion of the environment. Later on, Lamarck adapted his principle of an equilibrium balancing life against mass to provide evolution with a mechanism. It was analogous to erosion, no doubt because Lamarck was writing his uniformitarian treatise on geology at the time when he first put forward his evolutionary view of life.[15] The property of life fluids is to wear away new channels, new reservoirs, and new organs in the soft tissues, and thereby to dif-

[14] Ibid., pp. 150-152.
[15] It cannot be too much emphasized that Lamarck saw his own work as a single body of thought. His original plan was to follow his "physique terrestre" with the *Hydrogéologie* (1802), a *Météorologie*, and a *Biologie*. It was material which he had originally reserved for the latter that he drew on for the *Philosophie zoologique* and its predecessors. But he never did draw his mineralogical writings together.

ferentiate structures and specialize functions.[16] The individual organism silts up after a time and dies, but it leaves more highly complicated descendants.

Lamarck's first scientific essay was a chemical treatise of 1776. It contains an interesting note. In order to explain physically the origin and mechanism of the universe (and he aspired to nothing less), we need to understand, so he wrote, three things: the cause of matter, the cause of life, and the cause of that activity everywhere manifest.[17] Having dealt, therefore, with Lamarck's views on the origin and essence of matter and organism, let us turn briefly to the third problem of his trilogy, the problem of activity. In all his chemistry Lamarck attached primary importance to the element of fire. Later on he was to attack oxygen as a perfectly gratuitous postulate. Not only has it never been seen, but combustion is explicable as the action of fire, which can be seen in the act of burning or shimmering over a hot stove or a tile roof in the sun of a summer day.[18] But this was not simply a disagreement over the most common chemical reaction. For fire is the principle of activity. It exists in many states, of which Lamarck undertook, characteristically enough, a taxonomy.[19] In the fixed state, in coal, wood, or what will burn, fire is the principle of combustion. Conflagration is fire in its state of violent expansion, penetrating the pores of a burning body and ripping it to shreds. Evaporation occurs when fire in a state of moderate expansion surrounds molecules of water and bears them upward, so many tiny molecular balloons, to rejoin the clouds where the specific gravity of the water molecule encased in its light shell of fire balances that of air. (Lamarck also aspired to found the science of meteorology.) Finally—not to follow fire into all its states—there is a natural state, to which fire strives ever to return. And all the phenomena of light and heat, all the effects of sun and atmosphere, are manifestations of fire in its different states, forever striving to regain that which is natural.

Nor did Lamarck ever abandon his commitment to fire. It provided him with a physical basis of feeling and of life itself, and this will make clear the mistake of those who have taken him for a

[16] *Recherches sur l'organisation des corps vivants*, pp. 7-9. Cf. *Causes*, II, pp. 184-219.
[17] *Causes*, II, p. 26.
[18] Ibid., I, pp. 47-60.
[19] This is the subject of most of volume I of the *Causes*; see too *Réfutation de la théorie pneumatique* (Paris, 1796), pp. 31, 36.

vitalist. His dichotomy of organic and inorganic nature provides no escape into transcendentalism, and that has always been the door through which vitalists have slipped from science into mystery. Life is a purely physical phenomenon in Lamarck, and it is only because science has (quite rightly) left behind his conception of the physical that he has been systematically misunderstood and assimilated to a theistic or vitalistic tradition which in fact he held in abhorrence. In his view spontaneous generation was no continuing miracle. Life was activated by the stirring of fluids. Lamarck hinted that this process is quickened by fire, and on the mechanism of sentience he was explicit. Its physical basis is the nervous fluid, the same substance as the electrical fluid, which itself is only a special state of fire.[20] The pyrotic theory, therefore, embraces matter, life, and activity, and in that theory lay the common origin of the three aspects of the *Philosophie zoologique*—its psychology, its physiology, and its evolutionary view of species.

Moreover, the emergence of Lamarck's evolutionary views from his chemistry is more than a curious adventure of ideas. For chemistry was then the locus of the continuing scientific revolution. Lavoisier carried out Lagrange's dictum that the future of chemistry must lie in turning it into a kind of material algebra. Lamarck's notion of chemistry was different. Not only was his chemistry contemplative rather than experimental and analytical, but it contemplated a different sort of world, a world of which the ultimate seizable characteristic is continuity rather than that susceptibility to analysis which depends on objectification of discrete entities. Lamarck's nature is continuous both in structure and operation. Thus, matter itself is something infinitely plastic, however inert, and he escapes the consequences of the particulate views everywhere accepted by denying to the molecule, the ultimate particle, that permanence required by the doctrine of chemical fixity. So, too, when he considered the history of nature, his conception of process was uniformitarian even before the geological issue was fairly raised. In this question, only accident found him on the side of sobriety.

Mind itself, finally, is continuous with nature. For Lamarck, scientific explanation cannot assume the posture of standing outside the subject. Unless an explanation is causal and graphic, it is nothing, and the business of physics is to bring each problem back to some inherent principle or tendency to perfection, carried out by

[20] See, e. g., *Corps vivants*, pp. 163-164, 195.

the agency of a subtle fluid, which is distinguishable only in its effects, but which must exist lest the effects remain inexplicable. He criticized the chemists for holding that addition of one substance to another can form a new one. This presents the product as pre-existent in the reagents. If that is true, nothing has happened. If it is not true, nothing has been explained. To invoke "affinity" is to conjure up an occult force. Instead, the chemist must describe how active principles permeate and alter bodies in reaction.[21] Principles are what combine, and they do so in total mixture. For Lavoisier, on the other hand, materials are what combine, even though he did see caloric as a material substance. The issue was deeper than the argument over phlogiston, and Lamarck lumped Priestley with Lavoisier in the "pneumatic school." Before them all, he flung down the pyrotic theory and the old retort invoked against Galileo and Newton in their day, and to be invoked against Darwin in his: that to describe is not to explain. At the same time, he implied the newer retort of romanticism: that to analyze and quantify is to denature. His attack upon Lavoisier is of a piece with Goethe's *Farbenlehre* and with the writings of Marat, from whom Lamarck drew a certain inspiration, extending even to their mutual resentment of the claims of mathematics to speak as the language of science.[22]

Lamarck's philosophy, therefore, is no anticipation of Darwin, but a medley of dying echoes: a striving toward perfection; an organic principle of order over against brute nature; a life process as the organism digesting its environment; a primacy of fire, seeking to return to its own; a world as flux and as becoming. He is the last important scientist to give them back, these old echoes; and it is no biologist, it is Sainte-Beuve, who has best caught the spirit of this philosophy:

> M. de Lamarck was the last representative of that great school of naturalists and general observers who held sway from Thales and Democritus right down to Buffon. He was the mortal enemy of the chemists, of experimentalists and petty analysts, as he called them. No less severe was his philosophical hostility amounting to hatred for the tradition of the Deluge and the Biblical creation story, indeed for everything which recalled the Christian

---

[21] *Mémoires*, pp. 7-20; *Réfutation*, pp. 69-77.
[22] See the suggestive discussion of G. Bachelard, *La formation de l'esprit scientifique* (Vrin, Paris, 1938), pp. 226-228. For Lamarck's references to Marat, see *Causes*, I, pp. 343-368.

theory of nature. His own conception of things was simple, austere, and full of pathos. He constructed the world out of the smallest possible number of elements, and with the fewest crises and the longest duration imaginable. . . . Similarly in the organic realm, once he had admitted the mysterious power of life, in as minimal and elementary a form as possible, he supposes it developing on its own, building itself up, complicating itself little by little. Various organs were born of unconscious needs, of simple habit working in the different environments against the constant destroying power of nature. For M. de Lamarck separated life from nature. Nature in his eyes was cinder and stone, the granite of the tomb, death itself. Life intervenes only accidentally, as a strange but singularly industrious intruder, fighting a perpetual battle with some little success, achieving here and there a certain equilibrium, but always vanquished in the end.[23]

III

If Lamarck's lifework had been only the last of the Greek philosophies of nature, it might be resigned with other anachronistic vagaries to the keeping of the antiquarian of ideas rather than the historian. But it has, perhaps, a wider significance, not for its content, but as an epitome of the problem of the tension between science and the culture that creates it as its most dynamic and distinctive activity. From Newton to Darwin the preference of romantic thinkers for the sciences of life is as striking as the predilection of rationalistic thinkers for the physical sciences. So it was that Voltaire popularized physics and Rousseau botany. So it was that Paley referred a moral philosophy to astronomy and Bernardin de Saint-Pierre to natural history. It is no accident that the *Jardin des Plantes* was the one scientific institution to flourish in the radical democratic phase of the French Revolution, which struck down all the others.

In the interaction between life and matter which Lamarck saw as the dynamical process of the world, he asserted the qualitative primacy of life in the organism's monopoly of chemical creativity. And down to Darwin, who subjected life to nature, the idea of evolution held great appeal for the romantic mind. Goethe's view of nature, to take the most notable example, is in all essentials the same

[23] *Volupté* (Paris, 1927 edition), pp. 192-194.

as Lamarck's, ministering to the same needs and drawn from the same moral and humane tradition: the unity of nature triumphs over the diversity of the world in universal metamorphosis.[24] For Goethe will replenish the dream, not just with the Faust legend, and magic, "this one book of mystery / From Nostradamus' very hand," but also—"Thee, boundless Nature, how make thee my own?"—with the inter-maxillary bone, the metamorphosis of plants, and a theory of colors as the "deeds and sufferings of light." And perhaps the humanist attempt to understand nature through self-knowledge, though never again to be the way of science, will always be the way of art. Not only of art, but of history, or rather of historicism, for the consonance of the Lamarckian and the Hegelian dialectic is obvious. More importantly, Herder's philosophy of history presupposes the same idea of nature as Lamarck's philosophy of evolution. Its reality is process and unfolding. Its laws are universal extensions of those which govern the birth, growth, and life course of the single organism. It transposes the correspondence of microcosm and macrocosm from space to time, and its model of order is the organism and not the machine, the wisdom of the whole and not the precision and predictable motion of the parts.

In a book on conservatism I recently came upon the following statement appreciatively underlined by a succession of student readers, fortunate young men of the American élite: "Without the creative principle of voluntary action and a healthy degree of self-organization the organic life of society perishes in the arms of an efficient despotism, even though it takes unto itself the sacred name of democracy. The purpose of government is not to concentrate but to diffuse power. Diffusion of power is the characteristic of organic life, just as the concentration of power is the characteristic of mechanism."[25] And (if one follows this lead from pure to political romanticism, where the nexus of authority running between the one and the many really matters) it is almost alarming to think for a moment of the vast structures of reasoning about the state and society which depend upon substituting the metaphor of organism for atomism and mechanism. Where is Burke left without it? Where is the whole conservative apologia of the 19th century if it read the wrong science when it assimilated the notion of the body politic to naturalism? What

[24] See René Berthelot, *Science et Philosophie chez Goethe* (Alcan, Paris, 1932).
[25] R. J. White, *The Conservative Tradition* (Kaye, London, 1950), pp. 8-9.

becomes of socialism if the idea of society as a collectivity which it lifted from romantic conservatism crumbles into atomistic (or individualistic) dust? Nothing would happen to political realities, of course, but at least political apologists would be deprived of the right to draw dogmas from the nature of things and thrown upon their own resources—though since that is where Voltaire threw them two centuries ago, there is no reason to expect them to remain any more content with their own resources than other people do.

Deep interests, then, have been bound up with the view of nature which Lamarck expressed, deep interests and deep feelings. What German, for example, could quite break loose from the spell of *Naturphilosophie*? Not at any rate the greatest of the Darwinians, for Ernst Haeckel celebrated the 50th anniversary of *On the Origin of Species* with an essay in syncretism.[26] And if one looks behind Lamarck into the 18th-century sense of nature, there too it will appear how his evolutionary theory is to be taken as a link between the Enlightenment and romanticism, and not as a way station between the Newtonian spirit and the Darwinian. I have recently had occasion to suggest, in connection with another argument, that Lamarck's writings are the last, though one of the most explicit, of a whole series of attempts, some sad, some moving, some angry, to escape the consequences for naturalistic humanism of Newtonian theoretical physics.[27] For the significance of Newtonian physics for human affairs is that it has none.

Since this argument calls into question the usual view of the Enlightenment as rising in the Newtonian triumph, I may, perhaps, be permitted to summarize it here. Lamarck's mentor, of course, was Buffon. But his significance for intellectual history is more neatly apparent in the adumbration of his evolutionary theory in the speculations of a greater and more sensitive mind than Buffon's or his own. For that theory was an elaboration, and ultimately an application to taxonomy, of the major themes in Diderot's philosophy of nature.

Diderot, too, has often been taken as a forerunner of Darwin and a prophet of some rightful leadership which biological science was

---

[26] Ernst Haeckel, *Das Weltbild von Darwin und Lamarck* (Kröner, Leipzig, 1909).
[27] For a full discussion of this topic with a documentation and further elaboration of Diderot's philosophy of science, see my paper, "The *Encyclopédie* and the Jacobin Philosophy of Science," *Critical Problems in the History of Science*, Marshall Clagett, ed. (University of Wisconsin Press, Madison, in press).

to assume in the 19th century. His famous allusion to the imminent decline of mathematics in science is forgiven as the momentary enthusiasm of one who would restore ordinary sight to eyes temporarily dazzled by the glamor of Newtonianism.[28] But as often when he spoke lightly, Diderot meant this seriously as a repudiation of abstract conceptualization. For it he would substitute at the heart of science that subjective sense of organism which merges nature into consciousness. Like Lamarck, like Marat, like Goethe (and before any of them), Diderot objected to the claims of mathematics to be the language of science. His grounds were ontological, mechanical, and moral. Mathematics idealizes and falsifies real situations. Mathematics deprives bodies of the perceptible qualities in which alone they have existence for an empirical, sympathetic science. Mathematics has turned mechanics into trivial description by mistaking the measurement of bodies for understanding the activity which animates them. Mathematics tempts the mind out into the meaningless infinite, whereas science, like everything related to man, must be limited by his interests and his good.

Diderot, therefore, will save nature from the blight of mathematics by investing matter with sentience and assimilating it to organism. In the *Dream of d'Alembert*, the mathematician is put into a delirium in order to speak out truths.[29] But it is the doctor, seeing nature in the perspective of human nature, who recognizes this stream of innocent consciousness as truth, and even anticipates its findings. He knew the answers all the time, and rejecting the conceptual obscurity of a Newton, Diderot makes scientific explanation into that illumination by nature which through ordinary understanding can penetrate the common breast. For nature is the combination of her elements, of which man is the chiefest, and not just an aggregate. Diderot, therefore, addresses himself to continuity, not divisibility. He, like Lamarck after him, writes of the transience of molecules, not of their existence. In genetics he rejects the atomistic as well as the providential implications of *emboîtement*. For nature knows no limits. Sex blends into sex, mineral into mineral, one living species into another. Individual animals are only moments of tightly organized activity, borne along a stream of seminal fluid, a great river of organic process flowing from the matrix womb of nature herself.

[28] Diderot, "*De l'interprétation de la nature*," in *Oeuvres philosophiques*, ed. Paul Vernière (Garnier, Paris, 1956), pp. 178-179.
[29] *Le Rêve de d'Alembert*, loc. cit., pp. 285-371.

"Tout change; tout passe; il n'y a que le tout qui reste," [30] and Diderot uses two metaphors to express this unity. In one the universe is a cosmic polyp. Time is the unrolling of its life. Space is where it lives. Continuity is its structure. This is the more frequently quoted of his figures, for it associates in one the ideas of evolution and universal sensibility in matter. This latter, of course, Diderot took from Buffon and Maupertuis. But in a very curious way, in a way reminiscent of what is said of Heraclitos, he treats development as a continuum consequent upon the impossibility of dividing time into discrete instants, and it is clear, therefore, that his time could never be a dimension, either of motion or of history, for it is the time of biological subjectivity (as, indeed, is his whole natural philosophy). Properly understood, therefore, Diderot gives us no reason to think of Darwin. And when we, on the contrary, do think of Hegel, or of Marx, it is the second of Diderot's metaphors which seems the more significant. This evokes the swarm of bees. Therein, the solidarity of the universe is social. It is a cosmic anthill, a cosmic hive, where the laws of community are laws of nature. And here, in this social naturalism and not in evolution, Diderot expressed a really prescient, and to the liberal an alarming, concordance between the one and the many, the whole and the parts.

It is clear that this is the same natural philosophy as Lamarck's, not only in itself but in its source. Diderot, too, drew it from chemistry, from the archaic chemistry of qualities and principles best epitomized in the work of Venel.[31] The object of this chemistry was to restore body to that matter which Newtonian mechanics deprived of every attribute but dimension and location—but body in the original sense of internal organization. For until Newton, or at least until Galileo, "body" implied organism in some sense, and this defense of chemistry, therefore, was a defense of organic against inorganic science, of life against the doom of physics. The question, then, as Venel himself wrote on one occasion, was the structure of nature: "The majority of the qualities in bodies which physics regards as modes, are in fact real substances which the chemist knows how to separate out and either restore or incorporate in other bodies. Such among others are color, inflammability, taste, odor, etc." And the congruence of Lamarck's natural philosophy with Diderot's may, therefore, suggest a final considera-

[30] Ibid., pp. 299-300.
[31] See his article "Chimie" in the *Encyclopédie*.

tion on its significance. For the whole point of Diderot's conception of science, as of much of the natural philosophy of the Enlightenment, was to abolish what he thought the unreal distinction between the physical and the moral worlds, to make virtue rise from nature as the object of science. And reciprocally, it is moral insight into nature which is the arm of science, not the conceptual objectification of nature which alienates man, the creator of science, from his own creation.

IV

What fundamental scientific generalization ever came into the world in so unassuming a guise as Darwin's theory of evolution? Is there any " great book " about which one secretly feels so guilty as *On the Origin of Species*? None in the history of science gives me, at any rate, such uphill work with students. Possibly there is a certain cruelty about student judgments, arising from the failure of skeptical young minds to perceive greatness where the scholar and the teacher say it resides. But to have to test our own enthusiasms according to our power to impart them may, perhaps, help maintain a sense of proportion, or at least force us to examine the grounds of what we say. If facts be faced, neither does a truly compelling interest shine spontaneously out of the law of falling bodies, or Newton's laws, or Lavoisier's theory of combustion. But there, at least, the teacher is assisted by the arresting force of Galileo's vision of science, by the daemonic quality of Newton's genius, or by the dramatic shadow of Lavoisier's destiny. Darwin gives one no such help. He claims for himself not power of abstract thought, but only the worthiest and dullest of intellectual virtues—patience, accuracy, devotion. His might be taken as the classic illustration of what Duhem meant when he described the English mind in science as weak and comprehensive. Nothing in the history of science is more familiar than his theory, or than the steps which led him to it by way of the Galápagos Islands and Malthusian political economy.

Like the law of falling bodies, the theory of natural selection is so widely taken for granted that its magnitude is not on the face of it apparent. And rather than rehearse it once again, it may perhaps set Darwin off to better advantage to consider briefly what his theory did not do and what it forbade others to do. For Darwin's opponents—the serious ones, not the theologians, who were only pathetic—did not deny the fact of evolutionary variation. But they did want things from biology which science cannot give without ceasing to be science and becoming moral or social philosophy. And this perspective will make apparent the justice of the judgment which attributes to Darwin the importance for biology that Newton has for physics, so that his rather numbing humility becomes, not the attribute of inferiority, but only the quality fitted to a science in which observation plays a larger role than abstract formalization.

Both Newton and Darwin, to begin with, were criticized for ingratitude to their predecessors. This is not just a question of scholarly manners: once the theory—gravitation, natural selection—is repudiated, then what is left is the evidence—the inverse square relationship, the fact of variation—in which intellectual property might indeed be claimed for predecessors. But in both cases the theory was rejected, not as mistaken, but as meaningless. For Newton and Darwin had a way of simply accepting the phenomena as given. They excluded reason and purpose, according to this complaint, not in any dogmatic or positive fashion, but simply as an abdication of judgment. Thus, they prevented philosophy from coming to grips with science.

Fontenelle, for example, dismissed Newton's geometric manner of proof on the same grounds that led Darwin's critics to deny merit to the concept of fitness in the organism. These are not scientific demonstrations, say the critics. Nor can they be because they come out precisely even. They are simply tautologies which circle through the phenomena right back to their starting point. What is causation in Newton and what in Darwin?—only a formless sequence of results extending backward or outward endlessly into a metaphysical limbo. Newton purports to unite his system with the principle of gravity, and Darwin with the principle of natural selection. But if either is asked what causes gravity, or what causes the variations that are selected, he does not know. Nor did the theory depend upon his trying to say. Indeed, its success hinged precisely on dropping that question. Of Darwin, too, it could be said, "Hypotheses non fingo," and in the same sense in which it is true of Newton: not as a sterile assertion of empiricism, but as a statement that theories (speculations are another matter) must just embrace the evidence. The Cartesians of the 18th century, however, and after them the romantics of the 19th century, wanted a science which would account at once for the behavior and the cause of phenomena, which would see nature steadily and see it whole. They wanted a science which would seize on the unity of nature, instead of fragmenting it into discrete events connected only by chance and circumstance, and not by reason or purpose. And in the case of Driesch—the most self-revealing of Darwin's critics—the heart of the position is that Darwin simply impoverishes biology, that he gives no rational insight into events, that he is a recorder posing as a philosopher, a chronicler rather than a historian of nature, and all this because he and his followers will not see that the laws of life are absolutely different from the laws of physics, and that in the organism purpose is all.[32]

Biological romanticism never made much impression in the world of English letters, where Samuel Butler and Shaw have been the most widely read of Darwin's critics. In their case, too, a comparison with certain themes of the 18th-century Enlightenment is instructive, for it makes clear that the question is no biological discussion, but simply the continuing expression of a moral resentment which wants more out of nature than science finds there. To read Diderot and Butler together is a curious experience, itself almost a vindication of Butler's *Unconscious Memory*. For one has the impression that this and Butler's other writings upon nature were products of his own rather painful and labored reflection, and yet how unoriginal they are! These, for example, are the four principles of Butler's *Life and Habit*: " The oneness of personality between parents and offspring; memory on the part of offspring of certain actions which it did when in the persons of its forefathers; the latency of that memory until it is rekindled by a recurrence of the associated ideas; and the unconsciousness with which habitual actions come to be performed." [33] Butler follows Diderot's route out of atomistic materialism: " It is more coherent with our other ideas, and therefore more acceptable, to start with every molecule as a living thing, and then deduce death as the breaking up of an

---

[32] For a discussion of the Cartesian response to Newtonianism, see my " Fontenelle and Newton," *Letters and Papers of Sir Isaac Newton concerning Natural Philosophy*, ed. I. Bernard Cohen (Harvard University Press, Cambridge, 1958). For an epitome and guide to the anti-Darwinian literature, see the very transparent Emmanuel Rádl, *History of Biological Theories* (tr. E. J. Hatfield, Oxford University Press, London, 1930), a book recommended as an unfortunate example rather than reliable guide. For the sense to be attributed to Newton's " Hypotheses non fingo," see Alexandre Koyré, " L'hypothèse et l'expérience chez Newton," *Bulletin de la Société française de Philosophie*, 50 (avril-juin, 1956), 59-97.

[33] Samuel Butler, *Unconscious Memory* (Fifield, London, 1910), p. 19.

association or corporation, than to start with inanimate molecules and smuggle life into them; and . . . therefore, what we call the inorganic world must be regarded as up to a certain point living, and instinct, within certain limits, with consciousness, volition, and power of concerted action." [34]

But from the point of view of one who admires the intellectual achievements of science, it is Shaw rather than Butler who by contrast to his pretensions seems drastically diminished in stature by his ventures into scientific criticism. The famous preface to *Back to Methuselah* [35] presents clichés with that air of lordly malice which Shaw knew how to assume as the right of a superior intelligence which did not mind pointing to its own perversity. But it was an intelligence which, far from transcending science, had never given itself the trouble to understand the force or limitations of scientific demonstrations, and in the perspective of history Shaw on Darwin will surely find a place side by side with Bellarmine and the papal jury setting the astronomers right about natural philosophy. It does not appear that Shaw ever thought to ask the biologists whether natural selection was true. It was simply "a blasphemy, possible to many for whom Nature is nothing but a casual aggregation of inert and dead matter, but eternally impossible to the spirits and souls of the righteous." Darwin is forbidden to banish mind from the universe: "For 'Natural Selection' has no moral significance: it deals with that part of evolution which has no purpose, no intelligence, and might more appropriately be called accidental selection, or better still, Unnatural Selection, since nothing is more unnatural than an accident. If it could be proved that the whole universe had been produced by such Selection, only fools and rascals could bear to live." [36] And the Shavian word on evolution, therefore, is in fact only a diatribe, another expression of the anti-vivisectionism —and in a certain sense the vegetarianism—of a personality whose Rousseauist attitude to nature involved more of sentimental hostility to intellect (as to any aristocracy) than is generally appreciated.

With this background in mind, one might almost have predicted that the latest thrust back to Lamarckism would have occurred in a Marxist context. And though the science of genetics will, unlike certain geneticists, survive its misadventures in the Soviet Union, this episode should stand as a warning that ideas have consequences,

[34] Ibid., p. 15.
[35] G. B. Shaw, *Back to Methuselah* (Brentano's, New York, 1921), p. xlvi.
[36] Ibid., pp. lxi-lxii.

and that to succumb to the very natural and often well-intentioned temptation to bend science to the socializing or the moralizing of nature is to invite its subjection to social authority, which is to say to politics. For your moralist knows what kind of nature he wants science to give him, and if it does not, he will either, like Shaw, repudiate it, or if like Lysenko he has the power, he will change it. Once again, as in Diderot, as in Goethe, as in Lamarck, resentment of mathematics, which expresses quantity and not the good, reveals the moralist beneath the natural philosopher—the Michurin school rejects in principle the mathematicization of biology in favor of the autonomy of organism. Lysenko's purported findings may, therefore, be taken as the nadir of the history of Lamarckism, and (one hopes) the end of the story. For in his demagoguery the humane view of nature is vulgarized by way of a humanitarian naturalism into the careerist's opportunity.[37] But there is nothing new about it. It is only the most recent expression of that pattern of resistance to science which has attended its entire history in reaction against the objectification of nature.

V

So far as the intellectual and cultural significance of evolutionary theory is concerned, therefore, Darwin had no predecessor in Lamarck. Lamarck's theory of evolution belongs to the contracting and self-defeating history of subjective science, and Darwin's to the expanding and conquering history of objective science. In the concept of natural selection, Darwin put an end to the opposition between mechanism and organism through which the humane view of nature, ultimately the Greek view, had found refuge from Newton in biology. Lamarck's theory, on the other hand, originated as the transfer to natural history of that old view for which Lavoisier had made chemistry, the science of matter, uninhabitable. It is for this reason that Darwin was the orderer of biological science, as Newton was of physics and Galileo of mechanics. He was the first to frame objective concepts widely enough to embrace the whole range of phenomena studied by his science. And it may be worthwhile to

[37] On Lysenko, see Conway Zirkle, *Death of a Science in Russia* (University of Pennsylvania Press, Philadelphia, 1949) and "L'Affaire Lysenko," *The Journal of Heredity*, XLVII (1956), pp. 47-57; and Julian Huxley, *Heredity East and West* (New York, 1949).

consider the theory of natural selection analytically for a moment, in order to specify what were the elements of its success, and how it was that, schematically speaking, Darwinian evolutionary theory stands in the same relation to Lamarckian in the overall structure of the history of science as does Galilean to Aristotelian mechanics.

In mechanics, Galileo achieved objectivity by accepting motion as natural, and considering its quantity as something to be measured independently of the moving body. This he accomplished by treating time as a dimension, after which motion in physics is no longer taken as a substantial change. In Darwin—to begin drawing out the parallel—natural selection treats that sort of change which expresses itself in organic variation in the same way. Instead of explaining variation, he begins with it as a fundamental fact of nature. Variations are assumed to occur at random, requiring no further explanation and pre-supposing no causative agent for science to seek out. This is what opened the breach through which biology might follow physics into objectivity, because it introduced the distinction, which Darwin was the first to make, between the origin of variations and their preservation. Variations arise by chance. But they are preserved according as they work more or less effectively in objective circumstance. In Lamarck, on the other hand, the two problems are handled as a single question, which in effect is begged by its solution in the inheritance of acquired characteristics. Lamarck, therefore, could no more have distinguished the study of variations from the study of the organism as a whole than the impetus school could separate motion from the missile.

In another and even more impressive respect Darwinian evolutionary change is analytically analogous to Galilean motion. There is direction in it, whereas in Lamarck's formulation life simply circles endlessly through nature. H. F. Blum has recently advanced the interesting argument that time as it enters into thermodynamic processes may be considered as a coordinate of evolution.[38] This amounts to saying on the one hand that evolution is capable of vectorial description, and on the other that biological time is a dimensional component of a physical situation and ceases thereby to be a refuge of becoming or a locus of flux. Quite generally, in fact, Darwin's work, though not of course quantitative in result, was nevertheless quantitative in method and manner of thought.

[38] H. F. Blum, *Time's Arrow and Evolution* (Princeton University Press, Princeton, 1951).

Thus, that he began with the Malthusian ratio was of far more significance for his success than was the question of its validity. It was, indeed, of utmost significance. What selection does in Darwin is to determine the quantity of living beings which can survive in any given set of objective circumstance. This aspect of the approach is more evident, perhaps, in Wallace's essay than in Darwin's more diffuse account. For example:

> Wild cats are prolific and have few enemies: why then are they never as abundant as rabbits?
> The only intelligible answer is, that their supply of food is more precarious. It appears evident, therefore, that so long as a country remains physically unchanged, the numbers of its animal population cannot materially increase. If one species does so, some others requiring the same kind of food must diminish in proportion. . . . It is, as we commenced by remarking, 'a struggle for existence,' in which the weakest and least perfectly organized must always succumb.[39]

And even more striking is Wallace's passage on natural selection as accounting

> . . . for that balance so often observed in nature,—a deficiency in one set of organs always being compensated by an increased development of some others—powerful wings accompanying weak feet, or great velocity making up for the absence of defensive weapons; for it has been shown that all varieties in which an unbalanced deficiency occurred could not long continue their existence. The action of this principle is exactly like that of the centrifugal governor of the steam engine, which checks and corrects any irregularities almost before they become evident; and in like manner no unbalanced deficiency in the animal kingdom can ever reach any conspicuous magnitude, because it would make itself felt at the very first step, by rendering existence difficult and extinction almost sure soon to follow.[40]

It has sometimes been remarked as a paradox that it should have been Darwin (and Wallace), the old-fashioned naturalists, and not the embryologists or physiologists of the continental laboratories, who brought the revolution in biology. But the reason is clear, and does not lie in the nature of their empirical contributions. It lies

---

[39] The Linnaean Society, *Darwin-Wallace Celebration* (London, 1908), p. 101. This reprints both the Darwin and Wallace papers of 1858.

[40] Ibid., pp. 106-107.

in the nature of their reasoning, which was concerned with quantity and circumstance. This is why it was they who liberated biology from its limiting dependence on classification and dissection, with the gulf between bridged insubstantially by that metaphor of goal-directed organism which the evidence never could control.

Missing the point of Mendel's work, all the eminent and puzzled biologists of the late nineteenth century who had to grapple with Darwin's legacy put themselves in a position like that of the Cartesians trying to explain the cause and describe the operation of gravity by a single concept. Whether of a romantic and speculative disposition like Nägeli, or of sober and ingenious temper like Weismann, this was the essential fault in their approach. In the structure of theory, both Nägeli's ideoplasm and Weismann's germ plasm were analogous to that Lamarckian inheritance of acquired characters which they repudiated. For both concepts gave at once a theory of heredity and development. Even Weismann's " ids " never proved capable of objectification. The historian of science may be pardoned for wondering what might have been the influence on biology had they known the history of science, and whether they might then have noticed the interest of Mendel's work. Suppose they had thought to compare his ratios with Dalton's, by which the revolution in chemical theory was reduced to numerical terms. Suppose, like Maupertuis, who had the insight from physics but lacked the information from biology, they had known of the relationship of the corpuscular philosophy of the 17th century to the Newtonian synthesis. Might they not have saved themselves much unprofitable reasoning and advanced the progress of their science by some decades? For nothing is more notable in the comparison of the biological to the Newtonian revolution than the reduction of the concept of natural selection to material atomism in the science of genetics. Just as the discontinuity of matter in atoms-and-the-void liberates motion from subjectivity, so biological objectivity was firmly seated in the discontinuity of the hereditary patrimony where inheritance could be comprised in number. (And as if to prove that this was the right track, genetics immediately gave that offense to moralizers and socializers of nature which throughout the history of science each successive step in objectification seems to have been bound to give.) In this perspective, therefore, it may appear as a kind of wisdom in Darwin rather than as a failing that his theoretical work began as an application to biology of the individualistic assumptions of classical political economy. He had, after all, no other basis for atomism. And the outcome is a conception of biological order no different from the order assumed by contemporary atomic physics—an order of chance to be analyzed by the techniques appropriate to mathematical probability.[41]

Darwin and Lamarck, therefore, speak their parts in that endless debate between atoms and the continuum, the multiplicity of events and the unity of nature, which is what has given the history of science its dialectic since its opening in Greece. Who seeks unity in nature believes in the continuum. Nor is it simply the wrong side, for it has been espoused by men worthy of attention: by Plato and Descartes, by Goethe and Einstein. But we must make distinctions. Einstein's is the continuum of geometry, and Goethe's of personality. For Lamarck the rational continuum resided in life. This was the root of his opposition to atomizing chemistry and of his emanationist evolutionary theory. At a time when he still believed in the immutability of species, he nevertheless expressed opposition to the Linnean system, hostility to which was the touchstone of romantic or metamorphosizing tendencies in taxonomy as rejection of atomism was in physics.

In one sense, therefore, the hiatus that one feels in Professor Lovejoy's work between the temporalizing of the chain of being and the foundation of evolution is inevitable. The latter is not the outcome of the former. For the continuum as the program of nature goes back to that aspect of classical philosophy which was a prolongation of cosmogony, back through the Stoics and Heraclitos to fire and the world as flux and process. But it is cosmology, the opposite of this, from which science derives; rather from the contemplation of being in the light of reason than of becoming in the light of process. And this resolves, perhaps, another apparent paradox: that providentialism and belief in fixity and divine design have in effect been more conducive to positive scientific work—in Newton, for example, in Linnaeus, or in Cuvier—than has belief in process. For though ultimately untenable, providentialism establishes a fixity in things for science to find. It posits the existence of entities which may serve as the term of metrical analysis. But in becoming, everything blends into everything and nothing may ever be defined. It is a mistake, therefore, to say with Cassirer that

---

[41] For a discussion of this, see Jean Rostand, "Esquisse d'une histoire de l'atomisme en biologie," *Revue d'histoire des sciences*, II, (1949), pp. 241-265, III, (1950), pp. 156-169, IV (1951), pp. 41-59, V (1952), pp. 155-170.

Darwin brought becoming within the pale of science.[42] What he did was to treat that whole range of nature which had been relegated to becoming, as a problem in being, an infinite set of objective situations reaching back through time. He treated scientifically the historical evidence for evolution, which had been marshalled often enough before him, but more as a travesty than an arm of science. Rightly understood, therefore, the question does come back to the starting point of Professor Lovejoy's treatment: the Darwinian theory of evolution turned the problem of becoming into a problem of being and permitted the eventual mathematicization of that vast area of nature which until Darwin had been protected from logos in the wrappings of process.

---

[42] Ernst Cassirer, *The Problem of Knowledge* (Yale University Press, New Haven, 1950), pp. 160-175.

ELEVEN

N EMBRYOLOGICAL ENIGMA IN THE ORIGIN OF SPECIES

JANE OPPENHEIMER

I

Charles Darwin was far from an embryologist. He devoted only one part of one short chapter of the *Origin of Species* to embryological considerations. Yet he spoke in it of the leading facts in embryology as "second in importance to none in natural history,"[1] and although his embryological remarks were brief, they were expressed in a tone that indicates that they came straight from his heart.

"Hardly any point gave me so much satisfaction when I was at work on the 'Origin,'" he wrote in his "Autobiography," "as the explanation of the wide difference in many classes between the embryo and the adult animal, and of the close resemblance of the embryos within the same class. No notice of this point was taken, as far as I remember, in the early reviews of the 'Origin,' and I recollect expressing my surprise on this head in a letter to Asa Gray."[2] Darwin's memory on this occasion, in contrast to some others, was correct: "Embryology is to me," ran his letter of September 10 [1860] to Gray, "by far the strongest single class of facts in favour of change of forms, and not one, I think, of my

---

[1] Charles Darwin, *On the Origin of Species by Means of Natural Selection,* ... A Reprint of the First Edition, ... Watts & Co., London, 1950, p. 382. The first edition was originally published by John Murray, London, 1859.

[2] Darwin, "Autobiography," in *The Life and Letters of Charles Darwin,* *including an Autobiographical Chapter,* ed. by Francis Darwin, D. Appleton and Company, New York, 1888, I, 72.

reviewers has alluded to this."[3] He had made a similar comment the previous year to J. D. Hooker: "Embryology is my pet bit in my book, and, confound my friends, not one has noticed this to me;"[4] and in 1857, two years before he had begun the final construction of the book, he wrote to Gray that "embryology leads me to an enormous and frightful range."[5]

Although it was only after the publication of the *Origin of Species*, and not before, that embryology's claim to relationship with evolution reached full voice, nonetheless the fact remains that there existed in pre-Darwinian writings many statements of embryological concepts akin to those expressed by Darwin. Embryology itself, furthermore, was undergoing in the years shortly prior to the completion of the *Origin of Species* the most drastic transformation ever to occur in its own history, as it developed away from *Naturphilosophie* in the direction of what is often called modern epigenetic theory. It should therefore prove rewarding to investigate what influence his acquaintance with the old or the new embryology might have exerted on the genesis of Darwin's ideas.

The reader of this essay will be spared detailed enumeration of all the eighteenth and early nineteenth century statements of the so-called laws, and their variations, relating embryos to ancestors. These are discussed in a number of accessible works; Kohlbrugge,[6] in particular, has presented a list of seventy-two authors, beginning with Goethe and Autenrieth in 1797 and concluding with Haeckel in 1866, who anticipated or expressed concepts of parallelism or of recapitulation, and it is superfluous either to abbreviate or to amplify his catalog here. We may leave it also to others to attempt to establish priorities for these ideas. Such speculations are idle, and since it is hardly either feasible or desirable to examine here the thought of more than six dozen of Darwin's embryologically-minded precursors, we might instead best proceed by searching into the works of the most distinguished and influential of the early nineteenth century embryologists, Karl Ernst von Baer, for facts and concepts in which Darwin might have been interested.

It is appropriate for a number of reasons to center the discussion

[3] Darwin, *Life and Letters*, II, 131.
[4] Darwin, *Life and Letters*, II, 39.
[5] Darwin, *Life and Letters*, I, 478.
[6] J. H. F. Kohlbrugge, "Das biogenetische Grundgesetz. Eine historische Studie," *Zoologischer Anzeiger* (1911), XXXVIII, 447-453.

on von Baer. His great embryological treatise [7] was published two or three decades before the *Origin of Species*, and Darwin might be expected to have known it. It was a climactic study, the culmination of all embryology that had gone before and the point of departure of all that was to follow; this was the book which transformed the embryology of *Naturphilosophie* to that of the laboratory of today. And von Baer's mind was of equal intellectual power with Darwin's, a judgment in which even Huxley, one of Darwin's most ardent admirers, concurred. "Von Bär was another man of the same stamp"[8] as Darwin, wrote Huxley, shortly after Darwin's death.

Von Baer himself, although he expressed himself as strongly opposed to Darwin in later years,[9] knew that he was somehow involved as one of his precursors. Now, von Baer had not only a uniquely great mind, but also a uniquely great personality; and like many another uniquely great personality, von Baer's was extraordinarily complicated. More of this on another occasion; the fact is mentioned here as an introduction to the statement that at different times he painted his earlier thoughts in different lights. "Know thyself," he began his first lecture on anthropology, "... I know of no inquiry which is worthier of free and thinking man than the exploration of himself,"[10] yet he was himself highly inconsistent

---

[7] Karl Ernst von Baer, *Über Entwickelungsgeschichte der Thiere. Beobachtung und Reflexion*, Gebrüder Bornträger, Königsberg, I. Theil, 1828; II. Theil, 1837. The second part was incomplete at publication, and its conclusion was published posthumously: II. Theil, Schlussheft, ed. by L. Stieda, Wilh. Koch, Königsberg, 1888.

[8] Leonard Huxley, *Life and Letters of Thomas Henry Huxley*, D. Appleton and Company, New York, 1901, II, 42.

[9] "Über Darwins Lehre," in Karl Ernst von Baer, *Reden gehalten in wissenschaftlichen Versammlungen und kleinere Aufsätze vermischten Inhalts*. Zweiter Theil. Studien aus dem Gebiete der Naturwissenschaften. H. Schmitzdorf (Karl Röttger), St. Petersburg, 1876, pp. 235-480. A much shorter version had been published in 1873: "Zum Streit über den Darwinismus," *Augsburger Allgemeine Zeitung*, No. 130, pp. 1986-88. For an excellent critique of von Baer's views on evolution see S. J. Holmes, "K. E. von Baer's Perplexities over Evolution," *Isis*, XXXVII (1947), 7-14. Holmes also discusses Darwin's knowledge and interpretations of embryology in "Recapitulation and its supposed causes," *Quart. Rev. Biol.*, XIX (1944), 319-331.

[10] "Erkenne Dich selbst! ... In der That weiss ich keine Untersuchung, welche des freien und denkenden Menschen würdiger wäre, als die Erforschung seiner selbst." Cited from Ludwig Stieda, *Karl Ernst von Baer, Eine biographische Skizze*, 2nd ed., Friedrich Vieweg und Sohn, Braunschweig, 1886, p. 202.

in his various portrayals of the development of his thinking, and this is plainly evident in what he said about his relationships to Darwinian evolution.

He wrote to Huxley in the year 1860 about the similarity of some of his own ideas to those of Darwin, and in his letter he specified particularly that only in the area of geographical distribution did his thought overlap that of Darwin. Huxley reported this to Darwin in a letter which merits full quotation:

August 6th, 1860
My dear Darwin,—I have to announce a new and great ally for you. . . .

Von Bär writes to me thus:—" Et outre cela, je trouve que vous écrivez encore des rédactions. Vous avez écrit sur l'ouvrage de M. Darwin une critique dont je n'ai trouvé que des débris dans un journal allemand. J'ai oublié le nom terrible du journal anglais dans lequel se trouve votre récension. En tout cas aussi je ne peux pas trouver le journal ici. Comme je m'intéresse beaucoup pour les idées de M. Darwin, sur lesquelles j'ai parlé publiquement et sur lesquelles je ferai peut-être imprimer quelque chose—vous m'obligeriez infiniment si vous pourriez me faire parvenir ce que vous avez écrit sur ces idées.

" J'ai énoncé les mêmes idées sur la transformation des types ou origine d'espèces que M. Darwin. Mais c'est seulement sur la géographie zoologique que je m'appuie. Vous trouverez, dans le dernier chapitre du traité ' Ueber Papuas und Alfuren,' que j'en parle très décidément sans savoir que M. Darwin s'occupait de cet objet."

The treatise to which von Bär refers he gave me when over here, but I have not been able to lay hands on it since this letter reached me two days ago. When I find it I will let you know what there is in it.[11]

---

[11] Darwin, *Life and Letters*, II, 122-123. Von Baer's French letter to Huxley might be translated as follows: " And besides, I find that you are still writing reviews. You have written on Mr. Darwin's work a criticism of which I have found only fragments in a German journal. I have forgotten the terrible name of the English journal in which your review appeared. Also in any case I cannot find the journal here. As I am much interested in the ideas of Mr. Darwin, on which I have spoken publicly and on which I shall perhaps have something put into print, you would oblige me infinitely if you would be able to have forwarded to me what you have written about these ideas. I have expressed the same ideas on the transformation of types or origin of species as Mr. Darwin. But it is only on zoological geography that I rely. You will find, in the last chapter of the treatise " Über Papuas und Alfuren "

The "something" which von Baer wrote to Huxley of having in mind to put into print about Darwin certainly was to include the long essay "Über Darwins Lehre" published in 1876. In this he confessed his belief that he had "supplied some material for [the] foundation [of Darwin's doctrine], even though time and Darwin himself have erected on the fundament a structure to which I feel myself alien," [12] and acknowledged that others of his ideas besides those concerned with geographical distribution anticipated those of Darwin. Before proceeding to his analysis of the doctrine, he took up his "own small scientific efforts, to explain in what relationship they stand to the transmutation doctrine," [13] first, his embryological studies, next, an essay entitled "Das allgemeinste Gesetz der Natur in aller Entwickelung," and finally the treatise "Über Papuas und Alfuren" mentioned in the letter to Huxley.

The "Papuas und Alfuren" may be considered first here. In this anthropological study von Baer expressed, to put it in Darwin's words, "his conviction, chiefly grounded on the laws of geographical distribution, that forms now perfectly distinct have descended from a single parent-form." [14] This essay was read to the Academy at St. Petersburg, according to Stieda,[15] on April 1, 1859. A notation accompanying the title in the volume in which it appears stated that the paper was read on April 8th, and a note elsewhere in the volume specified that this was published in September 1859, the month that the corrected proofs of the *Origin of Species* were returned to their publisher. Von Baer himself emphasized how few readers the treatise enjoyed.[16] And in fact, when Darwin finally got around to adding a reference to it to the historical sketch in the fourth

that I speak of it very positively without knowing that Mr. Darwin was concerning himself with the subject." The treatise to which von Baer refers is "Über Papuas und Alfuren. Ein Commentar zu den beiden ersten Abschnitten der Abhandlung *Crania Selecta*, . . ." *Mém. de l'Acad. Imp. des Sci. de St. Pétersbourg*, VI. Sér., Sci. Nat., VIII (1859), 269-346.

[12] "In der That glaube ich für die Begründung derselben einigen Stoff geliefert zu haben, wenn auch die Zeit und Darwin selbst auf das Fundament ein Gebäude aufgeführt haben, dem ich mich fremd fühle," von Baer, *Reden*, II, 240.

[13] "Ich will jetzt ganz einfach meine eigenen kleinen wissenschaftlichen Bestrebungen durchgehen, um darzulegen, in welchem Verhältniss sie zu der Transmutationslehre stehen," von Baer, *Reden*, II, 241.

[14] Darwin, *Origin of Species*, . . . 4th ed., John Murray, London, 1866, pp. xx-xxi. The statement was found also in the 5th and in the 6th editions.

[15] Stieda, *Karl Ernst von Baer*, p. 163.

[16] Von Baer, *Reden*, II, 247.

edition of the *Origin of Species* (1866), he referred not to the original article but to "Rudolph Wagner, '*Zoologisch-Anthropologische Untersuchungen*,' 1861."[17] Thus it exerted no influence on the pattern of Darwin's ideas on geographical distribution, which had already begun their development by the time Darwin left the *Beagle* in 1837.

It is equally clear that Darwin was without influence on von Baer with respect to theories of geographical distribution. It will be remembered that when Huxley wrote to Darwin about von Baer in the letter reproduced at the beginning of this essay, he stated that von Baer when in London had given him a copy of the article in question. Von Baer was in London during the late summer or early autumn of 1859.[18] He devoted only one short final chapter of his autobiography to his life after 1834, and there he gave no details of the visit to London beyond making the remark that the journey, which included also visits to Copenhagen, Stockholm, and Paris, was devoted primarily to anthropological interests.[19] In the essay "Über Darwins Lehre" he was, however, more specific:

> I must expressly point out that this treatise was not written under the influence of Darwinian theory. I already had it with me when I visited England in 1859 and I gave it to Messrs. Owen and Huxley along with another communication on distinctive skulls of different peoples from the St. Petersburg collections. On this occasion I first learned that Charles Darwin was occupied with a complete demonstration of the transmutation doctrine. The book itself had however not yet appeared. I became acquainted with it after my return to St. Petersburg after the end of the year.[20]

---

[17] Darwin, *Origin of Species*, 4th ed., p. xx. The reference remained the same in the subsequent editions.

[18] Stieda, *Karl Ernst von Baer*, p. 163.

[19] Von Baer, *Nachrichten über Leben und Schriften des Herrn Geheimraths Dr. Karl Ernst von Baer, mitgetheilt von ihm selbst*, . . . 2d ed., Friedrich Vieweg und Sohn, Braunschweig, 1886, p. 436.

[20] "Ausdrücklich muss ich bemerken, dass diese Abhandlung nicht unter dem Einflusse der Darwinschen Theorie geschrieben ist. Ich hatte sie schon mit, als ich im Jahre 1859 England besuchte, und theilte sie mit einer anderen Abhandlung über ausgezeichnete Schädel verschiedener Völker aus der St. Petersburger Sammlung den Herren Owen und Huxley mit. Bei dieser Gelegenheit erfuhr ich erst, dass Charles Darwin mit einer vollständigen Demonstration der Transmutationslehre beschäftigt sei. Das Buch selbst war aber noch nicht erschienen. Ich lernte es nach meiner Rückkehr nach St. Petersburg nach dem Schlusse des Jahres kennen," von Baer, *Reden*, II, 248.

According to his biographer Stieda,[21] von Baer delivered a lecture to the Geographical Society in St. Petersburg on October 10, 1859, and thus he must have returned there well before November 24th, the publication date of the *Origin of Species*.

In the second contribution in which, in the essay of 1876, von Baer considered himself a forerunner of Darwin, "Das allgemeinste Gesetz der Natur in aller Entwickelung,"[22] he considered among other things what would now be called the phylogeny of various animal groups. He went so far in this communication, originally a lecture delivered in Königsberg in January 1833 or 1834, as to conceive the possibility—he did not state it as a fact—that what are now separate species of a single genus, or at most those of a closely related genus, might have resulted from the development and propagation of a common type. "Yet I especially point out," he wrote in the essay published in 1876, "that I found no probability that all animals have developed through transformation."[23]

In the words of the original essay itself:

If under these circumstances it is easier to suppose that one form of antelope, or of sheep, or of goat, was created for the old world, and was transformed here into the now separate and permanent-appearing forms, than it is to suppose that many antelope, sheep and goats were created for the old world and none at all for the new world, where in contrast other genera resolve themselves into other species,—if it is even permitted to imagine that antelope, sheep and goats, that are related in so many ways, may all have developed from a common original form,—yet I can on the other hand find no probability that all animals have developed from one another through transformation.[24]

---

[21] Stieda, *Karl Ernst von Baer*, p. 164.

[22] Von Baer, "Das allgemeinste Gesetz der Natur in aller Entwickelung," in *Reden*, . . . Erster Theil, H. Schmitzdorf (Karl Röttger), St. Petersburg, 1864, pp. 35-74.

[23] "Doch bemerke ich ausdrücklich, dass ich keine Wahrscheinlichkeit gefunden habe, die dafür spräche, dass alle Thiere sich durch Umbildung entwickelt hätten," von Baer, *Reden*, II, 245.

[24] "Wenn es unter diesen Verhältnissen näher liegt, anzunehmen, dass Eine Form von Antilopen, vom Schaaf, von der Ziege für die alte Welt geschaffen wurde und hier in die jetzt getrennt und bleibend erscheinenden Formen umgewandelt wurde, als anzunehmen, viele Antilopen, Schaafe und Ziegen wurden für die alte Welt geschaffen und gar keine für die neue, wo dagegen andere Geschlechter sich in andere Arten auflösten, wenn es sogar erlaubt scheinen möchte, sich zu denken, dass Antilope, Schaaf und Ziege, die so vielfach verwandt sind, sich aus einer gemeinschaftlichen Urform entwickelt

It is unlikely in the extreme that Darwin knew this essay, which was originally published in 1834 in what must have been to Darwin an obscure journal;[25] in any case, it was not referred to by Darwin either in his correspondence or notebooks as so far published, or in the historical introduction which formed part of the last four editions of the *Origin of Species*.

The principal work, however, in which von Baer might have contributed to the basis of evolutionary doctrine was his monograph *Über Entwickelungsgeschichte der Thiere*. Von Baer wrote with extreme lucidity concerning the relevance of this embryological study to Darwinian theory:

> I believe that it was through my investigations of the manner of development of animals and the general speculations connected with them—however much these have been obscured through recent work of this kind—that I furnished some material for the currently prevailing opinions concerning the development of organic forms. Only I cannot subscribe to all the applications of this material. Although in the work which bears the title " Ueber die Entwickelungsgeschichte der Thiere," I have demonstrated the transformations of animal organisms during individual development, I believe I did not speak in support of a descendence theory in the sense of the more recent one. On the contrary, in the Fifth Scholion of the first volume I have emphatically expressed myself as opposed to the then dominant theory of transmutation.[26]

---

haben, so kann ich dagegen keine Wahrscheinlichkeit finden, dass alle Thiere sich durch Umbildung aus einander entwickelt hätten," von Baer, *Reden*, I, 55-56.

[25] " *Vorträge aus dem Gebiete der Naturwissenschaften und der Oekonomie*, gehalten vor einem Kreise gebildeter Zuhörer in der physikalisch-ökonomischen Gesellschaft zu Königsberg. Erstes Bändchen mit Vorträgen von . . . . herausgegeben von dem Prof. K. E. v. Baer. Königsberg 1834, bei Unger." This reference is given by von Baer in the introduction to the essay as published in *Reden*, I, 37.

[26] " Ich glaube allerdings durch meine Untersuchungen über die Entwickelungsweise der Thiere und die daran geknüpften allgemeinen Betrachtungen, so sehr sie auch durch die neuesten Arbeiten dieser Art verdunkelt sind, einigen Stoff zu den jetzt vorherrschenden Ansichten über die Ausbildung der organischen Formen geliefert zu haben. Allein ich kann nicht mit allen Verwendungen dieses Materials mich einverstanden erklären. In dem Werke, welches den Titel führt: "Ueber die Entwickelungschichte der Thiere," habe ich allerdings die Umwandlungen der thierischen Organismen in der Entwickelung der Individuen nachgewiesen, allein einer Descendenztheorie, in dem Sinne der Neueren, glaube ich nicht das Wort geredet zu haben. Vielmehr habe ich mich im fünften Scholion des ersten Bandes gegen eine damals

In fact, the Fifth Scholion was an eloquent disputation against "the prevalent notion, that the embryo of higher animals passes through the permanent forms of the lower animals." [27] Only in a footnote to it did von Baer deign to speak explicitly of the allied concept, with which he disagreed, "that all forms are immediately developed out of one," [28] and the whole argument served to refute on an embryological basis the theory that embryological sequences mirror the transformations of one animal type to another:

> *The embryo of the vertebrate animal is from the very first a vertebrate animal*, and at no time agrees with an invertebrate animal. A permanent animal form, however, which exhibits the vertebrate type, and yet possesses so slight a histological and morphological differentiation as the embryos of the Vertebrata, is unknown. *Therefore, the embryos of the Vertebrata pass in the course of their development through no (known) permanent forms of animals whatsoever.*[29]

"My opposition to the view of recapitulation" [to call it by its later name], wrote von Baer in the 1876 essay, "received fairly general recognition. Johannes Müller, who in the first edition of his

herrschende Ansicht von Transmutation nachdrücklich ausgesprochen," von Baer, *Reden*, II, 241.

[27] "Die herrschende Vorstellung, dass der Embryo höherer Thiere die bleibenden Formen der niederen Thiere durchlaufe," von Baer, *Entwickelungsgeschichte*, I, 199. The English translation in my text is not by myself, but, for reasons to be apparent below, one made by T. H. Huxley, which is to be found in "Article VII. Fragments relating to Philosophical Zoology. Selected from the Works of K. E. von Baer," in *Scientific Memoirs, selected from the Transactions of Foreign Academies of Science, and from Foreign Journals. Natural History*, ed. by Arthur Henfrey and T. H. Huxley, Taylor and Francis, London, 1853, p. 186.

[28] "Es war natürlich, ja nothwendig, dass man nun versuchte, die *einfachste* Form dieser Modificationen durchzuführen, die der unmittelbaren Entwickelung aller Formen aus einer," von Baer, *Entwickelungsgeschichte*, I, 201 fn. Translation in text by Huxley, *Scientific Memoirs*, ed. by Henfrey and Huxley, p. 189.

[29] "*Der Embryo des Wirbelthiers ist schon anfangs ein Wirbelthier*, und hat zu keiner Zeit Uebereinstimmung mit einem wirbellosen Thiere. Eine bleibende Thierform aber, welche den Typus der Wirbelthiere hätte, und eine so geringe histologische und morphologische Sonderung, wie die Embryonen der Wirbelthiere, ist nicht bekannt. *Mithin durchlaufen die Embryonen der Wirbelthiere in ihrer Entwickelung gar keine (bekannten) bleibenden Thierformen*," von Baer, *Entwickelungsgeschichte*, I, 220. Translation in text by Huxley, *Scientific Memoirs*, ed. by Henfrey and Huxley, p. 210. Huxley's italics, reproduced in the current text, conform to those of von Baer in the original work.

physiology had adopted the doctrine of Meckel and Oken, deleted it from the second edition. In general nothing was heard of it for a long while. Only most recently it is cropping up again here and there, yet without serious foundation." [30]

It is inexplicable that in von Baer's letter to Huxley about his partial agreement with Darwin he mentioned only the "Papuas und Alfuren," and neither the embryological treatise nor the essay on "Das allgemeinste Gesetz." It is also puzzling that in neither the letter of 1860 nor the essay of 1876 did he mention his extensive communication "Beiträge zur Kenntniss der niedern Thiere," in which already in 1826 he had refuted the law of parallelism: "It has been concluded by a bold generalization from a few analogies, that the higher animals run in the course of their development through the lower animal grades, and sometimes tacitly and sometimes expressly they have been supposed to take their way through all forms. We hold this to be not only untrue, but also impossible." [31]

In the main substance of the essay "Über Darwins Lehre" von Baer expressed strong opposition to Darwin's general concept of the processes responsible for evolution. As a teleologist, he regretted its ethical implications. On a strictly biological basis, he was at odds with what is perhaps its most central feature, namely, the assumption that the forces which acted in the past were similar to those operating in the present. Von Baer believed, in contrast, that "we have to admit that in a far distant earlier time, a much stronger formative force must have prevailed on earth than we know now, whether this operated through the transformation of already existing forms or through the creation of completely new series of forms." [32]

---

[30] "Mein Widerspruch gegen die Ansicht vom Durchlaufen niederer Thiere hat auch ziemlich allgemeine Anerkennung gefunden. Johannes Müller, der in der ersten Auflage seiner Physiologie die Lehre von Meckel und Oken angenommen hatte, strich sie in der zweiten Auflage. Ueberhaupt war lange nichts von ihr zu hören. Allein in der neusten Zeit taucht sie hie und da doch wieder auf, jedoch ohne ernstliche Begründung," von Baer, *Reden*, II, 243.

[31] "Man hat . . . mit kühner Verallgemeinerung aus wenigen Analogien geschlossen, die höhern Thiere durchliefen in ihrer Ausbildung die niedern Thierstufen, und bald ausdrücklich, bald stillschweigend, sie den Weg durch alle Formen gehen lassen. Das halten wir nicht nur für unwahr, sondern auch für unmöglich," von Baer, "Beiträge zur Kenntniss der niedern Thiere," *Verh. d. kaiserl. Leopold.-Carolin. Akad. d. Naturforscher*, XIII (1827), IIte Abth., p. 760. Translation in text by Huxley, *Scientific Memoirs*, ed. by Henfrey and Huxley, p. 184.

[32] "Wir müssen also . . . zugestehen, . . . dass in einer weit entlegenen Vorzeit eine viel gewaltigere Bildungskraft auf der Erde geherrscht habe,

When Georg Seidlitz, a strong Darwinist in Dorpat, where von Baer was then living in retirement, published in 1876 a detailed and polemic treatise [33] attempting to prove that von Baer was more of a Darwinist than he knew, that he misunderstood Darwin, and thus that his divergencies from him were of less significance than he believed, von Baer was highly indignant. First he wished not to read the book at all, but finally did so; then he wanted to answer it "because it was too vexatious for him." [34] But only the introduction had been dictated by the time he died.

II

No matter what von Baer considered his own position vis-à-vis Darwin, it is evident that their thought converged in a number of areas. If Darwin was not cognizant very promptly of the works of von Baer which touched upon zoological affinities and geographical distribution, he seems to have become acquainted, fairly soon in the period during which he was developing his concepts on the *Origin of Species*, with von Baer's embryological ideas. He stated that he opened his first notebook "for facts in relation to the Origin of Species" in July 1837; [35] and according to Charles Singer, in Darwin's notebooks for 1842 and 1844 he "devotes much space to embryological discussion relying on von Baer." [36]

The early editions of the *Origin of Species* failed, however, to acknowledge explicitly any influence von Baer's ideas may have played on the formation of Darwin's. It was only in the third and subsequent editions that von Baer's name was found at all. In the third edition, a reference to what Darwin called von Baer's standards of advance in organization was first introduced into the chapter on

als wir jetzt erkennen, möge diese nun durch Umbildung der bereits bestehenden Formen oder durch Erzeugung ganz neuer Reihen von Formen gewirkt haben," von Baer, "Das allgemeinste Gesetz der Natur in der Entwickelung," *Reden*, I, 57. This passage was repeated word for word in "Über Darwins Lehre," *Reden*, II, 245.

[33] Georg Karl Maria von Seidlitz, *Beiträge zur Descendenz-Theorie*. Engelmann, Leipzig, 1876.

[34] "Weil es ihm doch zu arg sei." In quotation marks in Stieda, *Karl Ernst von Baer*, p. 192.

[35] Darwin, *Life and Letters*, I, 56.

[36] Charles Singer, *A History of Biology*, revised ed., Henry Schuman, New York, 1950, p. 473.

natural selection.³⁷ In the section of the chapter on geological succession entitled "On the State of Development of Ancient compared with Living Forms," a specific quotation from von Baer was added in the same edition: "To attempt to compare in the scale of highness members of distinct types seems hopeless; who will decide whether a cuttle-fish be higher than a bee—that insect which the great Von Baer believed to be 'in fact more highly organized than a fish, although upon another type'?"³⁸ The most extensive allusion to the work of von Baer in this edition made use of a quotation from von Baer's embryological treatise, in connection with what Darwin designated the law of embryonic resemblance. Darwin's section on embryology began, in the third edition, as follows:

> It has already been casually remarked that certain organs in the individual, which when mature become widely different and serve for different purposes, are in the embryo exactly alike. The embryos, also, of distinct animals within the same class are often strikingly similar: a better proof of this cannot be given, than a statement made by Von Baer, namely, that "the embryos of mammalia, of birds, lizards, and snakes, probably also of chelonia are in their earliest states exceeding like one another, both as a whole and in the mode of development of their parts; so much so, in fact, that we can often distinguish the embryos only by their size. In my possession are two little embryos in spirit, whose names I have omitted to attach, and at present I am quite unable to say to what class they belong. They may be lizards or small birds, or very young mammalia, so complete is the similarity in the mode of formation of the head and trunk in these animals. The extremities, however, are still absent in these embryos. But even if they had existed in the earliest stage of their development we should learn nothing, for the feet of lizards and mammals, the wings and feet of birds, no less than the hands and feet of man, all arise from the same fundamental form.³⁹

---

³⁷ Darwin, *Origin of Species*, 3d edition, John Murray, London, 1861, p. 133. This reference is also included in the later editions.

³⁸ Darwin, *Origin of Species*, 3d ed., p. 365. This statement of Darwin's remains virtually unchanged, except for minor variations in punctuation and wording, in the later editions. The quotation from von Baer was not made by Darwin but is almost identical with that published by Huxley (*Scientific Memoirs*, ed. by Henfrey and Huxley, p. 196). "Ich glaube daher, dass in der That die Biene höher organisirt ist, als der Fisch, obgleich nach einem andern Typus," von Baer, *Entwickelungsgeschichte*, I, 208.

³⁹ Darwin, *Origin of Species*, 3d ed., pp. 470–471. The passage remains principally the same, except for minor changes in wording and punctuation, in

The first two of these allusions to von Baer's work had no counterparts in the first and second (1859 and 1860) editions. And as for the third reference, von Baer's ideas on embryological resemblance were attributed in the two earlier editions not to their rightful author but to Louis Agassiz, as will be familiar to all who know these versions:

> The embryos, also, of distinct animals within the same class are often strikingly similar: a better proof of this cannot be given than a circumstance mentioned by Agassiz, namely, that having forgotten to ticket the embryo of some vertebrate animal, he cannot now tell whether it be that of a mammal, bird, or reptile.[40]

Singer [41] has stated that in his notes Darwin has correctly ascribed the passage to von Baer, and if this is the case, the error as to authority in the *Origin of Species* can only be explained by a lapse of memory. Huxley considered "a great memory" to be one of Darwin's outstanding characteristics; he emphasized this to Romanes when he wrote to him about the latter's obituary notice of Darwin.[42] But Darwin was more critical of himself, and said in his "Autobiography": "My memory is extensive, yet hazy: it suffices to make me cautious by vaguely telling me that I have observed or read something opposed to the conclusion which I am drawing, or on the other hand in favour of it; and after a time I can generally

the later editions. While this paragraph begins the section on embryology in the 3rd edition, two paragraphs precede it in the 6th and one in the 4th and 5th editions. The quotation from von Baer was not original with Darwin but is identical with that published by Huxley, *Scientific Memoirs*, ed. by Henfrey and Huxley, p. 210. "Die Embryonen der Säugethiere, Vögel, Eidechsen und Schlangen, wahrscheinlich auch der Schildkröten, sind in frühern Zuständen einander ungemein ähnlich im Ganzen, so wie in der Entwickelung der einzelnen Theile, so ähnlich, dass man oft die Embryonen nur nach der Grösse unterscheiden kann. Ich besitze zwei kleine Embryonen in Weingeist, für die ich versäumt habe die Namen zu notiren, und ich bin jetzt durchaus nicht im Stande, die Klasse zu bestimmen, der sie angehören. Es können Eidechsen, kleine Vögel, oder ganz junge Säugethiere seyn. So überstimmend ist Kopf- und Rumpfbildung in diesen Thieren. Die Extremitäten fehlen aber jenen Embryonen noch. Wären sie auch da, auf der ersten Stufe der Ausbildung begriffen, so würden sie doch nichts lehren, da die Füsse der Eidechsen und Säugethiere, die Flügel und Füsse der Vögel, so wie die Hände und Füsse der Menschen sich aus derselben Grundform entwickeln," von Baer, *Entwickelungsgeschichte*, I, 221.

[40] Darwin, *Origin of Species*, 1st ed., 1950 reprint, pp. 372-373. Second ed., John Murray, London, 1860, pp. 438-439.

[41] Singer, *A History of Biology*, p. 468 fn.

[42] Huxley, *Life and Letters*, II, 42.

recollect where to search for my authority." [43] We can only conclude that when Darwin paraphrased the passage for the first edition he failed to exert his usual caution in searching for his authority.

No one could have been more meticulous with respect to the labelling of specimens than Darwin, as he made clear in offering to potential field collectors

> a few pieces of advice, some of which I observed with much advantage, but others, to my cost, neglected. Let the collector's motto be, "Trust nothing to the memory," for the memory becomes a fickle guardian when one interesting object is succeeded by another still more interesting. . . . Put a number on every specimen, and every fragment of a specimen; and during the very same minute let it be entered in the catalogue, so that if hereafter its locality be doubted, the collector may say in good truth, "Every specimen of mine was ticketed on the spot." Anything which is folded up in paper, or put into a separate box, ought to have a number on the outside (with the exception perhaps of geological specimens) but more *especially* a duplicate number on the inside attached to the specimen itself. A series of small numbers should be printed from 0 to 5000; a stop must be added to those numbers which can be read upside down (as 699. or 86.). It is likewise convenient to have the different thousands printed on differently coloured paper, so that when unpacking, a single glance tells the approximate number.[44]

That a labeller so conscientious as Darwin could forget whether it was von Baer or Agassiz who had committed the zoologist's unpardonable sin suggests that his memory must have had a compelling reason indeed for the fickleness of its own guardianship. Von Baer, as a matter of actual fact, did not in so many words speak of his defection in terms of a lost label designated as such, he merely said that he had forgotten to make a note of the embryos' names (see his original statement quoted in footnote 39 above).

It would be interesting to know what finally moved Darwin to correct his mistake. The error was presumably not noticed in the early reviews; we have said that Darwin remarked that these made no mention of his embryological discussion. Less than two months elapsed between the publication of the first edition (November 24,

[43] Darwin, *Life and Letters*, I, 82.
[44] Charles Darwin, *Charles Darwin and the Voyage of the Beagle*, ed. by Nora Barlow, Philosophical Library, New York, 1946, pp. 152-153.

1859) and that of the second (January 7, 1860), and while Darwin managed to make some changes during the short interval—he stated in his "Autobiography" that "during the last two months of 1859 [he] was fully occupied in preparing a second edition"[45]—the second edition differed little from the first. It "is only a reprint," he wrote to Asa Gray on December 21 [1859], "yet I have made a *few* important corrections."[46] Most of the changes that were made were inserted at Lyell's suggestion: "It is perfectly true," Darwin wrote to Lyell on January 10 [1860], "that I owe nearly all the corrections to you, and several verbal ones to you and others."[47] Who were the others?

The third edition was published on April 30, 1861, and Darwin wrote Huxley that he must begin work on it November 22, 1860.[48] He may of course have learned of his mistake earlier in 1860. On November 23, 1859, the day before the publication of the first edition, Huxley wrote to Darwin to present his compliments after reading a prepublication copy:

> I finished your book yesterday, a lucky examination having furnished me with a few hours of continuous leisure.
> Since I read Von Bär's essays, nine years ago, no work on Natural History Science I have met with has made so great an impression upon me, and I do most heartily thank you for the great store of new views you have given me. Nothing, I think, can be better than the tone of the book; it impresses those who know about the subject.[49]

Huxley went on to enumerate the chapters with which he was in total or partial agreement, and of Chapter XIII, which in the first edition covered classification, morphology, embryology, and rudimentary organs, he said that it contained "much that is most admirable, but on one or two points I enter a *caveat* until I can see further into all sides of the question."[49]

Huxley's mention of von Baer at this time was apparently not sufficient to prick Darwin's memory, otherwise Darwin would presumably have made his correction to the second, not the third, edition. When he replied to Huxley's letter two days later, Darwin said in a postscript, "Hereafter I shall be particularly curious to

---

[45] Darwin, *Life and Letters*, I, 73.
[46] Darwin, *Life and Letters*, II, 40.
[47] Darwin, *Life and Letters*, II, 59.
[48] Darwin, *Life and Letters*, II, 144.
[49] Huxley, *Life and Letters*, I, 188.

hear what you think of my explanation of Embryological similarity." [50] On Saint Valentine's Day [1860], Darwin commented to Hooker that Huxley "has never alluded to my explanation of classification, morphology, embryology, etc.," [51] which hardly suggests that Huxley had then yet replied to Darwin's plea of the previous November for embryological criticism. We know that several months later, in August, 1860, Huxley was writing to Darwin about von Baer's notions of geographical distribution and transformation of types (see page 295, this article); was it this, close to the time of preparation of the third edition, which first incorporated the correction, that stimulated Darwin to look again into the writings of von Baer or into his notes concerning them?

The fact that he mentioned von Baer to Darwin in several letters is not the only reason to connect Huxley with the correction of the error. When Darwin finally made the amendment in the third edition, substituting the name of von Baer for that of Agassiz, and quoting rather than paraphrasing the anecdote of the absent label, he did not himself translate von Baer's statement in his own words, but used a translation by Huxley that had been published in 1853. The simultaneous insertion of all three text references to von Baer into a single edition, with the two direct quotations both borrowed (without acknowledgment, by the way) from Huxley's translation, makes it appear possible that Darwin, once aware of his mistake, returned to reread von Baer not in the original but in Huxley's translation, and that, having finally checked the passage on embryonic resemblance, he also found elsewhere in the translation ideas which he could use to bolster up his argument in other spots.

As a matter of fact, Darwin never in the *Origin of Species* referred to the works of von Baer in the original. Huxley, we have shown, drew Darwin's attention to "Papuas und Alfuren" in August, 1860, shortly before Darwin began the work for the third edition. This was the edition to which the historical sketch was first added, but reference was made to "Papuas und Alfuren" in the sketch for the first time only in the following (fourth) edition, and even here it was not the original essay that was cited, but Rudolph Wagner's reference to it (see above, p. 297 this article). Perhaps Huxley never found his own copy, which he admitted was misplaced when

---

[50] Darwin, *Life and Letters*, II, 28.
[51] Charles Darwin, *More Letters of Charles Darwin*, ed. by Francis Darwin and A. C. Seward, D. Appleton and Company, New York, 1903, I, 140.

he originally wrote Darwin of von Baer's having given it to him. But is it meaningful that Huxley promised Darwin that, if he found it, he would tell him what was in it, and not that he would lend it to him?

It is no secret that Darwin was something less than fluent in German. One annotation suggesting his concern with this fact dates as far back as 1836, where an entry in a shopping list prepared aboard the *Beagle* read "German books—Spelling Dict." [52] He was frankly struggling with the language, and protesting about its difficulty, eight years later: he wrote to Hooker in 1844, "I am now reading a wonderful book for facts on variation—Bronn, ' Geschichte der Natur.' It is stiff German: it forestalls me, sometimes I think delightfully, and sometimes cruelly." [53] Similar grumblings about the difficulties of the language were scattered throughout his letters for the rest of his life. Even in 1881, almost exactly a year before he died, when he wrote to Romanes of having received a copy of Roux's *Kampf der Theile*, he complained: " It is full of reasoning, and this in German is very difficult to me, so that I have only skimmed through each page; here and there reading with a little more care." [54] The important sections of von Baer's *Entwickelungsgeschichte* had been full of reasoning, too! Darwin's son Francis portrayed his father's perplexities over the language more vividly than he perhaps recognized when he wrote the chapter of " Reminiscences " included in the *Life and Letters*:

> Much of his scientific reading was in German, and this was a great labour to him; in reading a book after him, I was often struck at seeing, from the pencil-marks made each day where he left off, how little he could read at a time. He used to call German the " Verdammte," pronounced as if in English. . . . He himself learnt German simply by hammering away with a dictionary; he would say that his only way was to read a sentence a great many times over, and at last the meaning occurred to him. . . .
>
> In spite of his want of grammar, he managed to get on wonderfully with German, and the sentences that he failed to make out were generally really difficult ones. He never attempted to speak German correctly, but pronounced the words as though they were English.[55]

---

[52] *Charles Darwin and the Voyage of the Beagle*, p. 252.
[53] Darwin, *Life and Letters*, I, 390.
[54] Darwin, *Life and Letters*, II, 419.
[55] Darwin, *Life and Letters*, I, 103-104.

What a model Englishman Darwin was! Yet he was not alone among English biologists in his deficiencies in the use of the German language. In 1853, when Huxley published his translations of excerpts from von Baer's Fifth Scholion and from the concluding section of his "Beiträge zur Kenntniss der niedern Thiere," he claimed that it "seemed a pity that works which embody the deepest and soundest philosophy of zoology, and indeed of biology generally, which has yet been given to the world, should be longer unknown in this country," and he added in a footnote the remark that "Dr. Carpenter (Principles of General Physiology), is, so far as we know, the only English physiologist who has publicly drawn attention to Von Bär's philosophical writings." [56] Huxley himself first read von Baer's essays in 1850, as he wrote to Darwin in 1859 (see p. 306, this article). Hence if Darwin was in fact, as Singer stated, writing notes about von Baer by 1842 and 1844, he was one of his early devotees in Great Britain.

The "Dr. Carpenter" to whom Huxley referred was William B. Carpenter, a friend both of Darwin and of Huxley; he was sufficiently intimate with Huxley to have attended his wedding. His *Principles of General and Comparative Physiology* passed through three editions before 1853, when Huxley referred to it; they were dated respectively 1839, 1841, and 1851. One more edition, entitled simply *Principles of Comparative Physiology* (1854),[57] was published before the *Origin of Species*. Huxley late in 1851 specifically mentioned reading the "last edition," [58] presumably the third. Darwin wrote to Carpenter in 1859 of "the admiration which he had long felt and expressed" for the "Comparative Physiology;" [59] was this merely an elliptical way of writing, or did he mean thus to designate that he was referring particularly to the fourth edition, the only one to be entitled simply *Comparative Physiology*? In any case, two months later he wrote to Carpenter in particular

---

[56] Huxley, *Scientific Memoirs*, ed. by Henfrey and Huxley, p. 176 and 176 fn.

[57] William B. Carpenter, *Principles of General and Comparative Physiology*, ... John Churchill, London, 1839; 2d ed., John Churchill, London, 1841; 3d ed., John Churchill, London, 1851; 4th ed., published under the title *Principles of Comparative Physiology*, John Churchill, London, 1854. The third edition was available to me only in the American edition, *Principles of Physiology, General and Comparative*, 3d ed., Blanchard and Lea, Philadelphia, 1851.

[58] Huxley, *Life and Letters*, I, 100.

[59] Darwin, *Life and Letters*, II, 18.

approbation of the latter's embryological knowledge,[60] so we may assume him to have been acquainted with Carpenter's references to von Baer.

Carpenter himself, while he referred to the ideas of von Baer even in the first two editions, apparently read von Baer's works in their original form only some time after the publication of the second edition of his *Principles* in 1841. At least, when writing in the fourth edition about von Baer's principle of development from the general to the special as related to "those resemblances which are sometimes discernible, between the transitory forms exhibited by the embryos of higher beings, and the permanent conditions of the lower," he added in a footnote:

> It is owing to the ignorance of Von Baer's writings which has generally prevailed in this country, that the credit has been recently assigned to others, of having first enunciated the true view of this subject. The Author may refer to the second edition of the present work, published in 1841, as having contained the doctrine stated above, which he was also accustomed to teach in his Physiological Lectures; and although his own acquaintance with Von Baer's works at that time extended but little beyond the references made to them by Dr. Martin Barry, yet these were sufficient to enable him to comprehend and apply the great developmental law which Von Baer had so clearly enunciated.[61]

Carpenter was correct in judging that he had comprehended von Baer's concepts by 1841. Even in his first edition he presented them with clarity:

> There is a greater variety of dissimilar parts in the higher organisms than in the lower; and hence the former may be said to be *heterogeneous*, whilst the latter are more *homogeneous*, approaching in some degree the characters of inorganic masses. This law is, therefore, thus concisely expressed by Von Bär, who first announced it in its present form. "*A heterogeneous or special structure arises out of one more homogeneous or general, and this by a gradual change*". . . .
>
> Allusion was just now made to the correspondence which is discernible between the transitory forms exhibited by the embryos of the higher beings, and the permanent conditions of the lower. When this was first observed, it was stated as a general law, that

---

[60] Darwin, *Life and Letters*, II, 57.
[61] Carpenter, *Principles of Comparative Physiology*, 4th ed., pp. 96, 98 fn.

all the higher animals in the progress of their development pass through a series of forms analogous to those encountered in ascending the animal scale. But this is not correct; for the *entire animal* never does exhibit such resemblances.[62]

In the second edition, which according to the author's preface enjoyed the advantage of Barry's revision of the chapter on animal reproduction, the concepts of von Baer were further elucidated by the addition of a passage beginning: " It is to be remembered that every Animal must pass through *some* change, in the progress from its embryonic to its adult condition; and the correspondence is much closer between the embryonic Fish and the foetal Bird or Mammal, than between these and the adult Fish." [63] In the third and fourth editions all these ideas were elaborated in very considerable detail.

Clearly Carpenter was profoundly influenced by von Baer. His whole treatment of comparative physiology was based on a developmental analysis for which von Baer's generalizations provided the principal unifying concept. It is permitted to wonder, then, why he waited for so long a time to read what von Baer had really written; and it is inevitable to conclude that he too, like Darwin, was at least in his early years poorly versed in the German language. This is partially borne out by the fact that, out of the forty-three works which he enumerated in the preface to the first edition as having relied upon or consulted in the preparation of the volume, seven were by German authors, and all seven of these were cited in either English or French translation.

But what of Barry, who seems first to have introduced von Baer's concepts to Carpenter? Huxley may have been accurate to the letter when he named Carpenter the " only English physiologist " to have publicly called attention to the philosophical works of von Baer before 1853; but Martin Barry, a Scottish physician, before then had not only read these in German but had discussed them publicly and extensively in the English language.

In 1837, the very year that the principal portion of von Baer's

[62] Carpenter, *Principles of General and Comparative Physiology*, 1st ed., pp. 170, 171. The quotation attributed to von Baer is taken not from the original works of von Baer, but, with one word omitted, from Martin Barry, " On the Unity of Structure in the Animal Kingdom," *Edinburgh New Philosophical Journal*, xxii (1836-37), 141, 345.

[63] Carpenter, *Principles of General and Comparative Physiology*, 2d ed., p. 196.

second embryological volume appeared in print, Barry published in the *Edinburgh New Philosophical Journal* a pair of essays "On the Unity of Structure in the Animal Kingdom;"[64] some brief excerpts from the second of these may suffice to demonstrate that if Barry did not literally translate von Baer's words, as Huxley was to do sixteen years later, he at least gave a reliable presentation of some aspects of his theories in English:

> A heterogeneous or special structure arises only out of one more homogeneous or general, and this by a *gradual* change.... The manner of the change, is probably the same throughout the animal kingdom, however much ... the *direction* (or *type*) and *degree* of development may differ, and thus produce variety in structure....
>
> It is not unusual, however, to hear of the "higher" animals *repeating* or *passing through* in their development, the structure of the "lower:" and though this is said in reference, of course, to no more than single organs, it is a mode of speaking calculated to mislead.
>
> Such expressions might not be improper, did there exist in the animal kingdom a scale of structure differing in *degree* alone. But there is no such scale. We must "distinguish between the degree of elaboration and the type of structure"....
>
> No structure peculiarly *characterizing* any one set of animals in the perfect state, makes its appearance even in the embryonal life of any other....
>
> Besides which, as Von Bär has truly said, were it a law of nature, that individual development should *consist* in passing through permanent but less elaborate forms, there is not a feature in embryonal life, nor a part then present, that we should not expect to find, somewhere at least, in the animal kingdom. Yet in what direction are we to look for an animal carrying about its food, as the embryo the yolk, or a pendant portion of intestine, like the vesicula umbilicalis? ...
>
> The same author has well remarked, that inasmuch as embryonal relations produce forms that are present in no grown animal ... it is also impossible that any embryo can repeat the state of many

---

[64] Martin Barry, "On the Unity of Structure in the Animal Kingdom," *Edinburgh New Philosophical Journal*, XXII (1836-37), 116-141; "Further Observations on the Unity of Structure in the Animal Kingdom, and on Congenital Anomalies, including 'Hermaphrodites;' with some Remarks on Embryology, as facilitating Animal Nomenclature, Classification, and the Study of Comparative Anatomy," *ibid.*, 345-364.

groups of animals. All embryos are surrounded with fluid; and consequently incapable of immediately respiring air. The real character of insects, therefore,—a lively relation to the air,—can never be repeated in an embryo. For the same reason, the embryo of mammals can never resemble perfect birds.[65]

The periodical in which Barry published his interpretations of von Baer's concepts was not obscure; the volume in which they appeared included also papers by Berzelius, Treviranus, Rathke, Ehrenberg, and Milne Edwards, among others. Neither was Barry an unknown figure in his own day. At the time he published the essays he had already been President of the Royal Medical Society in Edinburgh. He published in 1837, 1838, and 1839 a series of distinguished communications on mammalian embryology in the *Philosophical Transactions of the Royal Society*,[66] for two of which he received the Royal Medal in 1839. Surely Darwin knew of these latter articles. In speaking of his preparation for writing the *Origin of Species* he said in his "Autobiography" that "when I see the list of books of all kinds which I read and abstracted, including whole series of Journals and Transactions, I am surprised at my industry;"[67] certainly the transactions he covered included the publications of the Royal Society, in which he always showed such a vital interest. Furthermore, Darwin's own article on Glen Roy,[68] his only paper in the *Philosophical Transactions*, appeared in the same volume as the final article in Barry's series; and in the Glen Roy paper Darwin referred a number of times to a geological communication in the *Edinburgh New Philosophical Journal* of the previous year (1838).

Besides, there was another reason for Darwin, and Huxley too, to have known the work of Barry. In the first two editions of the *Principles of General and Comparative Physiology* Carpenter made, on the same page on which he referred to von Baer's law of the development from the homogeneous to the heterogeneous, a brief

---

[65] Barry, "Further Observations on the Unity of Structure," pp. 345-346, 347, 348-349.
[66] Martin Barry, "Researches in Embryology," First Series, *Philos. Transacs.*, cxxviii (1838), 301-341; Second Series, ibid., cxxix (1839), 307-380; Third Series, "A Contribution to the Physiology of Cells," ibid., cxxx (1840), 529-593.
[67] Darwin, *Life and Letters*, I, 68.
[68] Charles Darwin, "Observations on the Parallel Roads of Glen Roy, and of other parts of Lochaber in Scotland, with an attempt to prove that they are of marine origin," *Philos. Transacs.*, cxxix (1839), 39-81.

footnote reference to the *Edinburgh Philosophical Journal* for July 1837; in the third edition, on the page immediately preceding his statement of this law, he referred not only to the periodical but also to Barry by name. In a footnote in the fourth edition, on the same page as the beginning of the paragraph in which von Baer's principle is applied as an explanation of the resemblances of embryos to lower forms, he not only gave the author's name and the journal reference, but also spelled out the title of the essays.[69]

Hence it is puzzling that Huxley in 1853 implied that Carpenter was the sole individual to have published von Baer's ideas in English. But it is even more extraordinary that in 1894 he could still admit, when referring to his own " translation of ' Fragments relating to Philosophical Zoology, selected from the Works of K. E. von Baer,' ... published in ' Scientific Memoirs ' for February and May 1853," that he still believed that " up to that time . . . Von Baer's ideas were hardly known outside Germany." [70]

Great honor has accrued to Huxley for his part in convincing English biologists of the importance of German science. His son as his biographer, among others, has emphasized the significance of his role in so doing: " One characteristic of his early papers should not pass unnoticed," wrote Leonard Huxley. " This was his familiarity with the best that had been written on his subjects abroad as well as in England. Thoroughness in this respect was rendered easier by the fact that he read French and German with almost as much facility as his mother tongue," and then he quoted an article by P. Chalmers Mitchell that had been published in *Natural Science* in 1895: " ' It is true of course that scientific men read French and German before the time of Huxley; but the deliberate consultation of all the authorities available has been maintained in historical succession since Huxley's earliest papers, and was absent in the papers of his early contemporaries.' " [71]

It was not absent from the papers of Barry; not only the essays in the *Edinburgh New Philosophical Journal* but also the articles in the *Philosophical Transactions* provided abundant references to

[69] Carpenter, *Principles of General and Comparative Physiology*, 1st ed., p. 170 fn; 2d ed., p. 195 fn; 3d ed., p. 575 fn; 4th ed., p. 580 fn.
[70] T. H. Huxley, " Owen's Position in Anatomical Science," in Richard Owen, *The Life of Richard Owen*, John Murray, London, 1894, II, 299 fn.
[71] Huxley, *Life and Letters*, I, 160.

the appropriate French and German works. In 1841 Huxley as a boy of sixteen noted in his Journal

> Projects begun—
> 1. German ⎱
> 2. Italian ⎰ to be learnt.
> 3. To read Müller's *Physiology*.[72]

Four years earlier Barry had already published the following passage (which may be of interest to those who remember Darwin's wonderful analogy of the great Tree of Life found at the end of his chapter on natural selection):

> Naturalists have begun, just where they should have ended. They have attended to details, but neglected general principles. Instead of analyzing, their process has been one of synthesis. Their attention has been directed to the grouping of the *twigs*,—as if thus they were to find their natural connections, without even looking for assistance towards the branches, or the trunk that gave them forth....
>
> But what other course *could* naturalists have taken? Truly none: their "circumstance" allowed no other. It is only now that a way is beginning to be opened, by which it may by and by be possible to proceed in an opposite direction; viz. from trunk to branches and to twigs.
>
> This, if ever accomplished, must be by the means of the *History of Development* or *Embryology*, both human and comparative; a science almost new, and regarding which, there prevails in this country the profoundest ignorance and indifference. The French are in advance of us; but it is to *German* enterprise, industry, and perseverance, that we are indebted for almost every fact known to us on this subject; at least of those brought to light in recent times. It is to be hoped, however, that ere long this science will begin to obtain, even among ourselves, some degree of the attention which its importance claims.[73]

Huxley later, through his vast popularity, may have succeeded in accelerating a movement towards the recognition of German contributions that without his spurring might have proceeded more slowly. But surely Barry merits some credit from history for his early, exhaustive, and influential treatment in British periodicals of the work of the most important of the German investigators in his field.

[72] Huxley, *Life and Letters*, I, 11.
[73] Barry, "Further Observations on the Unity of Structure," pp. 362-363.

III

If the ideas of von Baer were becoming known to British investigators in general and to Darwin in particular during the late thirties and early forties, it remains to evaluate to what degree Darwin's thought was affected by what he learned of them. Fortunately we are able to investigate the early growth of Darwin's embryological thinking by examining two preliminary sketches which he drew up in 1842 and 1844,[74] many years prior to the framing of the *Origin of Species* in its final form.

Darwin himself laid great emphasis on the importance for the later success of the *Origin of Species* of having written these sketches. " In June 1842," he wrote in his " Autobiography," " I first allowed myself the satisfaction of writing a very brief abstract of my theory in pencil in 35 pages; and this was enlarged during the summer of 1844 into one of 230 pages. . . . The success of the ' Origin ' may, I think, be attributed in large part to my having long before written two condensed sketches, and to my having finally abstracted a much larger manuscript, which was itself an abstract." [75] Now if, as Singer said, Darwin's notebooks for 1842 and 1844 discussed von Baer's embryological ideas, to what degree was this fact reflected in the sketches written in these very years?

Without doubt some of von Baer's principal ideas concerning embryonic resemblances were echoed in them. The rather brief embryological passage in the essay of 1842 included the following statement:

> This general unity of type in great groups of organisms (including of course these morphological cases) displays itself in a most striking manner in the stages through which the foetus passes. In early stage, the wing of bat, hoof, hand, paddle are not to be distinguished. At a still earlier ⟨stage⟩ there is no difference between fish, bird, &c. &c. and mammal. It is not that they cannot be distinguished, but the arteries ⟨illegible⟩. It is not true that one passes through the form of a lower group, though no doubt fish more nearly related to foetal state.[76]

---

[74] Charles Darwin, *The Foundations of the Origin of Species. Two Essays Written in 1842 and 1844*, ed. by Francis Darwin, University Press, Cambridge (England), 1909.
[75] Darwin, *Life and Letters*, I, 68, 70.
[76] Darwin, *Foundations of the Origin of Species*, p. 42. Words in angular brackets supplied by Francis Darwin.

The essay of 1844 expanded the embryological section to eight pages, from which the following excerpts may be relevant:

> The unity of type in the great classes is shown in another and very striking manner, namely, in the stages through which the embryo passes in coming to maturity. Thus, for instance, at one period of the embryo, the wings of the bat, the hand, hoof or foot of the quadruped, and the fin of the porpoise do not differ, but consist of a simple undivided bone. At a still earlier period the embryo of the fish, bird, reptile and mammal all strikingly resemble each other . . . . From the part of the embryo of a mammal, at one period, resembling a fish more than its parent form; from the larvae of all orders of insects more resembling the simpler articulate animals than their parent insects; and from such other cases as the embryo of the jelly-fish resembling a polype much nearer than the perfect jelly-fish; it has often been asserted that the higher animal in each class passes through the state of a lower animal; for instance, that the mammal amongst the vertebrata passes through the state of a fish: but Müller denies this, and affirms that the young mammal is at no time a fish, as does Owen assert that the embryonic jelly-fish is at no time a polype.[77]

It is noteworthy that, although according to Singer Darwin was thinking about the ideas of von Baer in 1842 and 1844, he did not designate him by name in these sketches. The reference to Müller in the 1844 essay is of interest in that his *Elements of Physiology* [78] had been recently translated into English; in fact, this was one of the books by German authors, referred to above, which Carpenter referred to in English translation.

The tenor of these passages indicates that Darwin incorporated at least one of von Baer's concepts into his own thinking many years before the completion of the *Origin of Species*. He grasped that mutual resemblances of embryos could be explained by unity and community of descent, and to this degree he utilized the contribution of von Baer as a bulwark for his own beliefs. Unfortunately, however, in his adoption of the idea of von Baer he concluded

---

[77] Darwin, *Foundations of the Origin of Species*, pp. 218, 219.
[78] Johannes Müller, *Elements of Physiology*, translated . . . with notes by W. Baly. 2 vols., London, 1837-42. Reference from *British Museum Catalogue*, which does not name the publisher. Carpenter in the preface to the 1839 edition of his *Principles* made his acknowledgment to " Müller's Elements of Physiology, translated by Dr. Baly, London, 1838."

from them very different generalizations than had von Baer himself. This was already demonstrated in the essay of 1844. Here the chapter including the embryological discussion terminated with a paragraph entitled " Order in time in which the great classes have first appeared," and this drew, at least tentatively, inferences incompatible with the beliefs of von Baer:

> It follows strictly from the above reasoning only that the embryos of (for instance) existing vertebrata resemble more closely the embryo of the parent-stock of this great class than do full-grown existing vertebrata resemble their full-grown parent-stock. But it may be argued with much probability that in the earliest and simplest condition of things the parent and embryo must have resembled each other, and that the passage of any animal through embryonic states in its growth is entirely due to subsequent variations affecting *only* the more mature periods of life. If so, the embryos of the existing vertebrata will shadow forth the full-grown structure of some of those forms of this great class which existed at the earlier period of the earth's history: and accordingly, animals with a fish-like structure ought to have preceded birds and mammals; and of fish, that higher organized division with the vertebrae extending into one division of the tail ought to have preceded the equal-tailed, because the embryos of the latter have an unequal tail. . . . This order of precedence in time in some of these cases is believed to hold good; but I think our evidence is so exceedingly incomplete regarding the number and kinds of organisms which have existed during all, especially the earlier, periods of the earth's history, that I should put no stress on this accordance, even if it held truer than it probably does in our present state of knowledge.[79]

If, in his essay of 1844, Darwin was inconsistent in that he adopted some of von Baer's ideas, and deviated from others of them, he showed no improvement in this respect when he completed his argument fifteen years later. He did state specifically in the *Origin of Species* that " certain organs in the individual, which when mature become widely different and serve for different purposes, are in the embryo exactly alike. The embryos, also, of distinct animals within the same class are often strikingly similar," [80] and here is where he told the story of the missing label. But at this time too he said further what von Baer could never have accepted:

[79] Darwin, *Foundations of the Origin of Species*, p. 230.
[80] Darwin, *Origin of Species*, 1st ed., 1950 reprint, pp. 372-373.

As the embryonic state of each species and group of species partially shows us the structure of their less modified ancient progenitors, we can clearly see why ancient and extinct forms of life should resemble the embryos of their descendants—our existing species. Agassiz believes this to be a law of nature; but I am bound to confess that I only hope to see the law hereafter proved true.[81]

In other words, in the *Origin of Species* as in the preliminary draft he went beyond von Baer's conclusions regarding embryonic resemblances to draw the inferences which Haeckel was later to exploit. He explained embryonic resemblances on the basis of community of descent, with great profit to his own argument; but he wished to believe what von Baer had so vehemently denied, namely, that embryos could mirror the history of the race by being similar to adult, though extinct, forms.

It was indubitably from Agassiz that Darwin derived his support for his claims of parallel development in fossils and embryos. He was obligated to Agassiz for the example of the heterocercal and homocercal tails of fishes which he used in the paragraph of the 1844 essay dealing with the order of time in which the great classes appeared. Agassiz had written in the first volume of his great work on fossil fishes (1833-1843):

> Nothing will dispute that the form of the caudal fin is of high importance for zoological and geological considerations, since it demonstrates that the same thought, the same plan, which presides today over the formation of the embryo has also manifested itself in the successive development of the numerous creations which have formerly populated the earth.[82]

Within five years of Darwin's drafting of his 1844 essay, Agassiz made his unequivocal statement concerning the more general bearing of embryology on classification:

> [The] natural series again correspond with the order of succession of animals in former geological ages; so that it is equally

---

[81] Darwin, *Origin of Species*, 1st ed., 1950 reprint, p. 381.

[82] "Nul ne contestera que la forme de la caudale (*sic*) ne soit d'une haute importance pour les considérations zoologiques et paléontologiques, puisqu'elle démontre que la même pensée, le même plan, qui préside aujourd'hui à la formation de l'embryon, s'est aussi manifesté dans le développement successif des nombreuses créations qui ont jadis peuplé la terre," Louis Agassiz, *Recherches sur les Poissons Fossiles*, Petitpierre, Neuchatel, 1833-1843, I, 102.

true to say that the oldest animals of any class correspond to their lower types in the present day, as to institute a comparison with the embryonic changes, and to say that the most ancient animals correspond with the earlier stages of growth of the types which live in the present period.[83]

Is it not possible that Darwin's unconscious may have tricked him into naming Agassiz rather than von Baer in connection with embryonic resemblances in the first edition because in reality he favored Agassiz' conclusions over those of von Baer? Darwin, like other scientists, prided himself on this open mind. "I have steadily endeavoured to keep my mind free," he wrote in his "Autobiography," "so as to give up any hypothesis, however much beloved (and I cannot resist forming one on every subject), as soon as facts are shown to be opposed to it." [84] But like other scientists, he could unconsciously succumb to his own prejudices.

"I am rather sorry you do not think more of Agassiz' embryological stages," he wrote to Huxley in 1854, "for though I saw how excessively weak the evidence was, I was led to hope in its truth." [85] Two months later he thanked Huxley for an abstract of the lecture in which the latter had demolished Agassiz' argument with respect to homocercal and heterocercal tails, and also the generalizations drawn from it: "Thank you for your abstract of your lecture at the Royal Institution, which interested me much, and rather grieved me, for I had hoped things might have been in a slight degree otherwise." [86] Darwin did not repeat the example of the heterocercal and the homocercal tails in the *Origin of Species*, but of the truth of Agassiz' general conclusions he became more rather than less convinced during the course of the years. True, when he spoke of them in the chapter on geological succession in the first edition of the *Origin of Species*, he described them temperately, and even referred to Huxley's misgivings about them:

[83] Louis Agassiz, *Twelve Lectures on Comparative Embryology*, . . . Henry Flanders & Co., Boston, 1849, p. 26.
[84] Darwin, *Life and Letters*, I, 83.
[85] Darwin, *More Letters*, I, 75.
[86] Darwin, *More Letters*, I, 82. An abstract of Huxley's lecture, "On Certain Zoological Arguments commonly adduced in Favour of the Hypothesis of the Progressive Development of Animal Life," was published in *Proc. Royal Inst.*, II (1854-58), 82-85, and in *The Scientific Memoirs of T. H. Huxley*, ed. by Michael Foster and E. Ray Lankester, Macmillan and Co. Limited, London, 1898, I, 300-304.

Agassiz insists that ancient animals resemble to a certain extent the embryos of recent animals of the same class; or that the geological succession of extinct forms is in some degree parallel to the embryological development of recent forms. I must follow Pictet and Huxley in thinking that the truth of this doctrine is very far from proved. Yet I fully expect to see it hereafter confirmed, at least in regard to subordinate groups. For this doctrine of Agassiz accords well with the theory of natural selection.[87]

But by the time the final edition was completed in 1872, he had withdrawn his reservations, and here the passage reads:

Agassiz and several other highly competent judges insist that ancient animals resemble to a certain extent the embryos of recent animals belonging to the same classes; and that the geological succession of extinct forms is nearly parallel with the embryological development of existing forms. This view accords admirably well with our theory.[88]

Darwin's option in favor of Agassiz had its effects on subsequent embryology. While Haeckel in any event, even without Darwin behind him, might have reverted to Meckel's antiquated concept of parallelism, yet it was Darwin's thesis that fortified his position with respect to the biogenetic law. Thus Darwin, through Haeckel, looked in two ways in the history of embryology. In so far as Haeckelianism had its progressive side, by inspiring so many young men to enter biology and embryology, Darwin made a positive contribution by having encouraged Haeckel's beliefs and by having facilitated their acceptance. But Haeckel had his retrogressive influence in embryology, also. So powerfully dogmatic was his teaching of the outworn law of recapitulation that for years embryos were investigated primarily for what they might reveal of their ancestry; and the development of analytical and physiological embryology had to await the subsidence of his surge of ideas. For this Darwin too must share responsibility.

Although it is fruitless, it is nonetheless fascinating to speculate what direction embryology might have followed had Darwin in the *Origin of Species* placed less emphasis on Agassiz' conclusions

[87] Darwin, *Origin of Species*, 1st ed., 1950 reprint, pp. 286-287.

[88] Darwin, *Origin of Species*, 6th ed., 1902 reprint, D. Appleton and Company, New York, II, 120. The 6th edition was originally published by John Murray, London, 1872.

and more on von Baer's. And it is tempting to wonder whether he might not have better understood the contributions of von Baer, and have become more convinced of the cogency and significance of his criticisms of the old recapitulation theories, had he read his works in the original more easily, or more frequently, or more comprehendingly. In 1881, the year before he died, he still did not even own von Baer's works.[89]

What Darwin took from von Baer emanated from von Baer's speculations about the whole embryo, or at most about the whole organ. He never so much as mentioned the most important factual contribution of von Baer to embryology, his descriptions of the germ layers, though it was these which provided the phenomenological basis for von Baer's conclusions and for the new embryology to follow. Darwin may have believed, as he said he did, that his embryological considerations were vital to the successful development of his concepts of natural history. But what he accepted from von Baer to further them was drawn from von Baer's reflections more than from his observations. What von Baer described that was new Darwin mainly ignored. The power of von Baer's ideas was sufficiently strong that these could eventually triumph, no matter how they were treated by the more revisionistic apostles of Darwin. And the strength of the *Origin of Species* need not be measured alone by the tenuousness of some of its constituent arguments; in it as in the other creations of nature and of men, the significance of the whole transcends that of its parts.

[89] Darwin, *More Letters*, II, 27.

TWELVE

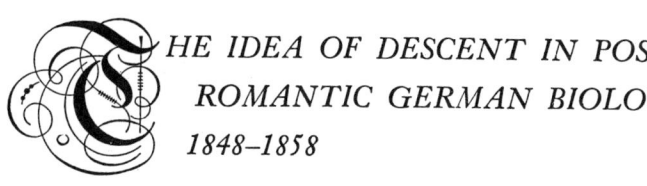
THE IDEA OF DESCENT IN POST-
ROMANTIC GERMAN BIOLOGY:
1848-1858

OWSEI TEMKIN

I

So great was the effect of Darwin's book on German scientists, philosophers, and educated laymen, that, in retrospect, the very doctrine of evolution seemed to have come as something new. Towards the end of his life, the philosopher Eduard von Hartmann (1842-1906) wrote:

> The time from 1830 to 1860 was a period of pure observation in biology. The students of the '50s had no idea that French scientists and German philosophers had envisaged such a thing as the evolution of nature, because at that time, a professor would have thought it unscientific even to mention such fantasies.
> Darwin's work "On the Origin of Species" (1858 English, 1859 German) struck this dull narrow-mindedness like lightning from a blue sky and brought about a veritable mental revolution.[1]

This picture is certainly wrong. The mere fact that Chambers' *Vestiges of Creation*, as Lovejoy has shown,[2] influenced a man like Schopenhauer and was translated into German in 1851,[3] is sufficient evidence that even in Germany the silence cannot have been com-

---

[1] Eduard von Hartmann, *Das Problem des Lebens*; Biologische Studien. Bad Sachsa im Harz: Haacke, 1906, p. 1 f. The importance of this reference evinces from its being cited in *Ueberwegs Grundriss der Geschichte der Philosophie*, vierter Teil, p. 315 f. (12th ed. Berlin, 1923).

[2] A. O. Lovejoy, "Schopenhauer as an Evolutionist," *The Monist*, 1911, 21: 195-222. See this volume, p. 415.

[3] See below.

323

plete. Darwin himself, in a later edition of the *Origin*, could cite a considerable number of those who had believed in descent before him. As early as 1862, A. Braun stated that the theory of descent did not reach an unprepared science.[4] By 1875 controversies were under way as to whether de Maillet, Robinet, and Buffon could be called " forerunners " of Darwin,[5] and Potonié, in 1890, cited about a dozen names of German evolutionists during the period with which we are concerned here.[6] Because many of the publications are hard to obtain even in Europe, I have not been able to analyze all of the works. On the other hand, I have taken the word German in the broad linguistic sense of German writing authors, thereby including the Austrian, Unger, and the Swiss, Nägeli. Though incomplete as far as coverage is concerned the following analysis may yet throw some light on the question: What was the relationship of German biologists towards the doctrine of evolution in the post-romantic decade before the appearance of the *Origin*?

The pertinency of this question is evident if we remember the role Darwinism was to play in German biology. It offered a solution to two problems: that of the origin of species and that of the cor-

---

[4] H. Potonié, " Aufzählung von Gelehrten, die in der Zeit von Lamarck bis Darwin sich im Sinne der Descendenz-Theorie geäussert haben," *Naturwissenschaftliche Wochenschrift*, November 9, 1890, *5*: 441-445; see p. 442. Braun's own attitude to transformism seems ambiguous; cf. Potonié, ibid., p. 444, and Walter Zimmermann, *Evolution*. Freiburg-München: Alber, 1953 (*Orbis Academicus* II/3), p. 380 ff.

[5] Georg Seidlitz, *Die Darwin'sche Theorie*. Zweite vermehrte Auflage, Leipzig: Engelmann, 1875, p. 31 ff.

[6] Potonié, op. cit. (ftn. 4). Of the subsequent books and articles listing German forerunners of Darwin I mention here E. Rádl, *Geschichte der biologischen Theorien*. II. Teil, Leipzig: Engelmann, 1909, *passim*; J. H. F. Kohlbrugge, " J. B. de Lamarck und der Einfluss seiner Descendenztheorie von 1809-1859," *Zeitschrift für Morphologie und Anthropologie*, 1914, *18*: 191-206; J. H. F. Kohlbrugge, "War Darwin ein originelles Genie? " *Biologisches Zentralblatt*, 1915, *35*: 93-111; J. F. Pompeckj, "J. C. M. Reinecke, ein deutscher Vorkämpfer der Deszendenzlehre aus dem Anfange des neunzehnten Jahrhunderts," *Palaeontologische Zeitschrift*, 1927, *8*: 39-42; Otto H. Schindewolf, "Einhundert Jahre Paläontologie (Paläozoologie). Ein Rückblick auf ihre Entwicklung in Deutschland." *Zeitschrift der Deutschen geologischen Gesellschaft*, 1948, *100*: 67-93 (Dr. Bentley Glass kindly drew my attention to this article); Schindewolf, " Einige vergessene deutsche Vertreter des Abstammungsgedankens aus dem Anfange des 19. Jahrhunderts, *Palaeontologische Zeitschrift*, 1941, *22*: 139-168. Eduard Uhlmann, "Entwicklungsgedanke und Artbegriff in ihrer geschichtlichen Entstehung und sachlichen Beziehung," *Jenaische Zeitschrift für Naturwissenschaft*, 1923, *59*: 1-116, is particularly important because it contains lengthy quotations from originals, including some not available to me.

relation between structure and function. It was no longer necessary to believe in the creation of species, and it became possible to explain the apparent finality of the organism and its parts by adaptation without recourse to plan or purpose. Both answers bolstered greatly the mechanistic and materialistic tendencies in German biology. Darwinism became the symbol of materialistic thought. Thus it seems worth while asking what role notions of biological descent played in the period before the appearance of the *Origin of Species* when materialistic tendencies were rising.

The materialistic tendencies of German biologists fell into two categories. The first aimed at a mechanistic explanation of life and denied the existence of a vital force. A decisive attack in this direction was made by Schwann in his *Microscopical Researches* of 1839. There followed Lotze's famous article on vital force of 1842. Shortly afterwards, three pupils of the great Johannes Müller (1801-1858) banded together in their fight against vitalism: Helmholtz, Brücke, du Bois-Reymond. They were joined by Carl Ludwig. The program of the school was formulated in 1848 by du Bois-Reymond in the preface to the first volume of his *Untersuchungen über thierische Elektricität*. Physiology was to be reduced to analytical mechanics. Ludwig's textbook of physiology was written in the spirit of this mechanistic philosophy which viewed even the cellular theory with some suspicion.[7]

The materialism of the second category was directed against the belief in the existence of an immaterial human soul or of any realities outside of "matter and force." The leaders of this faction were Carl Vogt, Moleschott, and Louis Büchner.[8] The biological mechanists, as we may characterize the former group, were not necessarily metaphysical materialists in the sense of the latter. Schwann held a Cartesian position and was a devout Catholic.[9] But materialism of a metaphysical shade was spreading among the German biologists of the fifties, and the controversy, which had a strong anti-clerical undertone, came to a head at the thirty-first meeting of the Society of German Scientists and Physicians at

---

[7] See Owsei Temkin, "Materialism in French and German physiology of the early nineteenth century," *Bull. Hist. Med.*, 1946, *20*: 322-327, and "Metaphors of human biology," in Robert C. Stauffer (ed.), *Science and Civilization*. Madison: University of Wisconsin Press, 1949, p. 173.

[8] On this and the following see Erik Nordenskiöld, *The History of Biology*. New York: Tudor, 1942, pp. 448-452.

[9] Temkin, "Materialism in French and German physiology," op. cit., (ftn. 7) p. 324.

Goettingen, in 1854, when Rudolph Wagner presented "an anthropological address" (as the subtitle called it) on "Human Creation and the Substance of the Soul."[10] The paper defended the belief in the descent of man from one pair (as against a multiple origin) and denied science the right to interfere in matters concerning religion. Although Wagner was not directly discussing animal descent, a passage in his published speech characterizes his attitude:

> As a matter of fact, only animals of one and the same species mix fruitfully. Animals of different, closely related species mix under special, mostly artificial, conditions. But the hybrids are sterile and die out. This fundamental law exists for the protection of the historical existence of species.[11]

The ensuing debate was stormy, with the materialists having the upper hand,[12] and the following years witnessed a literary feud between Carl Vogt (who was an old opponent of Rudolph Wagner) on the one hand,[13] and Rudolph Wagner and Andreas Wagner[14] on the other, in which both sides excelled in tactless invective. In the midst of this controversy appeared the popular book on "Force and Matter" by Louis Büchner, which became one of the basic works of German philosophical materialism and was reprinted again and again in subsequent years.[15]

The biological mechanists had one thing in common with the metaphysical materialists: both were opposed to the *Naturphiloso-*

---

[10] Rudolph Wagner, *Menschenschöpfung und Seelensubstanz*. Ein anthropologischer Vortrag, gehalten in der ersten öffentlichen Sitzung der 31. Versammlung deutscher Naturforscher und Aerzte in Goettingen am 18. September 1854. Goettingen: Wigand, 1854.

[11] Ibid., p. 13.

[12] For a description of the events see *Ueberweg's Grundriss*, (ftn. 1) p. 287 and Nordenskiöld, loc. cit. (ftn. 8).

[13] Carl Vogt, *Köhlerglaube und Wissenschaft*. Eine Streitschrift gegen Hofrath Rudolph Wagner in Göttingen. I have used the second printing of the fourth edition which appeared in Giessen: Ricker, 1856. In this pamphlet Vogt relates his earlier controversies with Wagner.

[14] Andreas Wagner, *Naturwissenschaft und Bibel* im Gegensatz zu dem Köhlerglauben des Herrn Carl Vogt, als des wiedererstandenen und aus dem Französischen ins Deutsche übersetzten Bory. Stuttgart: Liesching, 1855. The allusion is to Bory de St. Vincent (1780-1846), on whom see Henry Fairfield Osborn, *From the Greeks to Darwin*, second edition, New York: Scribner's, 1929, pp. 291-293.

[15] Louis Büchner, *Kraft und Stoff*. Empirisch-naturphilosophische Studien. In allgemein-verständlicher Darstellung. Frankfurt a. M.: Meidinger, 1855. He was a practicing physician who, for a short time, had taught at the University of Tübingen.

phie of the preceding era. Naturphilosophie in the spirit of Schelling was no longer a scientific power in the 1840's. But it was remembered as a destructive influence in German science. And it was still remembered that Naturphilosophie had encouraged the idea of descent. There is no doubt that evolutionary ideas had been rife among German romantics since the end of the eighteenth century. Yet in the individual case it is hard to ascertain whether the notion of evolution implied actual descent. Romantic thinkers were inclined to envisage all creation as an ever-ascending evolution of things from inanimate matter to man. But this concept of evolution did not imply that every higher species actually descended from lower ones. Many, if not the majority, believed that in creating higher species nature improved upon herself, but that, nevertheless, each species had to be created anew. They could, therefore, easily accept the doctrine of the destruction of prehistoric animals by catastrophes. Yet some among the German romantic biologists definitely believed that there was a transmutation of lower forms into higher ones.[16]

Both rightly and wrongly, Naturphilosophie, which had dominated German science during the romantic era, was tainted with the suspicion of upholding ideas of descent. The botanist, E. Meyer, in an article of 1854, wrote: " Among the deplorable [influences of Naturphilosophie] I reckon the deep shaking of the belief in the constancy of species. . . ."[17] Admittedly, a shaken belief in the constancy of species is not yet a positive belief in their historical transformation.[18] Isolated remarks which sound as if their authors affirmed descent must not be taken at their face value without careful investigation.

This holds equally true of the period under discussion. Thus the zoologist, Julius Victor Carus, is usually referred to as a forerunner

---

[16] See Jane Oppenheimer's contribution in this volume; also Owsei Temkin, "German concepts of ontogeny and history around 1800," *Bull. Hist. Med.*, 1950, *24*: 227-246, and the literature cited above, footnote 6.

[17] Quoted from Potonié, op. cit., (ftn. 4) p. 442 f. (E. Meyer's article is said to have appeared in the *Königsberger naturwissenschaftliche Unterhaltungen*, vol. 3, under the title: " Ueber die Beständigkeit der Arten, besonders im Pflanzenreich.")

[18] Walter Baron, "Die idealistische Morphologie Al. Brauns und A. P. de Candolles und ihr Verhältnis zur Deszendenzlehre," *Beihefte zum Botanischen Centralblatt*, 1931, *48*: 314-334; see p. 317. This article provides an excellent comparison beween the different points of view and aims of the "idealistic morphologists" and the followers of Darwin.

of Darwin, and this reference is based on a passage from his *System der thierischen Morphologie* of 1853. The passage in question reads:

> ... on the other hand, I may hope not to be misunderstood if, with regard to the present form of our classificatory endeavors and the *relationship* [*Verwandtschaft*] of certain forms of organic beings, I call attention to the fact that the first created forms, which from the admittedly oldest geological strata confront us as witnesses for an earlier creation which was closer at least to the first,—that these forms, apart from their organic character show the general character only of the group in which we place them. Consequently, in a sense limited of course by the absolute lack of possible demonstration, we can consider them the ancestors from which through perpetual generation and accommodation to progressively very different conditions of life, there originated the wealth of forms of the present creation.[19]

Taken by itself, this passage seems, if not clear, to be at least a clear reference to descent of present forms from earlier ones. But as far as I see, the bulk of the book leaves it undecided whether Carus has in mind concrete, or ideal, "evolutionary series in the animal world."[20] Towards the end of the book he expresses himself in a manner which seems to contradict the idea of descent. Morphology, he says, "not only demonstrates to us the immutable equality of the laws [i. e. of "formation"] but also their perfection from the very beginning. Often as the formative activity of the Creator manifested itself in the production of new animal worlds, still no forms proceeded from his hand which did not follow the same uniform plan."[21]

---

[19] J. Victor Carus, *System der thierischen Morphologie*. Leipzig: Engelmann, 1853, p. 5: "... darf aber auf der andern Seite wol hoffen nicht misverstanden zu werden, wenn ich mit Rücksicht auf die gegenwärtige Form unserer classificatorischen Bemühungen und auf die *Verwandtschaft* gewisser Formen der organisierten Wesen daran erinnere, dass die erstgeschaffenen Formen, welche uns aus den anerkannt ältesten geologischen Lagern als Zeugen einer früheren, der ersten wenigstens näher stehenden Schöpfung entgegentreten, ausser ihrem organischen Charakter nur den allgemeinen der Gruppe zeigen, zu welcher wir sie stellen, dass wir sie also, natürlich nur in einem durch den absoluten Mangel eines möglichen Beweises beschränkten Sinne, als die Urahnen betrachten können, aus denen durch fortgesetzte Zeugung und Accomodation an progressiv sehr verschiedene Lebensverhältnisse der Formenreichthum der jetzigen Schöpfung entstand." Part of that passage is quoted by Uhlmann, op. cit., (ftn. 7) p. 69 and misquoted by Ernst Haeckel, *Natürliche Schöpfungsgeschichte*, Sechste verbesserte Auflage. Berlin: Reimer, 1875, p. 98.

[20] "Entwickelungsreihe in der Thierwelt," discussed by Carus, op. cit., on pp. 275-278.

[21] Ibid., p. 505 f.

Seidlitz, in 1875, introduced the quotation of Carus' statement by the following remark: "The reserve with which this investigator, now an unconditional follower of Darwin, vents his transformistic thought in one single passage of the introduction is very significant for the public opinion of science at that time." [22] Seidlitz may have been right, but it must not be overlooked that, as the example of Virchow will show, ideas held before 1859 were sometimes given a new interpretation afterwards. It is for this reason that I have omitted a discussion of authors whose publications were not accessible to me, even where excerpts from their writings were available.[23]

To be sure, scientific opinion was preponderantly against the idea of transmutation. But there was no unity in the camp as to the alternatives. In view of the geological record, the eternity of all species could no longer seriously be defended, as Virchow made clear in his attack upon Czolbe.[24] Those who upheld a belief in creation in accordance with the biblical account may well have found their leader in the Swiss zoologist Louis Agassiz (1807-1873), who, in 1846, had emigrated to the United States.[25] In matters of descent,

---

[22] Seidlitz, op. cit., (ftn. 5) p. 52.

[23] As far as I can see, this omission includes:

W. Hofmeister (1849-1851), see Potonié, op. cit., (ftn. 4) p. 444 who refers to Sachs, *Geschichte der Botanik*, pp. 214-217. But in the English translation of this work (second impression, Oxford: Clarendon, 1906) Hofmeister is not cited as holding transformist views, although it is emphasized that his results were incompatible with the idea of constant species.

H. P. D. Reichenbach, who, according to Seidlitz, op. cit., (ftn. 5) presented a paper at the twenty-eighth meeting (at Gotha) of German Scientists and Physicians, published in 1854 at his own expense in Altona (where he practiced medicine) under the title *Ein kleiner Beitrag zur Anthropologie und Philosophie*.

Oswald Heer, who in his *Tertiäre Flora der Schweiz* of 1855, according to Seidlitz, op. cit., (ftn. 5) p. 53, spoke of "'Umprägung der Arten' . . . wie er die in relativ kurzer Zeit erfolgende Umänderung der Arten nennt."

F. T. Kützing, from whose "Historisch-kritische Untersuchungen über den Artbegriff bei den Organismen und dessen wissenschaftlichen Wert," *Programm der Realschule zu Nordhausen*, 1856, Uhlmann, op. cit., (ftn. 7) p. 71 ff. gives several excerpts.

Gustav Jaeger, who is credited by Seidlitz, op. cit., p. 53 with having "developed the theory of descent to its last consequences" in 1857 before a select group of young scientists. Its late publication in 1864 in the *Zoologische Briefe* is attributed to the reluctance to flaunt the opinion of scholars before 1859.

Although Schopenhauer's evolutionary ideas fall within our period, I have not referred to them in view of Professor Lovejoy's article.

[24] Rudolf Virchow, "Alter und neuer Vitalismus," *Archiv für pathologische Anatomie* etc., 1856, 9: 3-56; see below, p. 348.

[25] Louis Agassiz and A. A. Gould, *Outlines of Comparative Physiology, Touching the Structure and Development of the Races of Animals, Living and*

his views were uncompromising. Admitting that the fossil record showed " a manifest progress in the succession of beings," he insisted:

> But this connection is not the consequence of a direct lineage between the faunas of different ages. There is nothing like parental descent connecting them. The fishes of the Paleozoic age are in no respect the ancestors of the reptiles of the Secondary age, nor does Man descend from the mammals which preceded him in the Tertiary age. The link by which they are connected is of a higher and immaterial nature; and their connection is to be sought in the view of the Creator himself, whose aim, in forming the earth, in allowing it to undergo the successive changes which geology has pointed out, and in creating successively all the different types of animals which have passed away, was to introduce Man upon its surface.

Agassiz' thought was summed up as follows:

> To study, in this view, the succession of animals in time, and their distribution in space, is therefore to become acquainted with the ideas of God himself.[26]

The ease with which a fundamentally pious biologist like Schleiden who was opposed to materialistic tendencies of a philosophical kind could gloss over the Mosaic account of creation, demonstrates the tenuous hold of biblical orthodoxy on the minds of German scientists. It cannot be taken for granted that all those who neglected to express themselves on the origin of species, unquestioningly accepted their supernatural creation. Unless they dismissed the whole problem from their minds, another possibility was left to them, viz., a series of spontaneous generations. In their *Grundzüge der Botanik*, Endlicher and Unger, in 1843, wrote:

Extinct. For the Use of Schools and Colleges. Edited from the Revised Edition, and greatly enlarged, by Thomas Wright, M.D. London: Bohn, 1851. Agassiz' erstwhile student, Carl Vogt, in *Altes und Neues aus Thier-und Menschenleben*. 2 vols., Frankfort a. M.: Literarische Anstalt, 1859 (the preface is dated March 31), vol. 1, p. 353 quotes Agassiz's work on fossil fish, vol. 1, p. 171, for the belief in a personal creator who, twenty times or more, destroyed all life to replace it by better forms.

[26] Agassiz-Gould, ibid., pp. 417-418. There is a good deal of an older tradition in this whole line of thought. Walter Baron. " Zu Louis Agassiz's Beurteilung des Darwinismus," *Sudhoffs Archiv für Geschichte der Medizin und der Naturwissenschaften*, 1956, 40: 259-277, has pointed out that Agassiz was motivated by ideas of natural theology rather than of Naturphilosophie or sentimental religiosity. But it has to be added that the concept of "ideas of creation" is to be found in Herder and von Baer as well; see Temkin, " German concepts of ontogeny," op. cit. (ftn. 16).

The genesis of the present vegetation, like that of all earlier vegetations, can, however, only be conceived as a spontaneous generation in accordance with the idea of the plant organism eternally progressing towards perfection. It probably emerged, if a conclusion based on present experiences is admissible, from a slimy carbonaceous substratum [*Matrix*], out of which cells developed into shoots, and these into complete plants.

In these acts of spontaneous generation of which the earth does not seem constantly capable, but only after longer or shorter intervals, and only after previous violent changes on its surface and in the medium surrounding it, plant species were created in large numbers and according to certain contrasts. . . .[27]

This quotation reflects the status of the cellular theory in the 'forties, when the principle of *omnis cellula a cellula* had not yet been developed by Remak (1852)[28] and formulated by Virchow (1855). In 1859, Carl Ernst von Baer tried to limit the need for such spontaneous generations to a minimum. Like many other scientists, he felt uneasy about them, admitting that spontaneous generation was merely another word for creation minus the stigma of supernatural intervention.[29] But it is well to remember that Pasteur's first publication on the subject appeared in 1860, after Darwin's *Origin*. And even then spontaneous generation was not eliminated as a postulate, least of all by speculating Darwinists like Haeckel and Huxley. The difference was that after 1859 spontaneous generation was banished to a semi-mythical time at the beginning of evolution. Before 1859, many nontransformists as well as transformists assumed that it might have initiated every new " creation." This, I think, explains why, to the mechanistically minded biologist, spontaneous generation was more important as a principle than was descent. " Lest we take refuge in miracles and the incomprehensible, we must admit the origin of the first organized creatures on the earth by the free creative power of matter itself . . ." wrote Her-

[27] Stephan Endlicher und Franz Unger, *Grundzüge der Botanik*. Wien: Gerold, 1843, p. 461.

[28] Bruno Kisch, *Forgotten Leaders in Modern Medicine*: Valentin, Gruby, Remak, Auerbach. Philadelphia: The American Philosophical Society, 1954 (*Transactions of the American Philosophical Society*, New Series, vol. 44, part 2, 1954), p. 258.

[29] See Seidlitz, op. cit., (ftn. 5) p. 55 f. Zimmermann, op. cit., (ftn. 4) p. 503: " Bis zur Mitte des vergangenen Jahrhunderts nahmen auch namhafte Biologen, wie H. G. Bronn (1800-1862), K. E. v. Baer, K. v. Nägeli (für Pilze) plötzliche, sich in unseren Tagen abspielende Urzeugungen als erwiesen an."

332 FORERUNNERS OF DARWIN

mann Burmeister (1807-1892),[30] who dismissed the notion of descent, although he was not unsympathetic to it. And Rudolf Virchow, in 1856, insisted on the need of admitting spontaneous generation. Though he is not clear as to details, his way of expressing himself does not point to the limitation of spontaneous generation to a single event at the dawn of organic life.[31] He needed a principle that avoided eternity of species and supernatural intervention, yet upheld the idea of progress.[32] Against this background the various arguments for transformism have now to be viewed.

II

The beginning might well be made with the geologist Bernhard Cotta's (1808-1879) "Letters on Alexander von Humboldt's Kosmos." The first volume appeared in 1848, and the dedication, dated April 2 of that year, shows the author under the perturbing influence of the revolution; the occupation with the book has helped to quiet his mind, and he hopes "that the result of this occupation may affect others similarly."[33] As is to be expected from his desire for equanimity, the tone of the book is not polemical, although Cotta declares his antagonism to the speculative philosophy in Germany. The true scientist, he says, does not believe the world to be the result of his thought, but rather he believes his thought to be the end-product of the world, of his observation of the world. A few lines later Cotta offers a hypothetical explanation: "Indeed, from our limited earthly point of view we might even be tempted to

---

[30] Hermann Burmeister, *Geschichte der Schöpfung*. Eine Darstellung des Entwicklungsganges der Erde und ihrer Bewohner. Für die Gebildeten aller Stände. Fünfte verbesserte Auflage, Leipzig: Wigand, 1854, p. 325.

[31] Virchow, op. cit., (ftn. 24) p. 25: "Aber selbst wenn wir die Erfolglosigkeit aller Versuche, einzelne Zellen zu erzeugen, in unserer Zeit zugestehen müssen, so können wir darin keinen Grund für ihre Ewigkeit finden. Das Gesetz, nach dem ihre Bildung erfolgte, muss nothwendig ein ewiges sein, so dass jedesmal, wenn im Laufe der natürlichen Vorgänge die Bedingungen für seine Offenbarung günstig werden, die organische Gestaltung sich verwirklicht." See also below, p. 348.

[32] Ibid. "Die Zeiträume, welche der Mensch überblickt, sind, selbst wenn wir den grössten Rahmen ausspannen, indem wir die Jahrtausende nach Millionen rechnen, immerhin sehr beschränkt, und doch gestatten sie uns, die befriedigende Ueberzeugung zu gewinnen, dass ein freilich sehr langsamer und häufig unterbrochener Fortschritt in der Welt ist."

[33] Bernhard Cotta, *Briefe über Alexander von Humboldt's Kosmos*. Erster Theil, Leipzig: Weigel, 1848, p. III.

assert that man with his intelligence is the final result of a long evolutionary series [*Entwicklungsreihe*] of organic beings, a series which, through all its generations, has absorbed an immeasurable sum of external impressions (unconscious observations of nature and of its laws) and has thereby worked its way up to a thinking, rational being." [34] A rash interpretation of the sentence as referring to descent will be checked by the realization that it is the " evolutionary series " that works its way up.

Cotta offers a good example of how important it is to distinguish between evolution and transmutation (or descent). " In the 24th letter I wrote to you something about the sequence in which the organic forms originated one after the other on the earth. In this we recognized a *gradual evolution [Entwicklung] from the lower to the higher.*" [35] Cotta's italicization enhances the impression that he has evolution in the modern sense in mind, i. e., transmutation. But the immediately following sentence shows that evolution is used in a broader sense: " Yet how the single organisms, i. e. the first individuals of each species originated is still veiled in deep obscurity." [36]

Nevertheless, in the 24th letter as well as later he discusses transmutation in unmistakable terms. He presents it as a hypothesis advanced by others, even though his own inclination towards it is obvious. " Here I must point out to you right away," he writes in the 24th letter, " that much esteemed scientists have made the assertion that the diversity and the change of organic forms on the earth are altogether not caused by the origin of entirely new species, but only by an ever more diverse and higher evolution [*Entwicklung*] of the originally formed. According to this opinion, in the course of immeasurable periods, and corresponding to the respective external conditions there arose from one organic form all others. Accordingly we must consider man too as the highest evolutionary step of a first organic seed [*Keim*]. This explanation of the diversity in organic nature has a good deal for it. . . ." [37]

Even in his main discussion of organic life on earth Cotta does not forget to emphasize the hypothetical character of an explanation by descent.

The opinion *that the different organic forms developed gradually from one another* has already often been expressed, and

---

[34] Ibid., p. vi. This idea is taken up again at the end of the volume, p. 340, and given a rather clear materialistic sense.
[35] Ibid., p. 262.   [36] Ibid., pp. 262-263.   [37] Ibid., p. 140.

defended by argument, by true natural philosophers. Quite recently an anonymous Englishman sought ingeniously to prove this view in the book *Vestiges of the Natural History of Creation,* and indeed, there is much to be said for it even though it will, in the very nature of the matter, long remain a mere hypothesis. The law of the organization of matter is as yet completely undiscovered; its results have, as already stated, not yet been drawn into the sphere of the inevitable. If, however, for the moment we assume that such a law governing matter exists, that certain substances in certain circumstances unite and become organic forms, then the special diversity of these substances and circumstances will determine the diversity of the forms and be reflected in them. In the simplest external circumstances, the simplest, most suitable, organic forms will have arisen; and as the circumstances changed and became diversified, the organic forms gradually changed and became diversified also.[38]

From the above and a few other passages it is possible to form an impression as to why Cotta hesitated to associate himself frankly with transformism, why, nevertheless, he defended the idea, and along what line he looked for a possible explanation. Apart from our general ignorance of the laws of organization, he sees a main obstacle in the sharp separation of species. Never has an actual transition been observed between really different species. They differ, down to their minutest parts, "though *perhaps* we must assume that they developed historically from one another."[39] On what can he base the latter possibility? Cotta has great faith in the effect of long periods of time. While in the shortness of historical time we have not observed the formation of truly new species, yet table fruits and domestic animals can be considered as "almost" new formations of species.[40] Unbounded time could achieve what cannot be observed in relatively short periods. But this is at best a defense of transformism. What is Cotta's argument *for* it? In the first place, aversion to the acceptance of divine acts of creation: "If for every special form you refer to the creator of the world, that means cutting through the knot rather than untying it. You put an *assumption* instead of an *explanation* just because you lack the latter."[41] Apart from suggesting an explanation of the variety of species, the idea of transmutation also avoids a multiplicity of spontaneous generations and supports a progressive view

[38] Ibid., p. 263.
[39] Ibid., p. 262. Italics mine.
[40] Ibid., p. 264.
[41] Ibid., p. 263.

of man. As we shall see, Cotta, in these points, meets with the botanist Unger.

Since Cotta does not expressly make the idea of descent his own, it is not surprising to find relatively little thought given to its possible mechanism. Nevertheless, though he does not mention the name as far as I see, enough is said to characterize his attitude as Lamarckian. Greatly impressed by the harmony between structure and environment, Cotta cites as an especially impressive instance Charles Darwin's account of *Birgus latro*, a large crab living on the Keeling islands, that eats coconuts. " Had a species of crab previously living in the sea found pleasure in living on land and enjoying coconuts and had, therefore, through thousands of generations gradually changed its whole organization so as to become able to eat the nuts with relative ease? Or, was an animal, adapted for this special end, produced by a special act of creation? " Such questions cannot be answered directly; however, there are " the thousands of less conspicuous yet analogous cases in which the organization of animals manifestly changes gradually for a definite purpose through exercise, as the greyhound has taken on a form quite different from the dachshund or butcher's dog, and the race horse a form different from the draught-horse." [42]

In the same year, 1848, the botanist Schleiden too published a popular work, *The Plant*.[43] Schleiden argued for transformism but, in contrast to Cotta he did so unreservedly. Schleiden has no doubt that the world is God's creation, but to him that does not imply accepting the Mosaic story. He tries to fit the origin and development of plant life into the history of the earth, which makes him think that

> In some period of this gradual shaping of the earth, the first of organic existence originated, through forces, which may indeed still be in action, but under conditions and co-operation of those various forces such as now appear no longer possible upon our earth. The ocean was probably the birth-place of these organisms. . . .[44]

---

[42] Ibid., p. 266.
[43] M. J. Schleiden, *The Plant*; a Biography. Translated by Arthur Henfrey. Second edition, with additions. London: Baillière, 1853. However, judging from the quotations given by Uhlmann, *op. cit.* (ftn. 6), ideas of descent were already expressed in the first German edition of 1848.
[44] *The Plant*, p. 277 f.

A little later, Schleiden adds: "In subsequent periods, organisms also originated upon the dry land . . ,"[45] which sounds as if he too believed in repeated spontaneous generations. But then he seems to deny such an interpretation because he ascribes the whole subsequent process to gradual descent from a single cell.

> This view, that the whole fulness of the vegetable world has been gradually developed out of a single cell and its descendants, by gradual formation of varieties, which became stereotyped into species, and then, in like manner, became the producers of new forms, is at least quite as possible as any other, and is perhaps more probable and correspondent than any other, since it carries back the Absolutely Inexplicable, namely, the production of an Organic Being, into the very narrowest limits which can be imagined.[46]

This summary shows Schleiden's reason for believing in transmutation. As to the "how," he first proposed a hypothesis for the formation of varieties by referring to tropical conditions:

> The principal causes producing a luxuriant and varied world of forms in the tropics, are moisture and warmth, the causes of their multiformity appear to lie in the richness of the soil in readily soluble inorganic matters, which in the first place give rise to a variation in the chemical processes of the plant, and thus to a greater or less deviation in the form. These two conditions *meet* in the tropics, because they are dependent one upon the other, since the more luxuriant vegetation called forth by the moist, warm atmosphere, prepares by its death and rapid decay, a soil richer in soluble inorganic matters, for the succeeding generation.[47]

Once formed, varieties "when they have continued to vegetate under the same conditions for several generations, pass into subspecies, that is into varieties which may be propagated with certainty by their seeds. . . ." Assuming now that the same influences continue to act for a very long period,

> will not, at last, as the variety thus becomes a sub-species, so also this become so permanent, that we shall and must describe it as a species [?] Then, if the first cell be given, the foregoing points out how the whole wealth of the vegetable kingdom may have been formed by a gradual passage from it through varieties, sub-species and species, and thus onward, beginning anew from each

---

[45] Ibid. [46] Ibid., p. 293. [47] Ibid., p. 289 f.

species—in a space of time, indeed, of which we have no conception, for which, however, since there is nothing *real* wanting, we may provide at pleasure in our dreams; for it may be mentioned here that all the recent distinguished Geologists come ever more and more to the opinion, that very much, in the formation of the crust of our earth, which was formerly ascribed to violent, convulsive and sudden revolutions, has rather been the product of forces acting slowly through the course of enormous periods of time.[48]

Schleiden presupposes a very long geological time (which would have easily been granted to him by German scientists of the period); he also presupposes the principle of "actuality" in geology as defended by von Hoff in Germany and Lyell in England, and a modifying influence of the environment upon plant form. It should be noted that the environmental factors which he cites do not necessarily require adaptational tendencies on the part of the plant. They work through chemical and physical forces. Schleiden was opposed to any "vital force," though he was equally opposed to metaphysical materialism. Indeed, in 1863 he published a severe attack upon the materialistic tendencies in modern science.[49]

Environmental changes, though of a very different nature, are also invoked by Karl Heinrich Baumgärtner (1798-1886), from 1824-1862 professor of clinical medicine in Freiburg i.B.[50] In his younger years he had worked on the segmentation of the ovum, and to this work he referred again in the "Textbook of Physiology" of 1853, where he enunciated his theory of transmutation through influences upon surviving ova of creatures destroyed in earlier phases of geological history.[51] Two years later, in 1855, he issued in short theses the ideas prepared in the "Textbook" and subsequently elaborated.[52] They may here be sketched in the presentation he gave them in his *Schöpfungsgedanken*, "Physiological Studies for the Educated," as Baumgärtner called the book he pub-

---

[48] Ibid.
[49] See Temkin, "Materialism," (ftn. 7) p. 324.
[50] E. Th. Nauck, "Die ersten Jahrzehnte des psychiatrischen und neurologischen Unterrichts in Freiburg i. Br.," *Ber. Naturf. Ges. Freiburg i. Br.*, 1956, *46*: 63-84; see p. 64 f.
[51] Karl Heinrich Baumgärtner, *Lehrbuch der Physiologie mit Nutzanwendungen auf die ärztliche Praxis*. Stuttgart: Rieger, 1853, pp. 71 ff. and 190.
[52] K. H. Baumgärtner, *Anfänge zu einer physiologischen Schöpfungsgeschichte der Pflanzen- und der Thierwelt, und Mittel zur weiteren Durchführung derselben*. Stuttgart: Rieger, 1855.

lished in 1859.[53] The first part, giving his evolutionary thought, had appeared before;[54] so that any suspicion of an early influence by Darwin's *Origin* may be dismissed.

Baumgärtner had discovered—or believed to have discovered—what he called "the law of the ascending germ-metamorphosis in the periods of creation."[55] He argues that the chicken must have existed before the egg and the cock before the semen. Neither they nor their seed can have originated by spontaneous generation. Nor can they have originated from free organic substances. Likewise, the transformation of a complete organism into a different animal is very unlikley. Therefore, animals must have developed from germs; i. e., the higher animals must have developed from the germs of lower organisms.[56]

In particular, evolution took the following course. "Through organizing influences," as Baumgärtner calls them, in the first creative process, the simplest organism originated from nitrogen, carbon, hydrogen, and oxygen.[57] The original cell formed through spontaneous generation and became differentiated in analogy to the differentiation of animal and vegetal poles in the egg. A separation into primitive animals and plants may have been the direct result. The succeeding creative period began with a destruction of all the existing adult creatures and a change in their surviving seed. General cosmic phenomena accompanied by electrical polarizing forces were responsible for the catastrophe as well as the evolutionary step.[58] The new polarizations in the surviving seed brought about cleavages in new directions and therewith possibilities for the development of new, higher species. Baumgärtner does not assume that all the animals of the new creation were immediately available. They might start out as particles of the dividing ovum, gradually changing into the final forms. In particular, Baumgärtner depicted the evolution of man during the last period of creation as a transformation of low embryonic forms into the adult form. Evolution proceeded in many series, the various races of men might well have been the results of different evolutionary series. Spontaneous generation was not limited to the very first creation, but occurred again in each new period.[59] All this assured to evolution a great variety

[53] K. H. Baumgärtner, *Schöpfungsgedanken.* Physiologische Studien für Gebildete. Freiburg: Wagner, 1859.
[54] Ibid., p. iii-iv.
[55] Ibid., p. 307.
[56] Ibid., pp. 310-316.
[57] Ibid., p. 318.
[58] Ibid., p. 326 ff.
[59] Ibid., p. 336 f.

of possibilities and to the author an opportunity of indulging in peculiar speculations. For instance, beasts of prey and their victims are thought to be adapted to one another because they originally formed from the same egg. Again, some men were still leading a larval existence in the water while others more perfect watched them as they walked along the shore.[60]

Baumgärtner's arguments for descent are not convincing. If, as he believes, spontaneous generation can take place at the beginnings of all creative periods and if new species of animals lead a long life in a larval state, why then is it necessary to assume a mutation of earlier germs at all? One cannot help suspecting that Baumgärtner's motives for favoring descent were in part metaphysical ones. In his *Schöpfungsgedanken*, at least, the metaphysical element is obvious. All the changing organizations lead to the idea of a cosmic organism which, in turn, presupposes a directive power. "Thus," he writes, "physiological research leads us to the idea of God, viz. to the conviction that there is a thinking power to which the natural laws themselves and the final reason of all things must be traced back."[61] Moreover, the "law of transformation of germs and of the progressive evolution of mental life shows us that at any rate the movement marches on and that, therefore, our hopes for a higher destination are no empty illusion."[62]

In contrast to the popular books of Cotta, Schleiden, and Baumgärtner, Unger's "Attempt at a History of the Vegetable Realm" of 1852 is a work of a vastly different character. Himself professor of botany at the University of Vienna, Unger presents a solid paleontological investigation. And while we could quote a passage of 1843 where he (and Endlicher) clearly asserted repeated spontaneous generations, Unger has now changed his mind.[63] The assumption of spontaneous generation contradicts contemporary experience and the idea of descent recommends itself just because it helps to avoid such an awkward hypothesis. Going back in time, Unger finds algae as the most primitive of all later forms. In theory, at least, he dares to go even further and assume the previous existence of an "Urpflanze" (we are immediately reminded of Goethe),

---

[60] Ibid., p. 346.   [61] Ibid., p. 373.   [62] Ibid., p. 370 f.
[63] Among the German followers of the theory of transmutation during the period under discussion, Unger and Schaaffhausen (with the exception of Schopenhauer) are probably best known. Both are mentioned in Darwin's "Historical Sketch," and their writings are discussed by Zimmermann, op. cit., (ftn. 4).

or even a primary cell, "which is at the bottom of all vegetal existence." [64] Comparing the flora of past ages with that of more recent ones, he finds that the number of new species increases with the periods. If, therefore, all species originated spontaneously, spontaneous generation should have occurred more and more often, and most frequently in our own period. This is contrary to all experience. Accordingly, descent is the alternative—descent, however, which is due to inner forces. External conditions may have played a minor role, but they could not have been decisive because of the regularity noted in the development of new plant forms. "Nothing has been added in this regulated evolutionary process of the vegetal world that had not been previously prepared and indicated, so to speak. Neither genus, nor family, nor class of plants has manifested itself without having become necessary in time." [65] This allows Unger to consider the totality of plants a "truly coherent organism." [66] But how does this agree with the alleged stability of species? Unger avoids a clear-cut answer. He says that uniformity does not exclude variations. Old species did not simply change into new ones. Rather, some individuals metamorphosed while the old type still remained existent for some time.

His idea of an "inner development" allows Unger to avoid the question as to the origin of the *Urpflanze* and, at the same time to place this development into a grandiose cosmic frame. The unity of plant life stands in contrast to inanimate nature. Though we do not know how the first cell originated, we can at least accept it as "the origin of all organic life and, therefore, the true bearer of all higher evolution." [67] There is no reason to believe that the productive activity will stop. Our present period is still young, our world is full of imperfections and thus man strives after accomplishment of the ideal he has in his mind. The same is true of animals and plants. The formation of new races is but the urge to perfection, the attempt to be ready for entry into the next cosmic period.[68]

In opposing transmutation to spontaneous generation Unger did not stand alone. Cotta had advanced the argument that the occurrence of spontaneous generation could not be decided by direct observation. Anyhow, Cotta thought, "if it is true that the single organic forms developed from one another through the influence

---

[64] F. Unger, *Versuch einer Geschichte der Pflanzenwelt*. Wien: Braumüller, 1852, p. 340.
[65] Ibid., p. 344.
[66] Ibid., p. 345.
[67] Ibid., p. 340.
[68] Ibid., p. 349.

of changed external circumstances of life, this is again a reason against the spontaneous generation of single individuals." [69] Likewise, in Cotta's opinion gradual development points to further progress. At present, man is the highest stage of organization, "but we cannot know what more gifted being will result from many thousands of years of further development." [70] Humanity as a whole advances in abilities, knowledge, and morality. Indeed the dependence of morality on development is the great justification for the scientist's endeavor "to replace the personal activity of a creator by evolution through scientific law." [71]

Here is the place to point out the role which the optimistic belief in progress played in biology before Darwin. Virchow, in the above mentioned essay of 1856, frankly stated his preference for a view that recognized progress in the history of living forms as against a view that considered species eternal and immutable.[72] I believe that at that time, Virchow, temporarily at least, acquiesced in the assumption of repeated occurrences of spontaneous generation. But the underlying sentiment was the same as Unger's, though Unger expressed it in terms of a theory of descent and an idealistic philosophy.

In the face of such views it is hard to decide whether even Unger was led to accept descent as a purely logical consequence of his paleontological and geographical research, or postulated it under the influence of metaphysical predilections. The form he gave to it indicates that his general views on man, the world, and their progress played a considerable part. It is, therefore, all the more interesting that Schaaffhausen's famous article "On the Constancy and Transformation of Species," which appeared but one year later, in 1853, is an acceptance as well as a rejection of Unger's evolutionary ideas. Schaaffhausen states that not only von Baer, but even Vogt and Burmeister recoiled from the idea that man originated from the orangutan.[73] Opinions vary, from Agassiz who believed in twenty renewals of creation, to Unger who propounded that "the evolution of the vegetal realm began with an *Urpflanze* and that

---

[69] Cotta, op. cit., (ftn. 33) p. 267.
[70] Ibid.
[71] Ibid., p. 268.
[72] Virchow, op. cit., (ftn. 24) p. 24 f. Erwin H. Ackerknecht, *Rudolf Virchow: Doctor, Statesman, Anthropologist*. Madison: University of Wisconsin Press, 1953, p. 200.
[73] Hermann Schaaffhausen, "Ueber Beständigkeit und Umwandlung der Arten," *Anthropologische Studien*, Bonn: Marcus, 1885, pp. 134-164; see p. 135.

the various types appeared gradually in ever-increasing number, having emanated due to an inner formative urge of the plant itself." [74] In the main, Schaaffhausen is concerned with the problem of the constancy of species. He combats this idea with all the material and arguments at his disposal. At the head of his summary there stands the sentence: "The immutability of species which most scientists regard as a natural law is not proved, for there are no definite and unchangeable characteristics of the species, and the borderline between species and subspecies [*Art* and *Abart*] is wavering and uncertain." [75] Schaaffhausen stresses the need for more and more exact observation, especially regarding variations. This is a very sober attitude, though more is involved than is at first apparent. Beside expecting an ever-increasing discovery of missing links, he also confronts the opponents of evolution with the alternative "that the oak or horse originated from the elements." [76] And in contrast to Unger's inner urge, he puts his belief in the influence of, and adaptation to, external influences.[77] But he does not explain the mechanism of such influences and adaptations. Moreover, his subsequent attitude shows that his work on species was not quite free from general ideas. In a paper of 1860, he opens a short polemic against Büchner and Virchow with the words: "In our days, when a new materialistic school even denies the order in the world at large because it leads to the assumption of the concept of God, when one admires the chaos rather than the cosmos, it is all the more necessary to call to mind the doctrine of the unity of life and to show how well founded it is." [78] And he closes the polemic by saying: "But if we see the organic forms emanating one from the other and life obtaining its vigor from the sources of telluric nature only, and the earth receiving air and warmth from another star whose course again follows a higher law, then with this insight we approach the eternal mind whose creative thought also was creative deed, by which he made and at every moment makes anew, the world and its wonders." [79]

At this point one may wonder why strictly scientific arguments were not allowed to stand by themselves. Unfortunately, scientific results were sometimes more confusing than enlightening. In 1851,

[74] Ibid., p. 136.
[75] Ibid., p. 163.
[76] Ibid., p. 161.
[77] Ibid., p. 138.
[78] "Ueber die Gesetze der organischen Bildung," op. cit., (ftn. 73) pp. 294-326; see p. 325.
[79] Ibid., p. 326.

Johannes Müller had observed snails developing within echinoderms. This alleged discovery threw him and many others into utter confusion.[80] Snails and echinoderms belong not only to different species, but to different phyla! If such a metamorphosis were possible, then anything was possible!

Among the effects of the puzzling phenomena of *Generationswechsel* and parasitism seems to have been a predilection to view descent as a series of mutations rather than continuous changes. Such a predilection was noticeable in Baumgärtner, we shall find it again in Büchner, and it looms large in the evolutionary thought of the botanist Carl Nägeli (1817-91), which we analyze on the basis of his paper on " Individuality in Nature " presented in 1856.[81]

The paper starts out with a reference to the dispute between materialists and spiritualists. It soon becomes clear that Nägeli, though a biological mechanist, occupies, philosophically, an idealistic position. Scientific empiricism, he insists, moves within the narrow bounderies marked by imperfections of our sensory perceptions, by the eternity of space and time, and by their infinite divisibility. It cannot transcend the threshold to the world of the spirit.[82] This agnostic criticism is balanced by an idealistic evaluation of progress. As he himself says, the main part of the paper attempts to show that "material nature, through individual formations which come into being and pass away, rises to the highest point of development in a continual evolutionary urge and necessary gradation."[83] All this is similar to the tendency of mankind towards perfection through a change of individuals. The organic species, as he writes at the end of the paper, "itself is an individual which develops through continuous change, which finds a limitation through this change, and in this limitation creates new species." Material forms perish, the idea remains, ". . . for the material phenomena are but the transitory points, empty of content, of a movement which ceaselessly strives towards a better goal."[84]

---

[80] Emil du Bois-Reymond, " Gedächtnisrede auf Johannes Müller," *Reden.* 2 vols., Leipzig: Veit, 1886-87; vol. 2, p. 262 ff.

[81] Carl Nägeli, " Die Individualität in der Natur mit vorzüglicher Berücksichtigung des Pflanzenreiches," *Monatsschrift des wissenschaftlichen Vereins in Zürich*, 1856, *1*: 171-212. I was led to this publication by Uhlmann, op. cit. (ftn. 6). The paper is a semi-popular presentation and it is possible that other works by Nägeli prior to 1859 deal with transformism from a different point of view.

[82] Nägeli, op. cit., (ftn. 81) p. 176.
[83] Ibid., p. 171.   [84] Ibid., pp. 211 and 212.

Within this idealistic framework, Nägeli discusses the reasons for giving preference to the hypothesis of transmutation of species over that of new creations. Even in the short interval of historical time, hereditary tendencies combined with external influences lead to the formation of new races. The parents bequeath to the offspring a tendency to resemblance, as well as divergence. The artificial cultivation of plants illustrates how the tendency to change in a certain direction can be utilized. If such a tendency went on and on over long periods of time, it would lead to new species. If in reality species appear constant, this does not mean that they do not undergo internal change. Their chemical condition is modified up to the point where it leads to a sudden change in the offspring, as illustrated by the alternation of generations.[85] Thus Nägeli believes in a saltatory evolutionary change rather than a gradual one, although he admits the existence of both.

Nägeli does not ascribe apodictic certainty to his views. "The way in which they [i. e. species] originated, remains a matter of surmise."[86] Nevertheless there are reasons for assuming descent. Since the history of the earth is the result of "natural laws acting by necessity," we like to assume the same for the changes of organic nature. "But it is more natural that plant originated from plant and animal from animal than that they were created anew from inorganic substances." Nägeli adds an argument that sounds very modern:

> To be sure it appears as if thereby the difficulty is only moved back, since organic life must have begun at some time. However, the difficulty is also diminished; for it is more probable that out of inorganic compounds organic substances originated, that from the latter the simplest unicellular plants became organized, and that out of these the higher plants developed in a gradual sequence.[87]

It is surprising to find that in spite of this argument Nägeli still believed in the actual spontaneous generation of certain fungi.[88]

Nägeli does not doubt that evolution proceeds from the lower to the higher, his illustration being the old one of the ladder with

---

[85] Ibid., p. 209, footnote.
[86] Ibid., p. 203.
[87] Ibid., p. 206.
[88] Ibid., p. 207. However it is in line with this belief that Nägeli still clings to the formation of cells from amorphous matter, ibid., p. 188, footnote 2.

higher rungs above those below. The law of recapitulation is cited in support of this belief.[89] But when we consider that the idea of progress was an integral part of the evolutionary thought of the time, we may also agree that the assumption of a transformation from "lower to higher" was not merely thoughtless anthropomorphism, which took it for granted that man stands at the top of creation. Rather it appears as a projection into the realm of nature of the general idea of progressive perfection.

In view of the place that the notion of "progress" occupied in the evolutionary thought of men like Cotta, Baumgärtner, Unger, and Nägeli, one may well inquire more closely into the attitude of such "progressive" biologists as Vogt and Virchow towards transformism.

This question leads to the German translation of Chambers' *Vestiges of the Natural History of Creation*. The English original when published anonymously in 1844 had created a scandal. The author had affirmed his belief in the spontaneous generation of organisms and the transmutation of species. This belief was part of his radical deistic position. God had created nature and her laws, which then did the rest without his active intervention. The German translation of the book of 1851 became well known. Baumgärtner and Schaaffhausen both cite it.[90] However, we are here concerned less with Chambers' work than with the translator, who was none other than Carl Vogt, "monkey Vogt," as he was to be popularly known later when he had started preaching the gospel of Darwinism. But in 1851 Vogt was as yet a confirmed opponent of descent. Why then did he translate the work and what did he have to say about it?

Although Vogt had studied medicine, he turned to zoology and for some time was a co-worker of Agassiz. In 1847, he was appointed professor in Giessen, and at the beginning of the next year the translation was finished and printing had begun. Then Vogt, a radical republican, became involved in the revolution, lost his position in Giessen, and the preface of his translation is dated from Bern, October 1849. Vogt, as we have already seen, was one of the most militant free-thinkers of the time. His "Physiological Letters," published in 1845-1847, was one of the starting points of German philosophical materialism.

[89] Ibid., p. 207.
[90] Cotta, as we have seen above, p. 334, depended on the English original.

But why did Vogt translate Chambers' work at a time when he was still opposed to descent? In his preface, written after the unsuccessful revolution of 1848, he makes the following remark:

> To the Constitutional Party of Germany, whose influence should soon be limited to the innocent reading of innocent books, I recommend this book in all good will. It will find here a constitutional Englishman who constructed a constitutional God, who at first made laws as an autocrat but then of His own accord gave up autocracy and, without direct influence on the governed, allowed law to rule in His stead. A splendid example for princes! [91]

Vogt, an extremely sarcastic person, must have written this with his tongue in his cheek. It can hardly have been the whole reason for undertaking the translation. However that may be, he takes up the same thought in a footnote stating his objection to Chambers.

> The dispute between the theory of succession which presumes a gradual transformation of creatures, as does our author, and the theory of revolution which continually allows new fauna to appear on the earth, is as old as the study of petrifactions itself; and from the theoretical point of view, no solution is possible. It is a question of judging the most detailed facts and particulars of the delimitation of the variations which a species may undergo, and the decision will only be known when we have shown to what extent the specific characteristics of every shellfish can be modified.

Thus far he is the objective scientist who postpones the decision until every detail is known, which means a postponement *ad Kalendas Graecas*. But then he continues in a different vein:

> In a more general discussion of the question from the theoretical point of view we must, however, take into consideration that the assumption of successive, different creations does not, as our author believes, in any way imply the concept of a Creator and opposes that of a natural law. Similarly, on the other hand, the assumption of a gradual modification does not necessarily postulate the assumption of a natural law without creative intervention. We believe also that no species passed from one formation into

---

[91] *Natürliche Geschichte der Schöpfung des Weltalls, der Erde und der auf ihr befindlichen Organismen, begründet auf die durch die Wissenschaft errungenen Thatsachen.* Aus dem Englischen nach der sechsten Auflage von Carl Vogt. Braunschweig: Vieweg, 1851.

another, but that with every geological revolution there was a complete destruction of organisms and a renewal of them; yet we do not on this account assume a Creator, either at the beginning or in the course of earth history, and find that a self-conscious Being existing apart from the world and creating the world is just as ridiculous if it changes the world and its organisms twenty-five times or still oftener till it hits on the right arrangement, as when, after the creating the world and making the natural laws, it pensions itself off and retires, as our author would have it do.[92]

In other words, Vogt objects to reference to any divine agency, conceived theistically or deistically. But he does not tell us how he envisaged the origin of species, though it would be intriguing to know, in view of the fact that he is highly sceptical of spontaneous generation.[93] In most of the footnotes he limits himself to factual corrections—which Chambers' work certainly needed. But one remark at least we cite because it confirms what we said at the beginning about the contemporary attitude to evolutionary ideas: " The opinion of the author [i. e. Chambers] resulting from Naturphilosophie, that the species in the course of the geological epochs gradually transformed into others. . . ."[94] Whether Chambers received his evolutionary bias directly or indirectly from German Naturphilosophie need not concern us here. But it is pertinent to state that to a German around 1850, transformation was immediately related to Naturphilosophie. It does not sound convincing that Vogt, who improved upon Cabanis by likening the production of thought by the brain to the secretion of urine by the kidney,[95] rejected descent because he recoiled from the idea of descent of man from the orangutan, as Schaaffhausen claimed.

It seems more likely to assume that Vogt had an ambivalent attitude towards transmutation. He was in sympathy with it and yet rejected it, because acceptance seemed to conflict with his ideal of scientific method. The following remark in a work that appeared in 1859 seems to bear this out.

> Stripped of the comical trimmings which the natural philosophers and their conscious and unconscious hangers on managed to give this theory, it still retains something which seems to us

---

[92] Ibid., p. 124, footnote.
[93] Ibid., p. 132 ff. and Vogt's comments.
[94] Ibid., p. 98, footnote.
[95] Temkin, " Materialism," (ftn. 7) p. 324.

to be of great importance. As far as that is possible in the limited scope, namely, it brings the history of earlier creations into harmony with the prevailing general, physical, and organic laws and banishes completely the fiat of the rational personality, a creator, assumed by many other scientists. This theory of the transition of one species into another and of the gradual evolution of the organic form-type under the influence of external agents, would undoubtedly offer much greater inner credibility, if the hitherto known facts did not thwart and hinder. To these, however, we must defer until perhaps the faulty observations contained in them are discovered, and thereby the fact itself is rectified.[96]

To put it pointedly, Vogt regrets that he cannot accept the theory of descent. A scientist has to bow to facts, but he is allowed to hope that some day the facts will be corrected.

I do not know whether in the preceding years, too, Vogt had secretly hoped for a removal of the obstacles to the theory of descent. But it seems very probable that Rudolf Virchow had had this hope from at least 1856. At that time, as we have seen, Virchow emphasized the necessity for assuming spontaneous generation, and left open the possibility of a series of such events to explain the formation of progressively higher species. He did not declare himself a transformist, but he did not hide his sympathy either:

> As long as it is still not possible in the sense of the school of natural philosophers to show a continuous development of plants and animals from the simplest form to the most highly developed organisms in such a way that genus is transformed into genus. . . .[97]

As Ackerknecht has pointed out, Virchow was not an anti-Darwinist, although he warned against uncritical acceptance and generalization of Darwinistic ideas.[98] In an essay " On the Mechan-

---

[96] Karl Vogt, *Altes und Neues aus Thier- und Menschenleben.* 2 vols. Frankfurt a. M.: Literarische Anstalt, 1859; see vol. 1, p. 351.

[97] Virchow, op. cit., (ftn. 24) p. 24: " So lange es noch nicht möglich ist, im Sinne der naturphilosophischen Schule eine fortschreitende Entwickelung der Pflanzen und Thiere von der einfachsten Form zu den höchst entwickelten Organisationen in der Art zu zeigen, dass Gattung sich in Gattung umbildet; so lange man, wie Czolbe selbst anführt, mit Linné schliessen muss, dass alle Gattungen schon von (ihrem) Anfang an als solche existirten: so würde es nothwendig sein, um den Gedanken einer Epigenese der Gattungen, einer 'Schöpfung' zurückzuweisen, dass man die Ewigkeit aller organischen Gattungen aufstellte."

[98] Erwin H. Ackerknecht, op. cit., (ftn. 72) p. 199 ff.

istic Concept of Life" that appeared in 1862, he claimed to have insisted on transmutation as a necessary link in a mechanistic view of life, even before the appearance of Darwin's book. This essay purported to be the speech he had made at the 1858 meeting of German Scientists and Physicians.[99] In the published essay we read:

> But life must have had a beginning, for geology leads us into epochs of the earth's development in which no life was possible, where no trace, no remnant, of life is extant. But if there was a beginning to life, then it must also be possible for science to fathom the conditions for this beginning. As yet this is an unsolved problem. Indeed, our experiences do not even give us the right to consider the invariability of species which at the moment seems to be so assured, as a fixed rule for all time. For geology teaches us to recognize a certain series of steps by which the species succeeded one another, higher ones succeeding lower ones, and however much the experience of our own time conflicts with it, I am bound nevertheless to confess that rather it seems to me to be a necessity for science to return to the possibility of transition from species to species. Only then does the mechanical theory of life attain real security in this respect.[100]

Virchow apparently felt that his readers might suspect him of having written under the influence of the *Origin* which had appeared in the meantime. Thus he added the following footnote:

> Charles Darwin's book (*On the origin of species by means of natural selection*, London, 1859) which became famous so quickly, had not yet appeared when the foregoing was written.[101]

Since Virchow emphasized that the speech was not read from a manuscript, it is quite likely that no manuscript existed in 1858. However, there appeared a short official report which included the following passage:

> For it is already very difficult to decide whether there are rigid differences between plant and animal cells. For the sequence of organic development from the simpler to the higher forms, as taught by Naturphilosophie, no proofs have as yet been found,

---

[99] Rudolf Virchow "Ueber die mechanische Auffassung des Lebens. Nach einem frei gehaltenen Vortrag aus der dritten allgemeinen Sitzung der 34. Versammlung deutscher Naturforscher und Aerzte. (Carlsruhe, am 22. September 1858)," in *Vier Reden über Leben und Kranksein*. Berlin: Reimer, 1862, pp. 1-33. Note that the speech was not read from a manuscript!
[100] Ibid., p. 31.   [101] Ibid., p. 31, footnote.

and one is inclined to constancy of form. There seems to be a limited, preestablished route by which the principal characteristics of organisms are reproduced by inheritance and remain constant. Always the same form arises from earlier form without particular development. However ingratiatingly the indicated sequence of development may present itself to us, however desirable its proof may be to the individual investigator, the natural scientist who sacrifices himself and his pet inclinations to calm knowledge is obliged to designate this view as a fantasy. We are bound to assume the immutability of genera and species as a necessary consequence of organic laws. This compulsion is valid, however, only for the present, for the state of our knowledge today, and permits us to hope that more information will be found later.[102]

The two versions differ not only in letter, which is to be expected, but also in emphasis. According to the official report, transmutation is an unproved idea of Naturphilosophie; the true scientist views it with scepticism; he must admit the immutability of genera and species at the present time, though he reserves judgment as to what future research may reveal.[103]

These notions agree reasonably well with the article of 1856, and they also agree with a remark Virchow made in 1877 when, in the face of the great success of Darwin, he wrote in a reminiscent vein:

For those of us who were still acquainted with the old Nature-philosophy, it was certainly somewhat surprising to see how the genius of a single man restored to its rightful place, after its long and, alas, not entirely unjustified banishment, an idea already

---

[102] The report is reprinted in Karl Sudhoff, *Rudolf Virchow und die Deutschen Naturforscherversammlungen*. Leipzig: Akademische Verlagsgesellschaft, 1922, pp. 5-7; see p. 6 f. The report continues: "Der Redner rechnete es sich als ein gewisses Verdienst an, dieses ihm widerstrebende Gesetz auch in Krankheiten nachgewiesen zu haben, indem er zeigte, der Körper tue nichts, wozu ihn nicht seine Bildung im voraus berechtigt; hieraus folgt, dass man die generatio aequivoca ablehnen muss. Der Naturforscher vermag auf die drängenden Fragen keine bestimmte Antwort zu erteilen."

Comparison with the printed version of 1862 indicates that Virchow thought above all of present conditions. In the version of 1862, he goes so far as to postulate the origin of life at some time in the geological past, allowing transmutation to take over from that event. The official excerpt is not clear as to Virchow's attitude to spontaneous generation in the past.

[103] In view of our uncertainty regarding the authorship of the official report and its denial of spontaneous generation, I hesitate to take it as manifest contradiction of the version of 1862. But it seems fair to say that the report throws doubt on Virchow's unequivocal support of transformism at the meeting.

given official status as an *a priori* necessity by the Nature-philosophers, not only reactivating it but making it the basis of a general conception of the history of the organic world.[104]

It thus seems likely that both Vogt and Virchow, before the appearance of the *Origin*, realized that a theory of transmutation would fit better than any other into a mechanistic biology. Both, however, believed that the true scientist must sacrifice " pet inclinations " to facts. It would, moreover, seem that, after 1859, Virchow persuaded himself that he had insisted on this theory with greater emphasis than may actually have been the case.[105]

Vogt and Virchow were professional scientists of repute and, therefore, supposed to put facts before theory whenever the two disagreed. Louis Büchner, on the other hand, was hardly more than a scientifically trained popularizer of materialistic metaphysics, and in his *Kraft und Stoff* of 1855 we find indeed an outright proclamation of the belief in descent as an almost scientific fact. To Büchner it appears " scarcely understandable how some scientists can object to the assumption of a law of gradual transmutation —for no other reason than that as a rule, under our present-day conditions, such a separation of the individual species of animals is observed, that like parents always create only like offspring." [106]

Admitting that recent research left little room for spontaneous generation, Büchner nevertheless deemed it " not unlikely that it is still possible for the smallest and most imperfect organisms." [107] Johannes Müller's alleged discovery of snails developing in echinoderms is proclaimed as incontrovertible proof " that conditions must once have been possible where such a process could take place in higher animals or in which a monkey, nay any other animal gave birth to a human being! " [108] We have no right to draw inferences from the short period of our experience and from prevailing conditions to the endlessly long prehistoric times and to those states of the earth " in which nature undoubtedly was younger and stronger and more powerful in producing organic forms." [109] There is a law

[104] Rudolf Virchow, " Standpoints in scientific medicine [1877]," translated by L. J. Rather, *Bull. Hist. Med.*, 1956, *30*: 537-543; see p. 540.

[105] In the following paragraph (ibid., p. 540), Virchow says that in the Karlsruhe lecture of 1858, he expressed " the variability of species as a necessary presupposition of the mechanical theory of life . . . in the most clear-cut manner."

[106] Louis Büchner, op. cit., (ftn. 15) p. 82.
[107] Ibid., p. 73.   [108] Ibid., p. 87.   [109] Ibid., p. 82 f.

of transmutation according to which the transmutation was not, as the old school of natural philosophy demanded, a very gradual one, but rather saltatory and vested in the embryonic development. Thus he summarizes:

> From the most improbable beginning, the simplest element of organic form produced by a union of inorganic materials in involuntary generation, from the sorriest plant or animal cell, it was possible, with the help of extraordinary natural forces and endless periods of time, for that whole, rich world to develop which, infinitely varied in its ramifications, we find surrounding us.[110]

III

The foregoing survey lacks completeness as to authors as well as works covered. However, none of the great biologists claimed as adherents of the idea of descent has been completely disregarded. It seems, therefore, permissible to attempt some general statements.

Our initial question related to the role of the idea of transmutation in German biology during a decade of rising materialism. The idea did not agree with the biblical account, but was otherwise philosophically neutral. If we use the terms "idealist" and "materialist" in the broad sense of the preceding pages, we find more numerous support among the idealists:[111] Schleiden, Baumgärtner, Unger, Schaaffhausen, and Nägeli. Among the materialists, only Büchner supported it outright. Cotta favored it without committing himself. Vogt, though sympathetic, opposed it, and however much it may have appealed to Virchow, his open acceptance is unproven.

This result serves first of all as a renewed reminder of the importance of distinguishing between the idea of transmutation and Darwinism as a particular mode of it. It is not our task to inquire into the reasons for the great success of Darwinism. But we shall try to find reasons for the weakness of the idea of transmutation before 1859. Whether the idea was intrinsically weak, i. e., whether it lacked logical compulsion and factual evidence, shall not occupy us here, since Professor Lovejoy has examined this question in a much broader framework. Rather, we shall see whether the argu-

---

[110] Ibid., p. 87 f.

[111] Rádl, op. cit. (ftn. 6), has indicated that evolutionary ideas were favored by a number of men who subsequently to 1859 became anti-Darwinists. In this connection the article by S. J. Holmes, "K. E. von Baer's perplexities over evolution," *Isis*, 1947, *37*: 7-14 is instructive.

ments offered were weak *qua* arguments and lacking in persuasive power.

Such weakness is demonstrated first of all in the failure of winning a large number of followers. I mention this obvious phenomenon only in order to emphasize my doubt that suppression by the prevailing scientific opinion is largely to blame. In the case of Vogt and Virchow this could scarcely be the case; their aggressive temperament would have induced them to fight for the principle of transmutation had they felt sufficiently sure of it. Conversely, their reserve towards a theory which had their sympathy indicates that, rightly or wrongly, German biologists before 1859 considered the evidence against transmutation to be too strong.

The weakness is further documented by the multiplicity of arguments and explanations among those upholding transformism. Remarkably enough, none of the authors, except Cotta, was satisfied with accepting the views of any contemporary or previous champion, and of Cotta it would not be strictly correct to speak in terms of acceptance. However much they were influenced by Lamarck, Naturphilosophie, the *Vestiges of Creation*, or any other of Darwin's forerunners, such influence was not considered decisive.[112] Instead, most of them elaborated explanations of their own, or, like Unger and Schaaffhausen, avoided entering into details. In particular, Cotta, Schleiden, Unger, and Schaaffhausen seem to have thought of gradual changes, whereas Baumgärtner, Nägeli, and Büchner thought rather of sudden mutations.

Moreover, of the explanations adduced, some were ingenious, some were vague, and others (like Baumgärtner's and Büchner's) were fantastic. None of them, however, carried Darwin's great gift to the physiologist, viz., freedom from the embarrassment of teleology. In view of the fact that some of the explanations offered, like those of Schleiden and Baumgärtner, might also be called "mechanistic," this seems significant. It suggests that a theory of descent that failed to explain the mutual adaptation between body structure and environment lacked a good deal of the appeal which Darwinism exerted for physiologists like Helmholtz and du Bois-Reymond, to whom the question of the origin of species was of secondary importance.[113]

---

[112] The only exception is Cotta who, however, went to the other extreme of not adopting openly the views which he reported.
[113] Helmholtz, in a paper of 1853 (*Popular Lectures on Scientific Subjects*, translated by E. Atkinson, New York: 1873, p. 48) mentions the view "that

There was agreement that supernatural influences had to be excluded. There was, moreover, a more or less strongly pronounced desire to exclude, or reduce to a minimum, the frequency of spontaneous generations. Nägeli stated the case most distinctly—but vitiated it by admitting spontaneous generation even in his own day. Unger was most consistent, but then carried over old concepts of *Urpflanze* and idealistic morphology. Most difficult is the evaluation of the relative weight of scientific and speculative arguments. With Baumgärtner and Büchner the impression prevails that they believed in descent and thus found arguments for it, while Schaaffhausen went perhaps furthest in postulating transmutation because the case for the constancy of species seemed very doubtful to him. Taking our authors in the aggregate, all we can say is that nowhere do we find a theory of descent defended in isolation from a *Weltanschauung*, even if exclusion of supernatural creative acts is classed as a premise of scientific reasoning.

On the other hand, it is manifest that the idea of descent was not shrouded in complete silence. Even if weak, it was alive and found outspoken supporters. Thus one cannot help wondering why it did not exert a larger appeal in a time and region which was so favorably inclined to historical thinking. The appeal of an idea does not, after all, necessarily rest on its logical strength and supporting facts. *Entwickelungsgeschichte* was fashionable among German historians and biologists. Why was there no stronger temptation to applying it to species, especially since the concept of progress spoke in favor of transformism? It cannot even be said that another idea of great psychological force, such as biblical orthodoxy, opposed it. Remarkably enough, the Mosaic account of creation plays a negligable role in the argumentation as compared with the constancy of species. The latter of course was the main argument opposed to transmutation. But, whatever its merits, its psychological power is open to doubt.

I think that a psychological explanation has to take account of the fact that, far from being forgotten, the idea of transmutation was

during the geological periods that have passed over the earth, one species has been developed from another, so that, for example, the breast-fin of the fish has gradually changed into an arm or a wing." Without committing himself he adds that probably the majority of observers do not incline to this view. For du Bois-Reymond see John Theodore Merz, *A History of European Thought in the Nineteenth Century*. Vol. II, 3rd ed. Edinburgh-London: Blackwood, 1928, p. 435.

only too well remembered from the days of Naturphilosophie and speculative science.[114] To the post-romantic generation of German biologists, transformism lacked the appeal of newness. This generation prided itself on its rigorously scientific attitude which implied resistance to mere speculation. A comparison with the parallel fate of the idea of a microorganismic etiology of infectious diseases may help our understanding of the situation. Both ideas, that of transmutation and that of the bacterial cause of disease, were widely accepted after 1859: the former suddenly and dramatically with the appearance of Darwin's book, the latter more slowly in the wake of the work of Pasteur. Both ideas were not only well known before 1859, but suffered the fate of being too well known.[115] Nothing essentially new had been added to them, and they might profitably be disregarded by progressive scientists, intent upon the discovery of concrete facts. Whether in either case supporting facts were so conspicuously lacking as to justify that attitude is a different question which cannot be answered within the narrow limits of the period under discussion.

---

[114] J. H. F. Kohlbrugge, "War Darwin ein originelles Genie?" op. cit. (ftn. 6), has emphasized the widespread interest in theories of transmutation before Darwin and has pointed out several of the factors that account for it. His arguments are, however, influenced by the desire to belittle Darwin's claims, which gives them a one-sided tendency.

[115] As Henry E. Sigerist, *The Great Doctors* (Translated by Eden and Cedar Paul, New York: Norton, 1933), p. 352, pointed out, Henle's *Pathologische Untersuchungen* of 1840, in which he cogently argued for the microorganismic etiology of infectious diseases, found little recognition because "people were willing to believe the evidence of their own eyes. But they were weary of speculation." Erwin H. Ackerknecht, " Anticontagionism between 1821 and 1867," *Bull. Hist. Med.*, 1938, 22: 562-593, with reference to Henle, says "that what to us appears a vanguard action, impressed Henle's contemporaries rather as a rearguard action. . . ." Yet even before Pasteur, the arguments for the *contagium animatum* were supported not only by speculation but by observable facts as well!

THIRTEEN

THE ARGUMENT FOR ORGANIC EVOLUTION BEFORE THE ORIGIN OF SPECIES, 1830–1858 *

ARTHUR O. LOVEJOY

In this year of the centenary of the *Origin of Species* it is worth while to raise two questions which have received less consideration than they merit. At what date can the evidence in favor of the theory of organic evolution—as distinct from the hypothesis of natural selection—be said to have been fairly complete: in other words, how early were the facts and principles from which the truth of that theory is now ordinarily inferred sufficiently known, to all competent men of science, to require the inference, even though it was not, in fact, generally made? And by what British writer was a logically cogent argument for the theory first brought together and put before the public? The interest attaching to these questions is more than merely historical. The answer to them will afford a sort of object-lesson in the logic of scientific reasoning. Here is a doctrine now accepted by all naturalists: at what point, in the century-long accumulation, through half a dozen separate sciences, of the evidences inclining to that doctrine, ought we to say that the balance of logical probability turned decisively in its favor. The inquiry will also be found, I think, to throw a somewhat instructive light upon the psychology of belief, and to show how far, even in the minds of acute and professedly unprejudiced men of science, the emotion of conviction may lag behind the presentation of proof.

* Part I is reprinted with revisions from *The Popular Science Monthly*, 1909; Part II has been added; and Part III is reprinted with extensive revisions and alterations.

By this time, no doubt, every schoolboy knows that Darwin did not invent the theory of evolution, and the general public is more or less acquainted with the names and works of at least some of the earlier protagonists of the doctrine: of the elder Darwin, namely, of Lamarck, of Geoffroy St. Hilaire, and of Herbert Spencer. It is less commonly remembered, but perhaps not universally forgotten, that among English-speaking naturalists the theory was a commonplace topic of discussion for two or three decades before 1859, and especially after the publication and immense circulation of the successive editions of Robert Chambers's *Vestiges*, of which the first appeared in 1844. Geological text-books of the period referred to the "theory of transmutation of species" as a matter of course, though usually only to reject it as an exploded hypothesis. Thus the *Elements of Geology* of Alonzo Gray and C. B. Adams, 1852, enumerates three theories which have been advanced respecting the origin of animal species: (1) Successive special creations (2) "transmutation, which supposes that beings of the most simple organization having somehow come into existence, the more complex and the higher orders of animals have originated in them by a gradual increase in the complexity of their structures," [1] (3) *generatio æquivoca* of individuals and species. The first is adopted, but the second is discussed at greatest length; on it the authors remark that "those who have adopted the theory of transmutation have generally detached it from Lamarck's theory of appetency, and not attempted to explain *how* the process of transmutation goes on." The argument for evolution is similarly discussed and "refuted" in *Geological Science*, a popular text-book by D. T. Ansted, F. R. S., 1854. To this refutation, indeed, the greatest of English geologists had devoted four chapters of his *Principles of Geology* [2] before 1835.

But though such facts as these are, as I have said, now fairly familiar, the notion still widely prevails, even among biologists, that no serious *proof* of evolution either existed or had been published before the appearance of the *Origin of Species*—or at all events, before the late 1850's.[3] Professor Joseph LeConte, indeed,

---

[1] I quote from the reprint of 1854, p. 87.
[2] The writer's copy of Lyell's *Principles* is the first American from the fifth London edition, 1837.
[3] This opinion has, for example, been expressed by Poulton in his *Charles Darwin and the Theory of Natural Selection*, 1902. "The paramount importance of Darwin's contributions to the evidences of organic evolution are [*sic*]

in his *Evolution and Its Relation to Religious Thought*,[4] made it a reproach against both Lamarck and Chambers that they had unscientifically embraced the hypothesis before the evidence for it was ripe; and he considered it fortunate for science that their notions died still-born, under the weight of the great authority of Cuvier and Agassiz. "I know," wrote LeConte, "that many think with Haeckel that biology was kept back half a century by the baleful influence of Agassiz and Cuvier; but I can not think so. The hypothesis was contrary to the facts of science, *as then known and understood*. It was conceived in the spirit of baseless speculation, rather than of cautious induction; of skillful elaboration, rather than of earnest truth-seeking. Its general acceptance would have debauched the true spirit of science.... The ground must first be cleared... and an insuperable obstacle to hearty rational acceptance must first be removed, and an inductive basis laid." This last, LeConte goes on to argue, was largely the work of Agassiz, opponent of evolutionism though he was. Now, it is, of course, undeniable that the premature adoption of a hypothesis is a sin against the scientific spirit, and that the chance acceptance by some enthusiast of a truth in which, at the time, he has no serious reason for believing, by no means entitles him to any place of honor in the history of science. But what constitutes prematurity in this particular matter? And *was* the evolutionist hypothesis "contrary to the fact of science, as known and understood" at any time after 1840?

The prevalent belief that it was is chiefly due to two things. The first is the fact that before 1859 few English naturalists of high standing accepted, and almost none publicly avowed, the theory of descent; whereas, after the publication of the *Origin*, such notable names as Huxley, Lyell, Hooker, and Asa Gray were speedily numbered among the disciples of the doctrine, and in the ensuing five years it was well on its way towards its eventual complete triumph. The other source of the supposition that Darwin presented the first adequate grounds for believing in evolution is the

often forgotten in the brilliant theory which he believed to supply the motive cause of descent with modification. Organic evolution had been held to be true by certain thinkers during many centuries; but not only were its adherents entirely without a sufficient motive cause, but their evidences of the process itself were erroneous or extremely scanty. It was Darwin who first brought together a great body of scientific evidence which placed the process of evolution beyond dispute, whatever the causes of evolution may have been" (p. 100).

[4] Second edition, 1905, pp. 33-35.

express testimony of Huxley, whose paper on the reception of the *Origin of Species*[5] has come to be the principal source of information on its subject. In that article, and in several letters and other writings, Huxley takes credit to himself for his rejection of the transformation-theory until he became acquainted with Darwin's work; and he never expressed any sentiment far short of contempt for Chambers's *Vestiges*. He wrote in 1887:

> I must have read the "Vestiges" . . . before 1846; but if I did, the book made very little impression on me, and I was not brought into serious contact with the "species" question until after 1850. . . . It seemed to me then, as now, that "creation," in the ordinary sense of the word, is perfectly conceivable. . . . I had not then, and have not now, the smallest *a priori* objection to raise to the account of the creation of plants and animals given in "Paradise Lost." . . . Far be it from me to say that it is untrue because it is impossible. I confine myself to what must be regarded as a modest and reasonable request—for some particle of evidence that the existing species of animals and plants did originate in that way, as a condition of my belief in a statement which appears to me highly improbable. And, by way of being perfectly fair, I had exactly the same answer to give to the evolutionists of 1851-58. . . . The only person known to me whose knowledge and capacity compelled respect, and who was, at the same time, a thoroughgoing evolutionist, was Mr. Herbert Spencer. . . . But even my friend's rare dialectic skill and copiousness of apt illustration could not drive me from my agnostic position. I took my stand upon two grounds: Firstly, that at the time the evidence in favor of transmutation was wholly insufficient; secondly, that no suggestion respecting the causes of the transmutation assumed, which had been made, was in any way adequate to explain the phenomena. Looking back at the state of knowledge at the time, I really do not see that any other conclusion was justifiable. . . . As for the "Vestiges," I confess the book simply irritated me by the prodigious ignorance and thoroughly unscientific habit of mind manifested by the writer. If it had any influence at all, it set me against evolution. . . . Thus, looking back into the past, it seems to me that my own position of critical expectancy was just and reasonable. . . . So I took refuge in that *tätige Skepsis* which Goethe has so well defined; and, reversing the apostolic precept to be all things to all men, I usually defended the tenability of received

---

[5] Published as Ch. xiv. of *The Life and Letters of Charles Darwin*, Vol. i.

doctrines, when I had to do with the transmutationists; and stood up for the possibility of transmutation, among the orthodox.

In this matter Huxley is assuredly a witness whose testimony should not lightly be set aside; for to his attainments as a naturalist he ordinarily joined logical acumen and fairness of mind. Yet I think it possible to show that the passage just quoted gives a thoroughly misleading view of the logical status of the argument for evolution, as it existed in the light of the science of the period; that the attitude which Huxley assumed from 1850 to 1858 was contrary to all sound ideas of scientific method; and that he does the reputations of both Spencer and Chambers serious injustice. I shall attempt to establish these conclusions mainly by showing that the arguments and facts chiefly relied upon by Huxley himself and other early champions of transformism were entirely familiar and well authenticated from fifteen to twenty years before 1859, and had virtually all been clearly noted and pertinently used in the published evolutionary reasonings of Chambers or of Spencer. The truth is—as the evidence to be adduced will make clear—that Huxley's strongly emotional and highly pugnacious nature was held back in his earlier years by certain wholly non-logical influences from accepting a hypothesis for which the evidence was practically as cogent for over a decade before he accepted it as it was at the time of his conversion. These influences did not in Huxley's case, as they did in so many others, proceed from religious tradition or temperamental conservatism. But Huxley had unquestionably been strongly repelled by the *Vestiges*. The book was written in a somewhat exuberant and rhetorical style; with all its religious heterodoxy, it was characterized by a certain pious and edifying tone, and was given to abrupt transitions from scientific reasoning to religious sentiment; it contained a number of errors in matters of biological and geological detail; and its author apparently believed in the possibility of spontaneous generation on some rather absurd experimental evidence. All these things were offensive to the professional standards of an enthusiastic young naturalist, scrupulous about the rigor of the game, intolerant of vagueness and of any mixture of the speculative imagination with scientific inquiry, a little the victim, perhaps, of the current scientific cant about "Baconian induction."

Full advantage, moreover, had been taken, by the eminent scientists who were also champions of religious orthodoxy, of the faults of Chambers's book; they contrived very successfully to put about the impression that to be a "Vestigiarian" was to be "unscientific" and sentimental and absurd. These were three qualities which Huxley would have been intensely averse to being charged with. Finally, he seems to have been exasperated most of all by a single loose piece of phraseology that now and then recurs in the *Vestiges*. Chambers, namely, was prone to speak of "laws" as if they were causes and, more particularly, as if they were secondary causes to which the "Divine Will" delegated its agency and control. To Huxley, from the beginning of his career, this hypostatizing fashion of referring to "laws of nature" was a *bête noire*; and in 1887 we still find him pursuing the author of the *Vestiges* with ridicule because of his "pseudo-scientific realism."[6] He, therefore,[7] in 1854, almost outdid the *Edinburgh Review* in the ferocity of his onslaught upon the layman who had ventured to put forward sweeping generalizations upon biological questions while capable of occasional errors upon particular points which were evident to every competent specialist.

[6] "Science and Pseudo-science," 1887. Huxley's criticisms are curiously beside the mark. He argues that, whether you suppose that the Creator operates uniformly but directly "according to such rules as he thinks fit to lay down for himself," or that "he made the cosmical machine and then left it to itself," in either case his "personal responsibility is involved" in every result into which this uniform operation works out. But Chambers, so far from denying this, was especially anxious to insist upon it. What he equally insisted upon, however, was the uniformity of this agency. When he spoke of the Creator as working "through" law, the expression, doubtless, was infelicitous; but his essential idea was plain and unexceptionable, viz., that neither organic nor inorganic phenomena "result from capricious exertions of creative power; but that they have taken place in a definite order, the statement of which order is what men of science term a natural law." These last words are Huxley's own, uttered in 1862, in an address before the Geological Society. It is, he added, logically possible to regard such a law as "simply the statement of the manner in which a supernatural power has thought fit to act"; the main thing is that "the existence of the law and the possibility of its discovery by the human intellect" be recognized. This was exactly the essence of the view for which Chambers was contending. Huxley was so unduly enraged by a bit of looseness of language that he actually overlooked the important idea which that language was manifestly intended to express.

[7] I have not had access to this article, published in the *Medical and Chirurgical Review*; but its character is sufficiently indicated in the correspondence of Huxley and Darwin. The former speaks of it as "the only review I ever have qualms of conscience about, on the ground of needless savagery." Darwin thought it "rather hard on the poor author"; and added a curiously mild intimation of his own belief: "I am perhaps no fair judge; for I am almost as unorthodox about species as the 'Vestiges' itself, though I hope not quite so unphilosophical" (*More Letters of Charles Darwin*, I, p. 75).

Yet the layman was, after all, sound in his main thesis; and, what is far more significant, his thesis was based upon sound and sensible arguments, substantially the same arguments that Huxley was destined before long to use in the same cause, though with greater skill as a debater. It will, I think, appear impossible to acquit the young Huxley of a certain measure of scientific Pharisaism in this episode. He was so shocked by minor breaches of scientific propriety in the *Vestiges*, that he forgot the weightier matters of the law of scientific method. In his irritation at Chambers's incidental slips in zoology, he became blind to the importance and suggestiveness of the general outline of that writer's reasoning. Quite other was Alfred Russel Wallace's reaction upon the little book. As early as 1845 he wrote:

> I have rather a more favorable opinion of the " Vestiges " than you appear to have. I do not consider it a hasty generalization, but rather as an ingenious hypothesis, strongly supported by some striking facts and analogies, but which remains to be proved by more facts and the additional light which more research may throw upon the problem. It furnishes a subject for every observer of nature to attend to; every fact he observes will make either for or against it.

By 1847 Wallace had become thoroughly convinced of the truth of transformism; and from that time forward his mind was occupied with the problem of explaining the cause and *modus operandi* of evolution. At this time, he writes:

> The great problem of the origin of species was already distinctly formulated in my mind. . . . I believed the conception of evolution through natural law, so clearly formulated in the " Vestiges," to be, so far as it went, a true one; and I firmly believed a full and careful study of the facts of nature would ultimately lead to the solution of the mystery.[8]

---

[8] Wallace, *My Life*, I, p. 254. Writing sixty years after, Wallace adds his final judgment of the *Vestiges*, "a book which, in my opinion, has always been undervalued, and which, when it first appeared, was almost as much abused, and for much the same reasons, as was Darwin's *Origin of Species* fifteen years later." Ralph Waldo Emerson (who, it is true, was not a naturalist) was at once converted by the *Vestiges*. In a letter of April 30, 1845, he comments favorably on that work, in spite of the " strictures of the journalists." (*Letters*, 1939, III, p. 283. Cf. also p. 290.) Another philosopher converted to the doctrine of the transformation of species, at least as early as 1854, by a reading of the *Vestiges* was Schopenhauer, as I have elsewhere shown. And a

Wallace thus escaped the fatal error in logical procedure into which Huxley fell. For Huxley, in the passage already cited, gives as one of his two reasons for refusing to accept, even provisionally, the evolutionary hypothesis, the fact that "no adequate suggestion respecting the causes of the transformation assumed" had then been made. But, that no causal explanation of a fact is at hand, is not a good reason for denying the fact, if serious evidence of its reality is presented. Wallace properly discriminated the two issues; becoming first convinced that there was an established balance of scientific probability in favor of the fact, he then set himself upon the quest of a hypothesis that would explain it. He had his reward; a decade later he appeared, with Darwin, as joint author of the doctrine of natural selection.

II

I now proceed to the proof of the contentions of this paper. It is necessary first of all to remind the reader of some distinctive features of the situation in the science of the period 1830-1859. It was a situation essentially different from that in which Lamarck had first carried on the propaganda of transformism. The difference was due chiefly to two changes that had taken place in the intervening decades. First, the science of geology had gone through a brilliant development. The two allied subsidiary sciences of paleontology and stratigraphic geology had been created through the work of Cuvier and William Smith, and by the 1830's presented an impressive body of observational data concerning animal species—most of them extinct—represented by fossil remains found in a series of superposed strata which, it was generally admitted, could not have been formed simultaneously, and must have required long ages for their production. Geologists of the two schools whose violent controversies had filled the early years of the century—the "Neptunists" and the "Vulcanists"—had, Lyell observed in 1830, erred primarily through "undervaluing greatly the quantity of past time"; they had "misinterpreted the signs of a succession of events, so as

naturalist of the highest standing, Richard Owen, in 1844 in a letter addressed to "the author of the 'Vestiges,'" wrote—after pointing out certain errors of Chambers—"upon the whole the zoology and anatomy of the work is correct and near upon the present level."

to conclude that centuries were implied where the characters imported thousands of years, and thousands of years where the language of nature signified millions." [9] When the recognized antiquity of the earth and of the existence of plants and animals upon it had been thus vastly extended (as Buffon had from more limited evidence pointed out almost a century earlier), the (logically) first step towards the theory of organic evolution had been taken. Enough time had now been provided for extremely gradual changes in the characters of successive generations of the descendants of any pair of organisms; and any plausible hypothesis of the natural descent of, e. g., extant species from ancestors from which they differed widely in form or functions, necessarily presupposed such gradualness. This immense lengthening of the time-span of terrestrial history was, of course, no evidence *for* evolutionism; but it removed what would have been an insuperable obstacle to the formulation and acceptance of the theory.

Historically, however, this was not at all the significance which most naturalists of this period saw in the evidence of the great antiquity of the earth that stratigraphic geology and paleontology had presented. What they saw in it was the embarrassing necessity of abandoning the chronology of the Book of Genesis—which had been precisely calculated by Archbishop Ussher in the 1660's, and was printed in the margins of some editions of the King James Version of the Bible. But though what had been regarded as the divinely revealed *date* of the creation could no longer be accepted, the belief in the direct creative action of the Deity in the formation of the earth and of all "living creatures" upon it could not be surrendered. Consequently, even the most orthodox geologists had, by the 1830's, been constrained by the "testimony of the rocks" to adopt a new and strange theory of the *modus operandi* of the Creator. It was no longer possible to believe in a single original creation of all things within six "days"; a number of repeated acts of "special creation" must be assumed, separated from one another by wide intervals of time, and (according to one increasingly prevalent view) confined to the production of organisms. The immediate effect, in short, of the progress of geological knowledge had been to *increase* the resort to supernatural agency in the current accounts of the genesis of organisms. In place of one great, obscure miracle at the origination of the universe, the revised version of the doctrine

[9] *Principles of Geology*, first American edition, 1837, pp. 89, 88.

of creation substituted a long succession of relatively petty and
definite miracles; it assumed, in Chambers' words, "an immediate
exertion of the creative power, at one time to produce zoophytes,
at another time to add a few marine mollusks, another to bring in
one or two crustacea, again to produce crustaceous fishes, again
perfect fishes, and so on to the end." [10] Creationism had been compelled, like the Ptolemaic astronomy before it, to interpolate some
very singular epicycles in its system. And while all these miraculous interpositions were taking place in order to keep the organic
kingdom in a going condition, the Creator was not for a moment
allowed, by most of these geologists, (including, as we shall see,
Lyell and his followers) to interfere in a similar manner in their own
particular province of the inorganic processes. Their attitude was
like that of the French authorities who, a century earlier, suppressed
the "miraculous cures" of the Jansenist abbé at the church of St.
Médard in Paris, and, in a famous lampoon, were represented as
posting the following proclamation on the church doors:

> De par le roi, défense à Dieu
> De faire miracle en ce lieu.[11]

So, in the opinion of most naturalists the only officially licensed
area in which miracles might be performed by the Creator was the
domain of organic phenomena. Here, as a measure of compensation,
the number of miracles scientifically sanctioned had been materially
increased.[12]

The second broad change in geological science since the beginning
of the century was the introduction of a new methodological principle (new, at least, in British geology)—"uniformitarianism." This,
though still resisted by the old guard of believers in "special creations," had been enunciated by Hutton and elaborated and defended
as early as 1830 in the most admired and most authoritative general
geological treatise of the period—Charles Lyell's *Principles of
Geology*, which had reached its fourth edition by 1835. The
author's professional reputation and the consequent extent of his
influence may in part be gathered from the fact that he was presi-

---

[10] Chambers, *Vestiges*, 1844, Ch. XI.

[11] "By order of the king, God is hereby forbidden to perform miracles
in this place."

[12] The reader will find amusing examples of this inconsistency in President
Hitchcock's *The Religion of Geology*, 1852, pp. 164-5, 339-40. Cf. also Gray
and Adams, *Elements of Geology*, 1854, pp. 16 and 89.

dent of the Geological Society, 1835-36, and was knighted in 1848. The distinctive purpose of his great work was to show, as against the "catastrophists," that "all former changes of the organic and inorganic creation are referable to one uninterrupted succession of physical events, governed by the laws of nature now in operation," [13] upon his title-page Lyell challengingly inscribed Hutton's thesis (as summarized by one of his disciples):

> Amid all the revolutions of the globe, the economy of nature has been uniform, and her laws are the only things that have resisted the general movement. The rivers and the rocks, the seas and the continents, have been changed in all their parts; but the laws which direct these changes, and the rules to which they are subject, have remained invariably the same.[14]

It will be noted that Lyell here included "the organic creation" as well as the inorganic in this generalization. Four chapters of the first volume of the *Principles* are devoted to "controverting the assumed discordance of the ancient and existing causes of change" as one of "the theoretical errors which have retarded the progress of geology."

But if later (including existing) species, in the long sequence of geological eras, were not produced discontinuously by *non*-physical causes of change no longer operative, how *were* they produced? The obvious answer might have seemed to be that later species must have been descended from earlier ones, the descendants having been gradually transformed by the cumulative action of natural causes upon successive generations. In short, the theory of organic evolution might, one would suppose, have been at once recognized as an evident corollary of uniformitarianism, for it alone "referred all the changes of the organic creation to one uninterrupted succession of physical events, governed by the laws of nature now in operation"; the "uninterrupted succession of physical events" was simply the generation of offspring together with the action upon them of environmental factors. And though "species" did not appear to change into other "species" before the eyes of naturalists,

---

[13] Op. cit., first American from the fifth London edition, 1837, p. 148.
[14] Cited from Playfair's *Illustrations of the Huttonian Theory*. On the subject of uniformitarianism, and the history of (chiefly British) geology as a whole in the first half of the century, the reader is referred to Professor Charles C. Gillispie's admirably documented and organized and brilliantly written book *Genesis and Geology*, Harvard University Press, 1951.

many existing forms, notably some breeds of dogs, were admittedly descendants of ancestors from which they differed immensely and very diversely, and these differences were heritable—only, in these cases, the opponents of the "theory of transformation" did not call them "species." True, the processes or "causes" by which "ancient" species were transformed into others could not be said, in 1830, to be well understood; but, upon uniformitarian principles, they *must* have been causes of a kind, or kinds, that always have been operative; it merely remained for further scientific inquiry to identify them. I am not suggesting here that Lyell's uniformitarianism alone amounted to a final proof of the theory of evolution; there were, in fact, still some apparent difficulties to be overcome. I am suggesting that, once uniformitarianism was accepted, evolutionism became the most natural and most probable hypothesis concerning the origin of species. If it were not accepted, special-creationism still appeared to be the only possible alternative. And the latter hypothesis was an alternative which could not really be reconciled with the uniformitarian premise that Lyell adopted and so vigorously defended.

Nevertheless, Lyell in 1830 and for three decades thereafter was unable to recognize any such logical relation between uniformitarianism and the theory of organic evolution. The former did not appear to him to lend any support whatever to the latter.[15] Since, however, it was held by a few contemporary naturalists, including so eminent a one as E.-Geoffroy Saint-Hilaire, Lyell felt it incumbent upon him to "explain the data and reasoning by which it may be refuted."[16] He accordingly proceeded, through some fifty pages, to "combat the notion that one species may be gradually converted into another by insensible modifications in the course of ages";[17] and in the final sentence of the first volume he set down, as the

---

[15] I am glad to find myself in agreement with Gillispie (*Genesis and Geology*, p. 239) in rejecting the suggestion of another writer that Lyell in the 1830's was "a reluctant rather than a firm adherent of the doctrine of special and successive creations." But though not reluctant, he was inconsistent in his acceptance of it.

[16] *Principles of Geology*, I, p. 482: "M. Geoffroy Saint-Hilaire has declared his opinion that there has been an uninterrupted succession in the animal kingdom, effected by means of generation, from the earliest ages of the world up to the present day, and that the ancient animals whose remains have been preserved in the strata, however different, may have been the ancestors of those now in being."

[17] Ibid., Preface, p. xii. Lyell is summarizing the principal contentions of his book.

conclusion of the whole matter, "that species have a real existence in nature, and that each was endowed at the time of its creation with the attributes and organization by which it is now distinguished."[18] While this was an entirely clear assertion that every species owed its origination to a separate act of "creation" by the Deity and not to the slow operation of natural causes, and that, once created, a species retained its original distinctive characters without alteration, the sentence was equivocal on one pertinent question. What *was* "the time of their creation?" Were all the species that have ever existed—those now extant as well as those long since eliminated by natural processes—created all at once by divine fiat at the beginning of the world (or at all events, at the beginning of "the organic kingdom"), or were they, as the special-creationists held, "introduced" at many separate moments along the vast stretches of geological time? To put the question concretely—though when so put it appears absurd—were cacti, oaks, and pond-lilies, pterodactyls and hippopotami, robins and earthworms, lions and lambs, elephants and moles, monkeys and men, all coexistent in the early Paleozoic Era, or were they not? In view of the paleontological facts established by 1830, anyone who affirmed that species were "created," and not gradually produced by natural processes, seemed obliged to accept one or the other of these alternative suppositions. But Lyell, in formulating his general thesis, quoted above, about the mode of origination of species, was carefully non-committal on this issue Similarly, he wrote in his Preface: "Whether new species are substituted from time to time for those which die out, is a point on which no decided opinion is offered; the data hitherto obtained being considered insufficient to determine the question."[19] But this profession of a cautious suspension of judgment on what was then a subject of violent controversy did not express Lyell's actual view on that subject, but was a plain contradiction of it. For on the very next page he enunciated an unmistakably "decided opinion" on precisely the same question:

> The fossil remains just alluded to have belonged for the most part to species which have ceased to exist upon the earth, and

---

[18] Ibid., p. 528. This is Lyell's answer to the question stated at the beginning of Bk. III (p. 481): "whether species have a real and permanent existence in nature? or whether they are capable, as some naturalists pretend, of being indefinitely modified in the course of a long series of generations?"

[19] Ibid., I, p. xii.

after studying the fossils of different strata, we find proofs that many distinct assemblages of animals and plants have flourished in succession on the globe.

And, *inter alia* in the second volume of the *Principles* Lyell offered a " proof that the entrance of man into the planet is, comparatively speaking, of extremely modern date," and quotes with admiration " a memorable passage written by Berkeley a century ago in which he inferred, on grounds which may be termed strictly geological, the recent date of the creation of man." [20]

Lyell, then, strangely failing to perceive the clear implication of his uniformitarianism, joined with most of the geologists and zoologists of the period in embracing the one doctrine with which uniformitarianism was wholly incompatible—the theory of numerous and discontinuous miraculous special creations. Of the actual occurrence of such supernatural events he, of course, offered no direct and positive proof, for there was none to offer. But it could, he assumed, be proved indirectly, by refuting what he supposed (erroneously) to be the only possible alternative to it that was entitled to any consideration. And this alternative was for him, not the theory of organic evolution in general, but the particular form of it which had been propounded by Lamarck; it is against the latter that the greater part of his long argument is directed.[21] But a refu-

[20] Ibid., II, pp. 77, 156-7. But in 1863, after his conversion to Darwinism, Lyell published a work devoted to disproving the " extremely modern date of the creation of man," and to defending the doctrine that species in general were not " created," *either* all at once *or* at scattered intervals, but slowly formed by the action of natural causes: *The Geological Evidences of the Antiquity of Man, with Remarks on the Theory of the Origin of Species by Variation*. Concerning the problem of individual psychology as to why Lyell in 1830 chose to state his position on the above question in the contradictory manner indicated, I refrain from conjecture.

[21] The theory of the evolution of species through natural selection was, of course, one possible alternative; and though Darwin's presentation of this was yet to come, such a hypothesis might naturally have occurred in the 1830's to any geologist as a conceivable explanation of the geological facts already known, and *should* have occurred to a consistent uniformitarian (see above, p. 378). It had, in fact, been suggested in a paper read before the Royal Society in 1813 by William Charles Wells, an American-born physician then practising in England; see Conway Zirkle, " Natural Selection before the Origin of Species," Proceedings of the *American Philosophical Society*, 1941, pp. 71-128; Charles A. Kofoid, " An American Pioneer in Science . . . ," *Scientific Monthly*, 1941, pp. 77-80; and Richard H. Shryock, " The Strange Case of Wells' Theory of Natural Selection," in *Studies and Essays in the History of Science and Learning Offered in Homage to George Sarton*, 1944, pp. 197-207. Dr. Shryock sums up Wells's contribution thus: " He put together two

tation of Lamarckianism was not equivalent to a disproof (which was what Lyell conceived himself to be presenting) of "the notion that one species may be gradually converted into another by insensible modifications in the course of ages." Lamarck (though not he alone) had, of course, asserted this thesis, and, at the time, was generally regarded as its principal representative. But he had also attempted to explain *how* such modifications are produced and how they eventually give rise to distinct and permanent species. There were, in short, two Lamarckian general theses, and there might be (and in fact, were) reasons for admitting the possible, or even probable, validity of the first, even though the second could be shown to be untenable or even absurd. Lyell, however, does not steadily distinguish the two. Most of his criticisms of Lamarck's explanation of *how* new species were produced were devastatingly effective; but they were irrelevant to the main issue. One of them, however, it is worth noting incidentally, offered an objection similar to one which was subsequently to be raised against the Darwinian theory, and was to be found troublesome by the defenders of the "all-sufficiency of natural selection." Lamarck, observed Lyell, "offered no positive fact to exemplify the substitution of some *entirely new* sense, faculty or organ, in the room of some other suppressed as useless. All the instances adduced go only to prove that the dimensions and strength of members and the perfection of certain attributes may, in a long succession of generations be lessened or enfeebled by disuse, or . . . be matured and augmented by active exertion, just as we know that the power of scent is feeble in the greyhound, while its swiftness of pace and exactness of sight are remarkable—that the harrier and staghound, on the contrary, are comparatively slow in their movements, but excel in the sense of smelling." [22] This was akin to the argument later urged by some critics of Darwinism, that the supposition merely of random slight

hitherto isolated ideas which had long been familiar to biologists—'natural selection' and the 'origin of species.' And it was just this combination which constituted the essence of Darwin's later hypothesis." The relevance of this to the present topic is that, since both ideas had been long familiar, it should have required no great originality or perspicacity to try the logical experiment of putting them together—if one accepted the principle that only "natural" causes of empirically verified phenomena must be sought for in the quest for "explanations" of those phenomena.

[22] Incidentally, it will be seen, Lyell is here conceding a good deal to Lamarck's belief in the importance of "use-and-disuse" of organs in the production of hereditary new characters—though he would not have been willing to call them "specific" characters.

variations in the offspring of a given pair of individuals does not account for the emergence of totally new *kinds* of senses or faculties (e. g., vision, memory) nor explain how, in their minimal, incipient stage, these could have any great value in the struggle for existence.

What is, perhaps, the most surprising thing in Lyell's attack on Lamarck's *Philosophie Zoologique* is his failure to see that—however unconvincing may have been the latter's notions about the "causes" of the genesis of new species—the French zoologist had been a loyal adherent and stout defender of the uniformitarian principle to the vindication of which the *Principles of Geology* was dedicated. All of Lamarck's supposed causal factors were believed by him to be "now operative"—and by Lyell were admitted to be so and to have a part in the production of new "races" or "varieties."[23] Arguing, as Lyell was to argue, against the theory of universal catastrophes followed by new creations, Lamarck had written:

> Unfortunately this facile method of explaining the operations of nature, when we cannot see their causes, has no basis beyond the imagination which created it, and cannot be supported by proof. Local catastrophes, such as those caused by earthquakes, volcanoes and other special causes are well known. . . . But why are we to assume without proof a universal catastrophe, when the better known procedure of nature suffices to account for all the facts which we can observe? Consider, on the one hand, that in all nature's works nothing is done abruptly, but that she acts everywhere slowly and by successive stages; and on the other hand that the special or local causes of disorders, commotions, displacements, can account for everything that we can observe on the surace of the earth, while still remaining subject to nature's laws and general procedure.[24]

Lamarck's reference here is primarily to the explanation of geological facts; he was expressing more lengthily—and more vigorously —the thesis which Lyell thus summarized in his Preface:

> It is presumed that the reader will understand enough to be convinced that the forces formerly employed to remodel the crust

---

[23] Cf. preceding note. One of Lamarck's crucial assumptions, the inheritance of acquired characters, is, of course, now generally recognized to have been erroneous. But it was not known to be so in 1813 or in 1830, and it was accepted by Lyell (I, pp. 508 ff.), though, as usual, with a denial that such characters could be specific.

[24] English tr., *Philosophical Zoology*, 1914, p. 46.

of the earth were the same in kind and energy as those now acting; or at least he will perceive that the opposite hypothesis is very questionable.[25]

But Lamarck saw that the same presupposition which the geologist should apply when reasoning about the formation of the "crust of the earth" should also be applied when reasoning about the diverse organic forms found in that crust, i. e., about the production of species; and he at least *attempted*, albeit unsuccessfully, so to apply it. Lyell, though he—in one sentence—seemed formally to admit that this presupposition held good for the organic as well as the inorganic "kingdoms," simply abandoned it when he came to deal with the problem of species, and gave his powerful support to those who, in zoology, totally repudiated it.

Not all of Lyell's arguments, however, were pertinent exclusively to Lamarck's form of the theory of evolution; some of them could be, and doubtless were, regarded by readers of the *Principles* as objections to any form of it. But these were familiar arguments, which continued to be current in the writings of anti-evolutionists of the next thirty years. Of these objections, and their logical status in the light of the geological and biological knowledge of that period, we shall take note shortly. What the foregoing analysis of Lyell's general position chiefly serves to make intelligible is the curiously double effect his influence had on the minds of his contemporaries.

(a) On the one hand, the fact that a naturalist so eminent—and, in the breadth of his knowledge of observed geological phenomena, so deservedly eminent—as Lyell had examined and scornfully rejected the "notion of the transmutation of species" and had adopted —and *not* on avowedly theological grounds—the doctrine of the special-creationists, naturally added immensely to the weight of scientific authority behind that doctrine. His pronouncement in its favor must have been, for many naturalists, and most laymen, almost analogous to a decision of the American Supreme Court, or the British Lord Chancellor, on a legal issue. It was, one cannot but suspect, his example and influence, more than the logical force of his arguments, that so long helped to sustain the prevalent belief that transformism was not a scientifically respectable theory; for the arguments were all either inconclusive or fallacious, as he himself was to point out in 1863.

[25] *Principles of Geology*, p. xc.

But (b) though the influence of Lyell's treatise thus, for the greater part of a generation, did much to retard the progress of biological (as distinct from purely geological) science, it also *subsequently* did much to produce the revolutionary change in both sciences which was to take place at the end of the 1850's. Lyell himself, of course—as is evident from what has already been said—did not intend his work to have any such consequence: he believed he had rendered that impossible. But his formulation and his persuasive advocacy of the "uniformitarian principle" was in the end to play an important part in converting nearly all naturalists of the time to the very doctrine which he had so confidently "refuted," and in bringing about the complete emancipation of biology from all theological considerations—and this in spite of the fact that his own statement of that principle (p. 366, above) said nothing explicitly about either the origin of species or theology. It was for him only an expression of his opposition to any form of "catastrophism" in geology, Cuvierian or other; the point which it was designed to emphasize was that the geologist, in seeking to account for past, i. e., no longer directly observable, phenomena must assume that the "causes of change" then were generically the same as those which are sufficient to account for analogous phenomena observably going on at the present time. But this assumption, as we have seen, had implications to which Lyell's own intellectual vision, when he was writing the *Principles of Geology*, was still blind. He simply did not *see* that a uniformitarian could not consistently accept special-creationism, and must therefore accept some form of evolutionism.[26]

[26] This seems to me the fairest way of stating in psychological terms the odd fact that Lyell was honestly able to combine zealous uniformitarianism with an equally zealous anti-evolutionism. The blindness may well have been due to some religious preconception, but if so, this is not clearly apparent in the *Principles*. That Lyell *was* intellectually honest in the composition of that work is implicitly denied by John W. Judd in his *The Coming of Evolution*, 1911. Judd, who was a friend and warm admirer of Lyell, asserted (op. cit., p. 83) that the latter was, from 1830 on, "convinced of the truth of the doctrine of the evolution of species" and "from the first, had seen that it would be impossible to avoid the conclusion that the principles which he was advancing with respect to the inorganic world must be equally applicable to the organic world" (p. 64). That Lyell, in the *Principles*, did not reveal this conviction Judd attributed to a laudable reluctance to shock the religious sensibilities of others and "a deep sense of the necessity of avoiding the *odium theologicum*." The quotations from Lyell's letters which Judd gives by no means justify this interpretation, and it is flatly contradicted by Lyell's statement in 1863 that, when writing the *Principles*, he "rejected transmuta-

Though the youthful Huxley had been one of the worst sufferers from this same blindness, he was apparently also one of the first to recover from it; he seems to have done so a little sooner than Lyell. For we find him, in a letter of June, 1859, earnestly endeavoring to bring his eminent senior—then over sixty—to see that he, as the great champion of the uniformitarian principle, was bound to recognize in evolutionism the working hypothesis logically demanded by that principle:

> I by no means believe [he wrote] that the transmutation hypothesis is proven, or anything like it. But I view it as a powerful instrument of research. Follow it out, and it will lead us somewhere; while the other notion is, like all the other modifications of "final causation," a barren virgin. . . . I would very strongly urge upon you that it is the logical development of uniformitarianism.[27]

In the self-same paper of 1887 in which we saw Huxley attempting to justify his own earlier refusal to adopt the doctrine of transmutation, even as a working hypothesis, there is also to be found the following passage:

> I have recently read afresh the first edition of the " Principles of Geology; " and when I consider that for nearly thirty years this remarkable book had been in everybody's hands, and that it brings home to every reader of ordinary intelligence a great principle and a great fact—the principle that the past must be explained by the present unless good cause can be shown to the contrary; and the fact that, so far as our knowledge of the past history of life on our globe goes, no such cause can be shown—I can not but believe that Lyell was, for others, as for myself, the chief agent in smoothing the road for Darwin. For consistent uniformitarianism postulates evolution as much in the organic as the inorganic world. The origin of a new species by other than

---

tion" and that the paleontological and other evidence then seemed to him to "favour the doctrine of the fixity of the specific character" (*Antiquity of Man*, p. 394). The fact that Lyell had devoted four long chapters to " combatting " and " refuting " the " notion of transmutation," to showing its " fallacies," even to ridiculing it (*Principles*, I, pp. 490-3) Judd carefully avoids mentioning. Never was a man so preposterously mispraised by a friend for a practice of elaborate hypocrisy and moral cowardice, of which he was certainly innocent. For further evidence that Lyell even in 1859 was still in need of conversion to evolutionism, see the next paragraph.

[27] *Life and Letters of Thomas Henry Huxley*, I, p. 174.

ordinary agencies would be a vastly greater "catastrophe" than any of those which Lyell successfully eliminated from sober geological speculation.

But however much Lyell may have "smoothed the road," Huxley and most of the biologists of those thirty years had declined, as did Lyell himself, to travel over it. It was left for an anonymous amateur, whom they thereupon with one accord fell to abusing, to point out the practicability of that highway. In the *Vestiges* Chambers had pressed home the argument from geological uniformitarianism to biological evolutionism with an effective use of concrete illustrations.

> If there is anything more than another impressed on our minds by the course of geological history, it is that the same laws and conditions of nature now apparent to us have existed throughout the whole time. Admitting that we do not now see any such fact as the production of new species, we at least know that, while such facts were occurring upon earth, there were associated phenomena of a perfectly ordinary character. For example, when the earth received its first fishes, sandstone and limestone were forming in the manner exemplified a few years ago in the ingenious experiments of Sir James Hall. . . . It was about the time of the first mammals that the forest of the Dirt Bed was sinking in natural ruin amidst the sea sludges, as the forests of the Plantagenets have been doing for several centuries upon the coast of England. In short, *all the common operations* of the physical world were going on in their usual simplicity, obeying the laws which we now see governing them; while the supposed extraordinary causes were in requisition for the development of the animal and vegetable kingdoms. There surely hence arises a strong presumption against any such causes.[28]

In his *Explanations*, 1846,[29] Chambers attacked the current assumption that the biological sciences are exempt from the rules of scientific method which the other sciences accepted and to which they owed their success:

> The whole question stands thus: For the theory of universal order—that is, order as presiding both in the origin and administration of the world—we have the testimony of a vast number of

---

[28] *Vestiges*, reprint in *Morley's Universal Library*, p. 114; italics, in original.
[29] This supplement to the *Vestiges* seems to be little known; it is in some respects superior to the original work.

facts in nature, and this one in addition—that whatever is reft from the domain of ignorance and made undoubted matter of science, forms a new support to the same doctrine. The opposite view, once predominant, has been shrinking for ages into lesser space, and now maintains a footing only in a few departments of nature which happen to be less liable than others to a clear investigation. The chief of these, if not the only one, is the origin of the organic kingdoms. So long as that remains obscure the supernatural will have a certain hold upon enlightened persons. . . . One after another the phenomena of nature, like so many revolted principalities, have fallen under the dominion of order and law; but here is one little province still faithful to the Boeotian government; and as it is nearly the last, no wonder it is so vigorously defended. As in the political world, however, men do not trust in the endurance of a dynasty which is reduced to a single city or nook of its dominions, so we may expect a speedy extinction to a doctrine which has been driven from every portion of nature but one or two limited fields.

This was to be a favorite line of argument of evolutionists—*after* 1859. Thus Tyndall, in his Belfast Address, 1874, pointed out that

> the basis of the doctrine of evolution consists, not in an experimental demonstration—for the subject is hardly accessible to this mode of proof—but in its general harmony with scientific thought. . . . We claim, and we shall wrest from theology, the entire domain of cosmological theory. All schemes and systems which thus infringe upon the domain of science must, in so far as they do this, submit to its control. . . . Acting otherwise has always proved disastrous in the past, and it is simply fatuous today.

So in 1882 Romanes put in the forefront of the arguments for evolutionism

> the fact that it is in full accordance with what is known as the principle of continuity—by which is meant the uniformity of nature, in virtue of which the many and varied processes going on in nature are due to the same kind of method, i. e., the method of natural causation. . . . The explanations of . . . phenomena which are at first given are nearly always of the supernatural kind. . . . Now, in our own day there are very few of these strongholds of the miraculous left. . . . No one ever thinks of resorting to supernaturalism, except in the comparatively few cases where science has not yet been able to explore the most obscure regions

of causation. . . . We are now in possession of so many of these historical analogies, that all minds with any instincts of science in their composition have grown to distrust, on merely antecedent grounds, any explanation which embodies a miraculous element. . . . Now, it must be obvious to any mind which has adopted this attitude of thought, that the scientific theory of natural descent is recommended by an overwhelming weight of antecedent presumption.

But this "overwhelming weight of antecedent presumption" against special-creationism and in favor of evolutionism had been clearly and repeatedly pointed out by Chambers in the middle forties; whereas Huxley in the next decade, as he himself has told us in a passage already cited, was not only still attacking "the evolutionists of 1851-58," but was doing so on the specific ground (among others) that there was *no* antecedent presumption against the notion of special-creation:—creation being "perfectly conceivable," it was not "impossible," etc.[30] This, of course, was a perverse misapprehension of the issue. It was not a question of conceivability, but of the relative probability of the only two available hypotheses. And the first criterion of probability in such a case must be the agreement of any proposed hypothesis with the general type of hypothetical explanations which the whole previous experience of men of science has found to be capable of fruitful application, and of the sort of verification which comes through fruitful application. By such a criterion, no hesitation between the two hypotheses was admissible. "Special acts of creative volition" had never been found by science to be a *vera causa* at all; the hypothesis was vague, sterile, impossible of verification, contrary to all the principles of method by the use of which the past successes of science had been achieved; "gradual development through natural descent" was, as a working theory, definite, suggestive of precisely formulable problems to which inductive tests could be applied, harmonious with the initial assumptions through which several other disciplines had already been converted from mere masses of information into sciences. Huxley after 1858 saw this clearly enough, and expressed it forcibly, though he never confessed the unreasonableness of his earlier position. The publication of Lyell's *Principles of Geology*, Huxley wrote in 1887, "constituted an epoch in the modern history of the doctrine

[30] See above, p. 359.

of evolution, by raising in the mind of every intelligent reader this question: if natural causation is competent to account for the not-living part of our globe, why should it not also account for the living part?" Every intelligent reader *should have* had this question in his mind after 1830. But Huxley himself, and almost every other naturalist of the 1840's saw no value whatever in the reasonings of Chambers and Spencer and (in the 1850's) a few others who were the only *writers* of the period, in Britain or America, with enough "ordinary intelligence" to raise explicitly the question described by Huxley as epoch-making—and to give the right (not to say, the obvious) answer to it. Lyell, undeniably, had by his polemic for uniformitarianism made it inevitable that this question would soon be raised by others, and it is for that reason that he has a place of importance in the modern history of the doctrine of evolution. But he himself did not, in the *Principles of Geology*, anywhere definitely propound it *as* a question—a separate and logically primary question—calling imperatively for consideration by any professed uniformitarian; and though his acceptance of the special-creation doctrine tacitly presupposed an answer to the question, it was an answer in the negative—that is to say, the wrong answer.[31]

III

Neither before nor after 1859 was the argument for the theory of organic evolution based solely upon the uniformitarian premise. That premise was, it is true, regarded by nearly all the new converts

[31] The judicious reader will have observed that the "uniformitarian principle" was somewhat equivocal. (*a*) Essentially it meant for Lyell, Chambers, Huxley, and the early Darwinians, simply that "natural" and "efficient causes" should be used to "explain" the phenomena of geology (Lyell's first period) and of biology (according to all the others), and that no "supernatural" and no "final causes" should be invoked by investigators in these sciences. But if uniformitarianism was construed (*b*) as implying that the *same specific kinds* of natural efficient causes operative in any past time, *and no others*, can be operative in any subsequent time, Hutton's and Lyell's "principle" was not true; the hypothesis of what is now called "emergent evolution" is logically possible, and I think, correct. On this see my "The Meanings of 'Emergence' and Its Modes," in *British Journal of Philosophical Studies*, 1927, pp. 167-81. But in so far as the evolutionists of 1830-1859 used the term "uniformitarianism" in their argument for organic evolution in the former sense, they were, as shown above, simply applying to the species-problem the usual and indispensable working postulates of scientific reasoning.

as sufficient by itself to justify the adoption of the theory, at least as a necessary working hypothesis; and one of them gave it a much higher logical rating. Joseph LeConte declared:

> We are confident that evolution is *absolutely certain.* Not, indeed, evolution as a special theory—Lamarckian, Darwinian, Spencerian—for these are all more or less successful modes of explaining evolution; . . . but evolution as a law of derivation of forms from previous forms; evolution as a law of continuity, a universal law of becoming. In this sense it is not only certain, it is axiomatic. It is only necessary to conceive it clearly, to see that it is a necessary truth.[32]

LeConte's reasons for this last assurance—which he admits may "seem paradoxical to some"—were, unfortunately, rather confused; he (in this passage, though not usually) first defined "evolution" as signifying, not merely the descent of animal or plant species from ancestral species by gradual modification, but as "the origin of things in every department of nature," and was thus enabled to identify "the law of evolution" in biology as "naught else than the law of necessary causation applied *to forms* instead of phenomena." But most of those who had previously been anti- or non-evolutionists but became converted in or after 1859, did not (so far as I am acquainted with their writings) regard the theory of the origin of species which they now accepted as *itself* a "necessary truth." It was for them simply the extension to organic phenomena of Lyell's uniformitarian principle, and the grounds for thus extending that principle were usually those set forth in passages already quoted. This, then, for the biologists of 1859 and after, was the primary and (as LeConte termed it) the "*general* evidence of evolution,"[33] adequate to prove the conclusion even if it had stood alone.

But it did not stand alone. There were numerous other evidences of a different character—not antecedent general postulates of scientific reasoning to be applied in the causal explanation of *all* observed

---

[32] *Evolution and its Relation to Religious Thought*, p. 65. LeConte was certainly among the early converts. Whether, from the time of his conversion, he already held this extreme opinion, expressed in his volume of 1888, I do not know.

[33] LeConte devotes a chapter to this (op. cit., pp. 53-66), and considers this "evidence" probative even if one does not agree with his elevation of evolutionism itself to the rank of an "axiom."

facts about organisms, but *specific* facts directly established—or at all events, accepted as established—by observation. They were facts which were in accord with and corroborative of the uniformitarian principle; but even to the plain man unacquainted with that generalization, some of them could hardly have made sense unless some extant species were assumed to be descended from other quite dissimilar species. Such facts were as well known and as well verified at least two decades before 1859 as they were at that date; and they played a past second only (and for some minds perhaps not second) to that of the uniformitarian presumption, in bringing about what may be called "the Great Awakening" among naturalists—their sudden and belated, but almost universal, conversion to evolutionism.

The argument drawn from these specific "evidences" was, however, less simple than that from the general uniformitarian premise. Not all of them were equally well verified inductively; certain of the "facts" now adduced as proofs of evolution had, during the preceding period, been considered—by the same men who now accepted them—not to be facts, or to disprove the theory of transmutation. And there still remained at least two seeming difficulties in the theory as a whole, which it was necessary for its new champions to face and overcome. They were therefore compelled to fight on the defensive as well as the offensive. But both on its defensive and on its offensive side, the logical status of the argument was the same before and after the publication of Darwin's great work. Not only were the indisputable relevant facts most favorable to the evolutionist theory already sufficiently proved, most of them before 1844; but also the apparent gaps or flaws in the evidence which could be plausibly exhibited by the opponents of the theory during the fifteen years preceding the appearance of the *Origin of Species*, were not removed in 1859 nor for a number of years thereafter. Whatever force the arguments for the transformist conception had after that year, they had before it; and whatever weaknesses they had before it, they still had after it. In presenting proof of this, I shall first indicate by direct citations the manner in which the several arguments were employed by the chief advocates of evolutionism in 1859 and for some years thereafter and then show the use of the same arguments by the evolutionists of the earlier period.

1. *Argument from the Sequence of Types in Paleontology*

Given the paleontological data known to naturalists in the period we are considering, two distinct questions could be asked. First, do those data, taken as a whole, exhibit a general *progression* in the order of time from simpler, less highly organized, "lower," types of plants and animals, possessing fewer powers or modes of action, to more complex, more highly organized, "higher," types? Second, do those data exhibit such close morphological similarities between types of fossils found in the same or in immediately successive formations, as to justify the inference that the organisms represented by the fossils in the later formations were *descended from* those whose remains are found in the earlier? There was nothing in the doctrine of special creations, as such, that was obviously inconsistent with an affirmative answer to the first of these questions; and the chief propagandists for the doctrine were, in fact, also "progressionists." Since the question was one of fact, the answer must, of course, be determined by the paleontological evidence; but the affirmative answer was doubtless also considered the more pleasing and the more edifying. Thus Sedgwick, whose *Edinburgh Review* article was the most ferocious of all the attacks upon Chambers's *Vestiges*, wrote in a Discourse of 1850:

> There are traces among the old deposits of the earth of an organic progression among the successive forms of life. [Examples are cited].... This historical development of the forms and functions of organic life during successive epochs, seems to mark a gradual evolution of creative power manifested by a gradual ascent towards a higher type of being. . . . There was a time when cephalopoda were the highest type of animal life, the primates of the world; . . . Fishes next took the lead, then Reptiles, . . . Mammals were added next, until Nature became what she is now, by the addition of Man.

But Sedgwick, of course, added that "the elevation of the fauna of successive periods was not made by transmutation, but by creative additions." [34] When thus adjusted to the sequence of types shown by paleontology, the theory of special creations, undeniably, took on a somewhat odd look, and must have suggested to some members of

---

[34] Quoted by Lyell, *Antiquity of Man*, pp. 395-6, from Sedgwick's *Discourse on the Studies of the University of Cambridge*, Preface to 5th ed., 1850. Lyell cites similar expressions of progressionism from Hugh Miller, Owen, and others.

the University of Cambridge, for whose benefit Sedgwick's *Discourse* was written, a theologically awkward question: Why was it necessary, or why was it rationally desirable, for the Creator to approach his final objective so slowly and through such strange detours? Why create the Reptilia—

> Dragons of the prime,
> That tare each other in their slime,

in the phrasing of another graduate of Cambridge, in a more celebrated work published in the same year—[35] as a prelude to *Homo sapiens*? To this question, it is true, Sedgwick did offer a seeming answer, in his striking phrase "*a gradual evolution of creative power, manifested by a gradual ascent towards higher types*"; [36] but it was a highly heterodox answer, since it implied that in the Mesozoic era God was not yet *able* to create any living beings "higher" than reptiles. Chambers had said nothing as heretical as this. But whatever the theological pitfalls latent in the combination of special-creationism with progressionism, the combination seems to have been adopted by most of the anti-evolutionists of the time. And they thus admitted a broad parallelism between antiquity of geologic strata and simplicity of the organisms represented by the fossils contained in them.

This, however, was not admitted by Lyell, in his anti-evolutionist period. He thought it "easy to show" that the "proposition,

---

[35] Tennyson, *In Memoriam*, Canto LVII. The knowledge of contemporary paleontology shown in this canto was in all probability derived from a reading of Lyell's *Principles*; cf. the poet's tragic doubts about the future of the human species: "And he, shall he, Man her last work, that seemed so fair, ... Be blown about the desert dust, Or sealed within the iron hills? No more? A monster, then, a dream, a discord." Lyell in chapter VIII-XI of Book III had dilated upon the innumerable "causes tending to the extirpation of species" and the immense number of species that have been thereby destroyed; and he "deduced as a corollary, that the species existing at any particular period must, in the course of ages, become extinct one after the other. 'They must die out,' to borrow an emphatical expression from Buffon, 'because Time fights against them'" (II, p. 93).

[36] The conception of an "evolving God," whose development is manifested in the progressive diversity and complexity of living beings, had been propounded and defended by Schelling and Oken in 1809-1810; see Lovejoy, *The Great Chain of Being*, 1953 edition, pp. 317-326. But though the phrase used by Sedgwick seemed to imply such a theology, it appears improbable that he clearly realized this. Yet he may, as he was writing, have had a momentary glimpse of the fact that this assumption was the only way of reconciling progressionism, and the admitted paleontological evidence which progressionism presupposed, with the doctrine of special-creations.

generally received, that there is a progressive development of organic life from the simplest to the most complicated, has but a slender foundation in fact." He did not deny " the absence of fossil bones of mammalia in the older rocks," but warned his readers that "we must not too hastily infer" from this "that the highest class of vertebrated animals did not exist in the remoter ages "; for most fossiliferous strata "were deposited beneath the sea" and "the casualties must always be rare by which land quadrupeds are swept far out into the open sea; and still rarer the contingency of such a floating body not being devoured by sharks and other predacious fish." [37] All that this sort of reasoning could prove was, at most, that so much of what *might* have been evidence on the question must have been destroyed, or is inaccessible, that the paleontologist has no means of determining in what geologic era any species, genus, or class first existed. Even in 1830 this was somewhat of an exaggeration, and certainly in the 1850's it could no longer be maintained Lyell himself eventually realized this, and in *The Antiquity of Man* candidly reversed his position, and accepted "the generalisation as to progression laid down" in the passage quoted above from Sedgwick, as "still holding good in all essential particulars." [38]

Thus the *general* fact of the gradual production of higher types in the course of geologic time, the existence of a broad correspondence between relative recency of the appearance of new species on the terrestrial scene and their relative rank in the taxonomic scale of being was admitted by most, though not all, paleontologists before as well as after 1859. And to many evolutionists after that year this fact seemed an impressive corroboration of the doctrine of the descent of species from other species through successive, slow, and cumulative modifications of the characters of the ancestors. The importance which Huxley attached to this argument in 1863 is shown by a passage in his *Lectures on the Phenomena of Organic Nature*:

> If you regard the whole series of stratified rocks . . . constituting the only record we have of a most prodigious lapse of time; if you observe in these successive strata of rocks successive groups of animals arising and dying out, a constant succession

[37] *Principles of Geology*, I, pp. 149, 157. But Lyell in the 30's, as we have seen, was not willing to apply this reasoning to all vertebrates; he held it to be possible to "prove" that one of them, man, appeared on the planet at "an extremely modern date." The inconsistency of this with the above is obvious.
[38] *Antiquity of Man*, 1863, p. 396.

giving you the same kind of impression, as you travel from one group of strata to another as you would have in travelling from one country to another; . . . when you look at this wonderful history and ask what it means, it is only a paltering with words if you are offered the reply, "They were so created." But if, on the other hand, you look on all forms of organized beings as the results of the gradual modification of a primitive type, the facts receive a meaning and you see that these older conditions are the necessary predecessors of the present. Viewed in this light the facts of paleontology receive a meaning—upon any other hypothesis I am unable to see, in the slightest degree, what knowledge or signification we are to draw from them. Again, note . . . the singular likeness which obtains between the successive faunae and florae, whose remains are preserved in the rocks; you never find any great and enormous difference between them, unless you have reason to believe that there has also been a great lapse of time or a great change of conditions.

Just so had Chambers already argued in his *Explanations* in 1846:

Fifty years ago science possessed no facts regarding the origin of organic creatures upon earth. . . . Within that time, by researches in the crust of the earth, we have obtained a bold outline of the history of the globe. . . . It is shown on powerful evidence that during this time strata of various thicknesses were deposited in seas; . . . volcanic agency broke up the strata, etc. . . . The remains and traces of plants and animals found in the succession of strata show that while these operations were going on the earth gradually became the theatre of organic being, simple forms appearing first and more complicated afterwards. . . . This is a wonderful revelation to have come upon the men of our time, and one which the philosophers of the age of Newton could never have expected to be vouchsafed. The great fact established by it is that the organic creation, as we now see it, was not placed upon the earth at once:—it observed a PROGRESS. . . . There is also the fact of an ascertained historical progress of plants and animals in the order of their organization. . . . In an arbitrary system we had surely no reason to expect mammals after reptiles; yet in this order they came.[39]

But the general fact of "progression," though it could fairly be said to establish (when combined with the uniformitarian principle) a presumption in favor of the hypothesis of evolution, fell

[39] Op. cit., pp. 21, 106.

far short of a "proof" of it; for it did not, by itself, justify an affirmative answer to the second of the two questions distinguished at the beginning of this section. It was *too* general. It did not show —what the theory implied—invariably close morphological similarities between the organisms now found in the upper and lower parts of the same or in directly superposed strata. And when the paleontological evidence then available was more minutely examined, it appeared to be by no means wholly favorable to the development hypothesis. This was so far the case that the theologically orthodox geologists were able, with some real plausibility, to turn this weapon against the evolutionists. One of the only two apparently serious reasons that could then be advanced for rejecting the transformist hypothesis lay in the observation that the facts of stratigraphic geology as then known failed to exhibit with any consistency, fullness and precision, the sequences that the hypothesis required. Much of the fighting, between the time of the *Vestiges* and that of the *Origin of Species*, took place around this issue; and the battleground was well chosen for the conservatives. The directly frontal assault, of course, came from those who did not accept even progressionism. Thus a competent geologist could declare in a manual published in 1855:

> Reptiles do not exhibit any growing progression in their typical forms . . . The Saurians, which are the most highly organized order [sc., of the Reptiles] were created towards the close of the palaeozoic epoch: the Protorosaurus . . . attests this fact. . . . Their numbers greatly diminished in the cretaceous stages; and in the tertiary strata we find the remains only of crocodiles and alligators. . . . In reptiles and fishes the orders most inferior in organisation were the last created of their class . . . The theory of progressive development receives no support from the facts unfolded by the history of fossil reptiles. The genera that were first created belong to the most highly organized types of the class. . . . This fact . . . exactly accords with what has been observed in the stratigraphical distribution of fossil fishes.[40]

But even the special-creationists who recognized a general progression from lower to higher classes or orders of plants and animals could without inconsistency, and did, join in attacking evolutionism on this ground. It would not have been at all to their purpose to carry progressionism so far as to admit that the paleontological

[40] G. F. Richardson: *Geology and Paleontology*, 2nd Ed., 1855, p. 304.

record showed in all cases, or even in any cases, the degree of similarity between temporally consecutive "*species*" that should be expected, if the theory of transmutation were true. And in point of fact, it did not.

The chief objections to that theory advanced in this period on paleontological grounds were five in number. There was, first, the general and undeniable fact of "missing links" in the chain of past organisms. Secondly, there was the fact of the apparently sudden appearance of groups of allied, and by no means absolutely primordial, species in the lowest fossiliferous strata then known. Thirdly, there was the sudden disappearance of whole groups of species at the end of certain geological periods, and their sudden replacement in the next period by species different in type from the former, and closely allied to one another. These two points—the second and third—were the special contribution of the Cuvierian school to the controversy. Out of Cuvier's doctrine of the abrupt extinction of faunas at the successive "revolutions of the globe," his disciples had elaborated the theory of the radical and world-wide discontinuity of the faunas and floras of the successive great periods, and had hence inferred the actual necessity of assuming a definite number of special creations of fresh organic worlds *en bloc*. D'Orbigny knew exactly how many such creations there had been:

> The first creation shows itself in the Silurian stage. After its annihilation through some geological cause or other, a second creation took place a considerable time after, in the Devonian stage; and twenty-seven times in succession *distinct creations* have come to repeople the whole earth with its plants and animals, after each of the geological disturbances which destroyed everything in living nature. Such is the fact, certain but incomprehensible, which we confine ourselves to stating, without endeavoring to solve the superhuman mystery which envelops it.[41]

Fourthly—to continue the enumeration of the paleontological difficulties—it was objected that, within the limits of single great geological formations, the arrangement of fossils in the strata did not exhibit the required order of progression from lower to higher types, but sometimes even reversed that order. This was Sedgwick's principal point in his *Edinburgh Review* article, as it was that of Hugh Miller in his *Footprints of the Creator*, 1849, the most widely

---

[41] D'Orbigny, *Cours Élémentaire de Paléontologie Stratigraphique*, 1849, II, p. 251; cited in Depéret, *Transformations of the Animal World*, 1909, pp. 18-19.

circulated of the replies to the *Vestiges*. Miller's argument may be summarized in his own words.[42] The latest discoveries in the Silurian and Cambrian series, he declared, do not show the

> sort of arrangement demanded by the exigencies of the development hypothesis. A true wood at the base of the old red sandstone, or a true Placoid in the limestones of Bala, very considerably beneath the base of the Lower Silurian system, are untoward misplacements for the purposes of the Lamarckian; and who that has watched the progress of discovery for the last twenty years and seen the place of the earliest ichthyolite transferred from the Carboniferous to the Cambrian system, and that of the earliest exogenous lignite from the Lias to the Lower Devonian, will now venture to say that fossil wood may not yet be detected as low in the scale as any vegetable organism whatever, or fossil fish as low as the remains of any animal? But though the response of the earlier geologic systems be thus unfavorable to the development hypothesis, may not men such as the author of the "Vestiges" urge that the geologic evidence, taken as a whole, and in its bearing upon groups and periods, establishes the general fact that the lower plants and animals preceded the higher, . . . that the fish preceded the reptile, that the reptile preceded the bird, that the bird preceded the mammiferous quadruped and that the mammiferous quadruped and the quadrumana preceded man? Assuredly yes! They may and do urge that geology furnishes evidence of such a succession of existences; and the arrangement seems at once a very wonderful and very beautiful one. Of that great and imposing procession of which this world has been the scene, the programme has been admirably marshalled. But the order of the arrangement by no means justifies the inference based upon it by the Lamarckian.

The reason why, according to Miller, it does not, constitutes the fifth objection urged against evolutionism from the side of paleontology; "superposition," as Miller put it, "does not mean parental relation," any more than the presence of gradually accumulated vegetable and animal refuse in a farmer's ditch means that the creatures whose remains lie at the bottom of the ditch begot those whose remains are found higher up.

The last argument does not call, and never did call, for serious consideration; it is a begging of the precise question at issue, con-

---

[42] Quoted from the American edition, 27th thousand, 1875, pp. 227-8; the edition has a eulogistic preface by Agassiz, 1851.

cealed by a specious but lame analogy. The other four arguments depended, with respect to their logical weight, upon the way in which they were applied. If it were assumed that the burden of offering specific proofs rested upon the transformationist, and that the paleontological evidence was put forward by him *as a proof*, the objections of the orthodox geologists were perfectly sound: the paleontological evidence was not clear nor complete. If, again, transformism were regarded as a hypothesis which implied the existence of certain specific geologic facts, then, also, the objections enumerated were pertinent—with one all-important and extremely obvious qualification: the implied facts in stratigraphic geology had *not* been verified—so far as inquiry into a record that will always and necessarily remain fragmentary had then extended. If, lastly, the objections were advanced as a positive *disproof* of the transformation of species, they were entirely incompetent, by reason of the necessity of adding the qualification last mentioned. The record being notoriously incomplete, it was impossible to infer from mere breaches of continuity, and from occasional failures in the general parallelism of geological antiquity with simplicity of organic type, that the order of appearance of species had not in fact been progressive, and the result of gradual modification through natural descent. The difficulties raised by the conservative paleontologists logically justified, at most, only a Scotch verdict of "not proven" —so far as this part of the testimony is concerned.

This continued to be the logical situation in 1859 and for a number of years thereafter. Darwin wrote to Quatrefages:

> My views spread slowly in England and America; and I am much surprised to find them most commonly accepted by geologists, next by botanists, and least by zoologists; . . . for the arguments from geology have always seemed strongest against me.

That the objection from the general absence of intermediate links between species was a pertinent one he acknowledged with characteristic candor:

> Geology assuredly does not reveal any such finely graduated organic chain; and this is perhaps the most obvious and serious objection which can be urged against the theory.

He recognized that, so far as geological knowledge then went, whole groups of species sometimes seemed to make their appearance abruptly; though he argued that the increase of such knowledge had steadily tended to diminish this semblance of abruptness. Wholly eliminated these sharp transitions have never been; an authoritative expositor of the general results of paleontology says of d'Orbigny that, though "his ideas" were "too absolute, his observations remain none the less exact in their broad lines, and the sudden replacing of marine faunas, when passing from one stage to another, or even from zone to zone, must be considered almost a general rule."[43] The same writer, who was, of course, a convinced evolutionist, observes:

> After all we can not forget that there exists an immense number of creatures without intermediate links, and that the relations of the great divisions of the animal or vegetable kingdom are much less strict than the theory demands. ... The keenest partisan of the descent theory must admit that the fossil links between the classes and orders of the two kingdoms exist in infinitesimally small numbers.

The second argument of the paleontological opponents of the theory Darwin regarded as not less deserving of serious consideration.

> The sudden manner in which several groups of species first appear in our European formations, the almost entire absence, as at present known, of formations rich in fossils beneath the Cambrian strata, are all undoubtedly of the most serious nature. The difficulty of assigning any good reason for the absence of vast piles of strata rich in fossils beneath the Cambrian system is very great. ... The case at present must remain inexplicable, and may be truly urged as a valid argument against the views here entertained.[44]

To all these objections, as to that drawn from the absence of a uniformly progressive sequence in the superposition of species of certain classes, Darwin opposed a single reply: "the imperfection of the geological record"—an imperfection due not only to the "inadequacy of geological exploration but to the inevitable absence of many chapters from the rock-history itself." Paleontology thus offered to neither side materials for a decisive proof of its case. Darwin's sixth chapter presented these considerations in a masterly

[43] Depéret, *Transformations of the Animal World*, 1909, p. 22; the following passage, p. 113.
[44] Citations are from *Origin of Species*, sixth edition, ch. x, passim; this was ch. ix of the first edition.

manner. But there was no time in the history of paleontology when they were not extremely obvious and familiar considerations. Chambers, in replying to his critics, had admitted that there are "some cases" in which "we cannot positively, from direct evidence, affirm the truth" of the hypothesis he was defending, but explained that "the absence of direct proof is accounted for by the rarity or obscurity of the phenomena, our deficient means of observing them, or the logical difficulties arising from the complication of the circumstances in which they occur; insomuch that, notwithstanding as rigid a dependence upon given conditions as exists in the case of any other phenomena, it was not likely that we should be better acquainted with those conditions than we are ";[45] i. e., if the hypothesis as a whole was true, it was to be *expected* that certain of the effects implied by it could not now be observed by us. This was still more clearly pointed out by Spencer in a brilliant article written in 1858 [46] and published in the *Universal Review*,[47] in July, 1859: "along with continuity of life on the earth's surface, there not only *may* be, but *must* be, great gaps in the series of fossils"; and that "hence these gaps are no evidence against the doctrine of evolution." He concluded:

> It must be admitted that the facts of Palaeontology can never suffice either to prove or disprove the Development Hypothesis; but that the most they can do is, to show whether the last few pages of the Earth's biologic history are or are not in harmony with this hypothesis.

In its later development, it is true, paleontology has been able to produce some striking supplementary evidences of evolution. In a limited number of cases, approximately complete and closely graduated series of forms of single orders or families can be exhibited in due stratigraphic superposition. But all the elaborate and impressive "form-series" have been worked out since 1859. Darwin himself made no original discoveries in this field; and as late as the sixth edition of the *Origin* the best evidence of the sort he presented from writers is, I believe, summed up in these two sentences:

> Several cases are on record of the same species presenting varieties in the upper and lower parts of the same formation.

---

[45] *Explanations*, 1845, pp. 146-7.
[46] *Life and Letters of Herbert Spencer*, II, p. 332.
[47] Reprinted in *Illustrations of Universal Progress*, 1868, pp. 361, 376.

Thus Trautschold gives a number of instances with Ammonites, and Hilgendorf has described a most curious case of ten graduated forms of *Planorbis multiformis* in the successive beds of a fresh-water formation in Switzerland.

Of the two instances cited the first is vague—the great studies of Waagen (1869) and of Neumayr (1871-5) on the Ammonites were still to come; and the observations of Hilgendorf seem already, by the time the sixth edition of the *Origin* was prepared for the press, to have been shown to be erroneous.[48] The best known example, to English readers, of a form-series is that of the Equidae. But Rütimeyer's *Beiträge zur Kenntnis der fossilen Pferde* appeared only in 1863; and Huxley's researches in this field, which were the consequence, not the cause, of his acceptance of the theory of descent, were first presented to the public in his presidential address before the Geological Society in 1870.

2. *Argument from Persistent Types*

If really good examples of graduated form-series were not available at the time of the *Origin*, the evolutionist of the period could still find in paleontology a sort of evidence decidedly unfavorable to one hypothesis opposed to his own—that of wholesale obliterations of faunas, and thorough-going new creations of the entire organic world. This evidence lay in the persistence of many orders and certain species through more than one geological epoch. The classic of the special-creation doctrine was the introduction to the third edition of Cuvier's *Recherches sur les ossements fossiles*; and the principal argument of that work was, in the words of one of Cuvier's disciples,[49] to the effect that "no fossil species, at least among the two classes of *mammalia* and *reptilia*, has any analogue among living species, or, in other words, that every fossil species is extinct." If this could be shown by positive evidence not to be the case, one of the principal supports of the special-creation hypothesis was taken away from it. Huxley made much of this line of attack in a paper of 1859 and in his address before the Geological Society in 1862. He pointed out, for example, that *lingula* and certain *mollusca* " have persisted from the Silurian epoch to the present day, with

[48] Cf. O. Schmidt, *Descent and Darwinism*, 1873, English tr., 1896, p. 96.
[49] Flourens, *Analyse raisonnée des travaux de G. Cuvier*, 1841; italics mine.

so little change that competent malacologists are sometimes puzzled to distinguish the ancient from the modern species." He noted that the " group of *crocodilia* was represented at the beginning of the Mesozoic age, if not earlier, by species identical in the character of their organization with those now living " and that, probably, even " certain types of the ancient mammalian fauna, such as that of the *marsupialia* have persisted with no greater change throughout as vast a lapse of time."

But the argument Huxley here used had not newly become available. Flourens' generalization had gone far beyond any evidence which he had offered, or which could in the nature of the case be offered. The proposition was, indeed, insusceptible of proof, save by a sort of reasoning in a circle. For when the special-creationists denied the survival of species from one epoch to another, they were using the word "species" in a sense different from that in which *they*, at least, usually employed it. In their zoology, the final test of specific difference between two forms was the sterility of the hybrid. But extinct forms can not be subjected to this test. In paleontology, therefore, differences of species had to be determined solely on grounds of morphological, mainly skeletal, dissimilarity; while it was, at the same time, recognized that in living animals an immense range of such dissimilarity might be consistent with identity of physiological species. If the pug-dog and the greyhound had been extinct, it is at least questionable whether paleontologists would have assigned them to the same species—especially if their fossil remains had been found at different geological horizons. Under such circumstances, it was open to the paleontologist to multiply species almost *ad libitum*; if he had adopted a theory which required that no species found in one stratum should be found in another, it was easy to make the most of slight variations of form. Differentiations of species thus made, however, were essentially subjective; all that could conceivably have been proved objectively was that no *form* remained the same through successive geological periods. Yet even of this no proof was forthcoming; the geologic record was not the sort of document that could furnish proof for a universal negative. It furnished, in fact, evidence on the other side. Even Cuvier's eulogist had been obliged (1841) to limit the generalization by adding "at least among mammals and reptiles "— and then to make further exception of two orders of mammals. And Hitchcock in his *The Religion of Geology* (1852), while exaggerating the discontinuity of then known types, could say no more than that, " of the thirty thousand species of animals and plants found in the rocks, *very few* living species can be detected." But a few were as good as a multitude as witnesses to the fact that there had been no such complete, simultaneous extinctions of faunas, and radical alterations of terrestrial conditions, as the Cuvierian theory supposed. And we find Chambers, in 1845, citing specific examples of persistency, as Huxley was to do seventeen years later.

> There is a badger of the Miocene which can not be distinguished from the badger of the present day. Our existing *Meles taxus* is therefore acknowledged by Mr. Owen to be " the oldest known species of mammal on the face of the earth." It is in like manner impossible to discover any difference between the existing wild cat and that which lived in the bone caves with the hyæna, rhinoceros and tiger of the ante-drift era, all of which are said to be extinct species. . . . There is a persistency of certain shells since the beginning of the tertiaries. . . . Several shells of the secondary formation straggling into the tertiaries are not less conclusive, in rigid reasoning, that all the tertiary species were descended from the secondary, though the wide unrepresented interval at that point allowed a greater transition of forms. In short, the whole of the divisions constructed by geologists upon the supposition of extensive introductions of totally new vehicles of life must give way before the application of this rule, and it must be seen that what they call new species are but variations of the old.[50]

But though the paleontological evidence of the long persistence of some species served to refute the Cuvierians, it was not, of course, an evidence of transmutation. The fact that a species can continue to exist unchanged through several geological eras tells us nothing about the mode of its origination. Thus, of the two special arguments from paleontology, the first, though it stimulated investigations which were before long to provide some sound support for the theory of descent, was not, in 1859, a really convincing proof of that theory; and the second was effective only as a destructive weapon against one part—what may be called the extreme right wing—of the forces of special-creationism.

[50] *Explanations*, p. 108 f. Chambers was not an authority on these matters, but Richard Owen was; he was generally regarded as the greatest comparative anatomist of the time.

Paleontology, then, down to and beyond 1859, could offer no cogent proof of the transformation of species; and though it also could offer no cogent disproof, it did seem to exhibit some facts which, as we have seen Darwin himself admitting, could be "truly urged as a valid argument against [his] theory." But these facts for many biologists of the period were not the worst of the difficulties which the partisans of that theory must surmount. The second great objection to transformism had no relation to paleontology, but was deduced from a proposition in genetics that was supposed to be capable of proof—and, in fact, to have been proved—by observation of now-existing organisms.

3. *Argument Concerning the Sterility of Hybrids; Rejection of Buffon's Definition of "Species"*

That some pairs of animals capable of sexual union but commonly classed as of different species produce, when mated, either no offspring or sterile offspring, everyone knew; the existence of mules was not a discovery of modern science. And if such sterility marked *all* cases of mating between different species, though matings between individuals of the same species are normally fertile, the greater part of "animated nature"—all living creatures except those that multiply by fission—would seem to be divided primarily into two sharply contrasted *types of classes* of organisms—those incapable and those capable of generating fertile offspring through cross-mating. For at least a century, i.e. since the publication of the second volume of Buffon's *Histoire Naturelle* [51] it had been assumed by most biologists that this dichotomy of the whole realm of organisms endowed with sex was subject to no exceptions—an empirically verified generalization, a "law of nature." And a very striking and significant generalization it seemed to these biologists to be. To express this distinction, a new biological terminology had been introduced. The word "species" had hitherto commonly been used to designate relatively large classes of organisms which were distinguished from others by easily observable differences of form, structure, and behavior. But under the influence of Buffon the term "species" was now given a new definition: it came to be used

---

[51] See above, p. 98. Buffon himself, however, in his later volumes sometimes wavered on this point.

to designate those kinds of organisms which cannot by mating with one another produce fertile offspring. Those which can do so were called "races" or "varieties." This, obviously, was merely a matter of definition, a statement of the connotation and denotation of certain words. But from a definition one cannot deduce any factual proposition—not even that entities having the properties mentioned in the definition exist—though in the case of Buffon's definition it was, of course, assumed to be a known fact that there are organisms having the property stated in the definition. But Buffon, as we have seen, combined with his new definition of the term "species" two factual propositions: (a) that organisms not of the same "species," in the sense defined, can have had no ancestors in common, and (b) that the so-called "specific" characters of a "species" are immutable, e. g., that all the descendants of gill-breathing marine animals are still marine animals, that no extant quadrupeds having solid hoofs can be among the posterity of any animals having toes or cloven hoofs. Neither of these general propositions—(a) and (b)—was included in or implied by the Buffonian definition of "species"; neither logically followed from the one relevant proposition that *was* true, viz., that matings between *some* (Linnaean) species produce sterile hybrids; and neither was itself true. Yet this illogical deduction of a supposed biological fact from a verbal definition dominated, and misled, the biology of the later eighteenth century; we have seen this illustrated in the case of Kant. And it still, even after 1858, appeared to many biologists to interpose a *caveat* against acceptance of the theory of organic evolution pending further extensive experimentation in cross-breeding. Thus Huxley, in his lectures of 1862, warned his audience of the one missing link in the chain of evidence—the fact that selective breeding has not yet produced species sterile to one another.[52] And Alfred Russel Wallace, later writing retrospectively, recalled that "one of the greatest, or perhaps we may say the greatest, of all the difficulties in the way of accepting the theory of natural selection as a complete explanation of the origin of species has been the remarkable difference between races and species when crossed." [53]

How was this difficulty dealt with by the evolutionists of 1859 and the following decade? Huxley disposed of it mainly by an act of faith. Convinced on other grounds that the transformation of

[52] Huxley's *Life and Letters*, I, p. 208.
[53] *Darwinism*, p. 152.

species must be accepted "as a working hypothesis," he not only refused to abandon that hypothesis merely because it seemed inconsistent with one striking fact which was generally held to have been experimentally verified; he boldly predicted that future experimentation would simply eliminate the difficulty—by showing that the production of sterile offspring is no evidence that the parents belonged to different and immutable "species," as Buffon had supposed. Huxley wrote Darwin in 1862:

> I have told my classes that I entertain no doubt whatever that twenty years' experiments on pigeons, conducted by a skilled physiologist, instead of by a mere breeder, would give us physiological species sterile *inter se* from a common stock, . . . and I have told them that when these experiments are performed I shall consider your views to have a complete physical basis.[54]

But this implied that until such experiments *had* been performed, with the result Huxley predicted, the evidence for Darwin's views would *not* be complete. And a mere prophecy of what would hereafter be shown by an investigation not yet carried out or even attempted, was likely to seem to the sceptical to be somewhat less than probative. Far more searching and effective was Darwin's own criticism of "the view commonly entertained by naturalists" about the implications of the sterility of hybrids. For the criticism was not based upon a prophecy of future discoveries, but upon a wide compilation and an acute analysis of relevant facts already well established.

While, as we have seen, he freely admitted the force of the objections raised by his opponents on paleontological grounds, and simply expressed his confidence that future discoveries in that science would invalidate those objections, he did not make a similar admission with respect to the anti-evolutionist argument from the sterility of hybrids. On the contrary, he devoted a long chapter of his book to demonstrating that this objection was without force in the light of facts *already* established and logical considerations which should have been obvious. He based this conclusion, not merely upon the fact that it had never been shown experimentally that *all* hybrids are sterile, but also (a) upon evidence that *some* offspring of crossed species—that is, of what were generally admitted to be different

---

[54] Ibid., p. 145.

species—are *not* sterile and (b) upon the observation that, in those cases in which the sterility of hybrids is empirically verified, it is presumably explicable as the result of natural causes affecting the reproductive organs of one or both parents or of their hybrid progeny, and preventing the normal conjunction or interaction of the elements contributed by the two parents. Darwin's evidence for the former contention was drawn largely from the experiments of botanists, but was not on that account the less relevant, since if Buffon's definition of "species" were accepted as an argument against the possibility of transmutation, it should hold good of plants as well as animals. Darwin reviewed and critically examined the recorded experiments in the hybridization of plants by botanists and became convinced that the weight of the evidence was against the assumption that crossed species are necessarily sterile. He pointed out, for example, that "the Rev. W. Herbert,"[55] a horticulturist of great skill, is emphatic in his conclusion that some hybrids are perfectly fertile—as fertile as the pure parent species; Herbert's experiments are cited in detail. "With regard to animals," Darwin wrote, "much fewer experiments have been carefully tried than with plants." Yet he cited several examples of the fertility of hybrids reported by other zoologists, e. g. "Mr. Quatrefages states that the hybrids from two moths" of different (Linnaean) species "were proved, in Paris, to be fertile *inter se* for eight generations. . . . The hybrids from two varieties of geese which are so different that they are commonly ranked in distinct genera have often been bred in this country with pure parents," and in India, according to "two eminently capable judges [whose names are given] whole flocks of these crossed geese are kept in various parts of the country; and as they are kept for profit, where neither pure parent-species exists, they must certainly be highly or perfectly fertile." Darwin's procedure here is characteristic; while other naturalists, even Huxley, were calling for further investigation of interbreeding and meanwhile provisionally suspending final judgment on the question, Darwin was patiently collecting and appraising the records of relevant experiments *already* performed by competent investigators in various parts of the world, and was thereby enabled to reach a definite judgment on the matter:

[55] Dean of Manchester, d. 1847. His monographs on plant breeding were published in 1832 and 1847. He was also a valued correspondent of Darwin.

Considering all the ascertained facts on the intercrossing of plants and animals, it may be concluded that some degree of sterility, both in first crosses and hybrids, is an extremely general result, *but it cannot, under our present state of knowledge, be considered as absolutely universal.*[56]

In short, Buffon's famous "definition of species" had, before the eighteen-fifties, been shown to be false if it was construed (as it usually had been) as a proved factual generalization, an assertion that sexual unions between admittedly different species never have fertile descendents.

In the three "special" arguments of which we have thus far been reviewing the history, the champions of evolutionism in the years from 1830 until the 1860's were, in the main, fighting on the defensive. They were seeking to rebut the reasonings of the adversaries of their doctrine, by showing that the premises on which those reasonings rested either had not been proved or were contrary to facts already verified. But such arguments in rebuttal fell short of a positive proof of the actual descent of existing species from more ancient and admittedly different species. The observed facts on which these arguments were based were drawn from the sciences of geology and genetics, the latter in the limited sense of the study of what may be called the gross externally observable characters of successive generations of individuals of now extant organisms.[57] We now turn to three further arguments for the evolutionist theory, two of them based wholly and the third partly on the comparative anatomy of the internal—i. e., *not* externally observable—structures found in organisms of identical classes (in the taxonomic sense of that term). Comparative anatomy, as we shall shortly see, was by no means a new science in the early 19th century; Buffon's associate Daubenton best deserves, I think, the title of "founder" of it. It was a subdivision but, as Darwin emphasized, an especially significant subdivision, of the science of morphology, which is, he declared, "the most interesting department of natural history, and may be said to be its very soul."[58] And in it he found exhibited—as had Chambers and a few others—three distinct *kinds of resemblance* between organisms which were manifestly not merely of different

---

[56] Quotations are from *Origin of Species*, first American edition, 1860; italics mine.

[57] See Professor Glass's article on Maupertuis in this volume.

[58] *Origin of Species*, first edition, p. 377.

species but of different classes or orders. These resemblances in internal structures seemed to him and his evolutionist successors—and precursors—to be obvious and absolutely irrefragable proofs of the descent of supposedly different "species" from common remote ancestors (of the same class or order). These positive "evidences of evolution," then, will now be summarized and the history of their discovery and of their use in the great controversy over the "species-question" be briefly related.[59]

4. *Argument from Rudimentary or Abortive Organs*

Not all the internal structures in adults of existing species are functional. Many of them, Darwin pointed out, are of no use whatever to the organisms in which they are found:

> Organs or parts in this strange condition are extremely numerous throughout nature. For instance, rudimentary mammae are very general in the males of mammals. . . . Nothing can be plainer than that wings were formed for flight, yet in how many insects do we see wings so reduced as to be utterly incapable of flight, and not rarely lying under wing-cases firmly soldered together. . . . The meaning of rudimentary organs is often quite unmistakable. . . . In reflecting on [these facts] every one must be struck with astonishment, for the same reasoning power which tells us plainly that most parts or organs are exquisitely adapted for certain purposes, tells us with equal plainness that these rudimentary or atrophied organs are imperfect and useless. In works on natural history rudimentary organs are generally said to have been created for the sake of 'symmetry,' or in order to complete the scheme of nature; but this seems to me no explanation, but simply a restatement of the fact.

---

[59] These three arguments were usually, not, indeed, confused—but combined and intermingled as if they were essentially the same, since they all are based on internal resemblances and all lead to the same general conclusion. But the resemblances are in one case between organs having different functions, in another between functional organs and those having no functions at all; also, in the first two arguments the resemblances are between structures observable in adults of different existing species, in the third argument the seeming resemblances are between the embryonic forms of still existing genera and the *adult* forms of both existing and extinct genera. In view of these logical differences, and of the fact that the first two arguments have had different historic fortunes from the third, it is essential to distinguish the three sharply from one another, as will be done in the following pages.

As this last sentence shows, the fact that there are rudimentary organs was not a discovery of Darwin's nor a recent discovery; it was a commonplace generally found in current books of natural history, a known fact for which the special-creationists had felt obliged to devise " explanations." Their explanations were like that which Molière's M. Jourdain gave of the soporific effect of opium: all organisms were supposed to be endowed with a *virtus simulativa* —though unlike opium, it had no potential utility for any purpose. The obvious explanation of these useless structures and the only rational explanation, Darwin maintained, was to regard them as instances of the familiar phenomenon of heredity, viz., that characters of ancestors are transmitted to their descendants. They the more plainly exemplified this phenomenon precisely because they *were* useless. They contributed neither to the pleasure of the creatures possessing them, since most such creatures (except modern biologists) were quite unaware that they possessed them; and since they did not function at all, they could not have any survival-value—a convincing argument for evolutionism, and an equally decisive argument again all would-be explanations of the characters of organisms by " final causes," or purposive " pre-arrangements " by the Creator.

But Chambers in the middle eighteen-forties had fully set forth the evidence concerning rudimentary organs and they were for him also evidences of the descent of species from other species.

> An organ will sometimes be seen developed to a certain extent, but wholly without use. This organ will, perhaps, be seen serving a certain use in a particular family of animals; but we advance into an adjoining or kindred family, and there we find a rudiment of the same organ, which, owing to the different conditions of the new set of creatures, is of no kind of service. Thus, some of the serpent tribes possess rudimentary limbs. In other cases, a portion of organization necessary in one sex, e. g. the mammae of the human female, also exist in the male, who has no occasion for them. It might be supposed that in this case there was a regard to uniformity for mere appearance ' sake; but that no such principle is concerned appears from a much more remarkable instance connected with the marsupial animals. The female of that tribe has a process of bone advancing from the pubis for the support of her pouch, and this also appears in the male marsupial, who has

no pouch and requires none. [Chambers apparently did not think this improved the appearance of the male marsupial.] [60]

Since Chambers liked to give an edifying turn to his presentation of scientific facts, rudimentary organs seem to have been a little troublesome for him, for there was nothing especially edifying in the existence of perfectly useless organs. The best that he could say for them, in that vein, was that they "appear but as harmless peculiarities of development, and as interesting evidences of the manner in which the Divine Author has been pleased to work." But it must be borne in mind that, as Chambers constantly insisted, the Author of Creation was pleased to work always through secondary causes acting in accordance with invariable laws. Such "interesting peculiarities of development" must therefore have been considered by Chambers as in accord with his "general hypothesis of the development of the vegetable and animal kingdom,"—and therefore as normally incidental to the gradual development of one species out of another by ordinary processes of reproduction. It was as evidences of the truth of this hypothesis that he adduced so many examples of rudimentary organs. But many of these examples must have been obvious to everyone. You did not need to study biology in order to learn that human males have functionless nipples or that some large birds have wings with which they cannot fly. We now turn to a class of facts which could only become known through the work of specialists in one of the biological sciences.

5. *Argument from the Homologies in the Internal Structures of Organisms of Different Species or Orders; the 'Unity of Type'*

The notable progress made by the science of comparative anatomy in the first half of the nineteenth century made evident a second indication of the common descent of organisms outwardly very dissimilar both in their forms and in their ways of performing certain essential functions, e. g., locomotion. This was especially, though not solely, apparent in the different classes, subclasses, or orders of vertebrates. Most birds and bats have wings which enable them to fly through the air (though in some birds, e. g., ostriches, these members have become merely vestigial), but birds and bats do

---

[60] *Vestiges*, 1857, p. 116. Several other examples of useless organs are cited.

not resemble one another in most of their other externally discernible characters. Fishes propel themselves by means of fins for pushing backward against the water, aquatic birds by means of webbed feet, but no taxonomist had ever placed such birds in the same class with fishes. Among four-limbed land animals, some, e. g., horses, can move forward by trotting (leaping on the foreleg of one side and hind-leg of the other side); others, e. g., kangaroos (not known to Europeans until after Captain Cook's voyage in 1770), move solely by leaping on immensely large hind-legs, the fore-limbs being virtually vestigial, i. e., non-functional. Countless other examples of such almost complete external dissimilarity in the organs used as means of exercising various similar or identical types of function, had long been familiar; no special knowledge of anatomy was needed to discover them. But the dissections of the comparative anatomists showed that the internal structures which operate these external structures, with their extremely diverse forms and modes of functioning, are *not* dissimilar, but, at least within a given class, such as the vertebrates, are generally composed of the same number of parts of the same kinds, arranged in the same manner. For example, the wing by which a bat flies and the arm by which a man throws a ball are moved by the same sort of structure inside those members. This fact—that there is an inner " identity of type " behind most such outward diversities, was chiefly made familiar to the English-reading public by the work of the most eminent comparative anatomist of the period, Richard Owen. It was he, also, who did most to introduce into English usage specific terms to designate the two ways in which organs may resemble one another. The terms were apparently invented by German naturalists early in the century, but Owen appended to his public lectures *On the Invertebrate Animals*, 1843, a Glossary in which the two needed terms were defined as follows:

> '*Analogue*': a part or organ in *one* animal which has the *same function* as *another* part or organ in another animal. '*Homologue*': the *same organ* in *different* animals under *every variety of form and function*.

These definitions were not, in fact, well expressed. For example, the wings of bats and the feet of quadrupeds or bipeds were classed by Owen as homologous, since they had the same general function, viz., locomotion, yet they had utterly different external forms. The significant distinction was that between the external forms and the internal forms employed for the performance of similar, i. e. "analogous," functions. But though the definitions might have been more precisely framed, they served the purpose of emphasizing the fact that species (or orders) which seem to the superficial observer too unlike one another to be descended from common ancestors, prove, when their inward parts are investigated, to show such similarity that community of descent seems the obvious explanation.

Owen, it is true, was not an evolutionist; after 1859 he violently attacked Darwin's doctrine, and sought to vindicate a rival theory of his own—which Darwin said that he had never been able to understand. But he at once saw that Owen had, however unintentionally, presented impressive evidence of "transmutation." "Why," Darwin asked in the *Origin of Species*, "should similar bones have been created in the formation of the wing and leg of a bat, used as they are for such totally different purposes? . . . What can be more curious, than that the hand of a man formed for grasping, that of a mole for digging, the leg of the horse, the paddle of the porpoise and the wing of the bat, should all be constructed on the same pattern, and should include the same bones, in the same relative positions? Geoffroy St. Hilaire has insisted strongly on the high importance of relative connexion in homologous organs: the parts may change to almost any extent in form and size, and yet always remain connected together in the same order. We never find, for instance, the bones of the arm and forearm or of the thigh and leg transposed. Hence, the same names can be given to the homologous bones in widely different animals. . . . Analogous laws govern the construction of the mouths and limbs of crustaceans. . . . Nothing can be more hopeless than to attempt to explain this similarity of pattern in members of the same class by utility, or by the doctrine of final causes." " On the ordinary view of the independent creation of each being, we can only say that so it is;—that it has so pleased the Creator to construct each animal and plant." [61] This, of course, seemed to Darwin no explanation at all; " these facts are inexplicable on the ordinary view of creation." But

the explanation is manifest on the theory of the natural selection of successive slight modifications. . . . In changes of this nature

---

[61] *Origin of Species*, first American edition, N. Y., 1860, pp. 377-80; order of citations somewhat rearranged.

there will be little or no tendency to modify the original pattern, or to transpose parts. The bones of a limb might be shortened and widened to any extent, and become gradually enveloped in thick membrane, so as to serve as a fin; or a webbed foot have all its bones, or certain bones, lengthened to any extent, and the membrane connecting them increased to any extent, so as to serve as a wing; yet in all this great modification there will be no tendency to alter the framework of bones or the relative connexion of the several parts. If we suppose that the ancient progenitor, the archetype, as it may be called, of all mammals, had its limbs constructed on the existing general pattern, for whatever purpose they served, we can at once perceive the plain signification of the homologous construction of the limbs throughout the whole class.[62]

Here, again, Chambers had anticipated Darwin. He wrote in the *Vestiges*: [63]

... while the external features of the various creatures are so different, there has been traced, throughout large groups of them, a *fundamental unity of organization*, as implying, with respect to these groups, that all were constructed upon one plan, though in a series of improvements and variations giving rise to the special forms, and bearing reference to the conditions in which each animal lives. Starting from the primitive germ, which, as we have seen, is the representative of a particular order of full-grown animals, we find all others to be merely advances from that type, with the extension of endowments and modification of forms which are required in each particular case; each form, also, retaining a strong affinity to that which succeeds. ... The ordinary observer is surprised to learn how much further the principle is carried. ... The wing of the bird contains bones representing those of our arm, though modified for so different a purpose. The paddles of the whale tribes and seals are other curious modifications of a member substantially the same. The bat, again, has the bones of its *hand* developed to an unusual extent, so as to become a frame for the membrane by which it flies: in the extinct pterodactyle, the same purpose was chiefly served by a development of the forefinger alone. ... The fundamental resemblance which lurks below various appearances is often startling. Thus, the giraffe, with its long neck, has, in that part, no

---

[62] Ibid., pp. 378-9.
[63] *Vestiges*, 1847, reprint from sixth edition, pp. 113-5.

more bones than are to be found in the neck of the elephant or pig, which hardly seem to have any neck at all. The cervical vertebrae are but seven in every one of the mammalian animals. . . . Unity of organization becomes the more remarkable when we observe that the corresponding organs of animals, while preserving a resemblance, are sometimes put to different uses. For example, the ribs become, in the serpent, organs of locomotion, and the snout is extended, in the elephant, into an instrument serving all the usual purposes of an arm and hand.

Chambers did not, of course, find in the homologies evidence for Darwin's hypothesis of natural selection, with which, naturally, in the eighteen-forties he was not acquainted. His aim was to prove, against the special-creationists, the doctrine of the gradual production of new species by natural causes operating through ordinary processes of reproduction—to vindicate the doctrine of organic evolution, not to propound a universal theory of the nature or *modus operandi* of its causes. It is to be noted, however—since he has sometimes been described as a Lamarckian—that he explicitly rejected the hypothesis of the author of the *Philosophie Zoologique*, concerning whom he wrote:

> Early in this century, M. Lamarck, a naturalist of the highest character, suggested an hypothesis of organic progress which deservedly incurred much ridicule, though it contained a glimpse of the truth. He surmised, and endeavored with a great deal of ingenuity, to prove, that one being advanced in the course of generations to another, in consequence merely of its experience of wants calling for the exercise of its faculties in a particular direction, by which exercise developments of organs took place, ending in variations sufficient to constitute a new species.

"It is possible," said Chambers, "that wants and the exercise of faculties have entered in some manner into the production of the phenomena we have been considering, but certainly not in the way suggested by Lamarck, whose whole notion is obviously so inadequate to account for the rise of the organic kingdoms, that we can only place it with pity among the follies of the wise." Though this dealt rather too harshly with Lamarck, it shows that Chambers had considered the only hypothesis which he knew on the problem to which Darwin was to offer an alternative solution, and had found it unconvincing for a valid reason. But in truth neither the Lamarckian nor the Darwinian hypothesis was pertinent to the question which to Chambers seemed primary; he—like Spencer—held that the main matter was to prove that the theory of natural descent of many diverse species from common ancestors was a *fact*, whatever the specific processes of the production and perpetuation of variations. And in the homologies, he maintained, were plain evidences of that fact, because they could not be explained by any other hypothesis.

6. *Argument from Comparative Embryology: the Theory of Recapitulation*

In the two decades after 1859, the best-known, most aggressive, and most influential elaborator and disseminator of this argument—and one of the earliest German champions of Darwinism—was Ernst Haeckel, in writings published in 1866 and 1875.[64] But I shall, to illustrate the continuing currency and general acceptance of the argument at the end of the century and beyond, quote from a later and widely read popular exposition of it by a distinguished American naturalist, whose severe strictures on Chambers's "premature" adoption of evolutionism have already been mentioned:

> It is a curious and most significant fact that the successive stages of the development of the *individual* in the higher forms of any group (ontogenic series) resemble the stages of increasing complexity of differentiated structure in ascending the animal scale in that group (taxonomic series), and especially the forms and structure of animals of that group in successive geological epochs (phylogenic series). In other words, the individual higher animal in embryonic development passes through temporary stages, which are similar in many respects to permanent or mature conditions in some of the lower forms in the same group. . . . Surely this fact, if it be a fact, is wholly inexplicable except by the theory of derivation or evolution. The embryo of a higher animal of any group passes *now* through stages represented by lower forms, because in its evolution (phylogeny) its ancestors *did actually have these forms*. From this point of view the ontogenic series (individual history) is a brief recapitulation, as it were, from memory, of the main points of the phylogenic series, or family history.[65]

---

[64] *Generelle Morpologie der Tiere*, 1866, and *Ziele und Wege der Entwicklungsgeschichte der Tiere*, 1875, English tr. of the latter by E. Ray Lankester, *The History of Creation*, 2 vols., London, 1876.
[65] LeConte, op. cit., 1888, pp. 130-31; italics in original. The passage is

This theorem—that "the laws of embryonic development are also the laws of geologic succession"—LeConte represented as "added" to biology by Agassiz, and he regarded it as "the only solid foundation of a true theory of evolution." It was for this reason that he declared that no one was reasonably entitled to believe in the transformation of species before the publication of the work of Agassiz, and hence that Chambers's evolutionism was a "baseless speculation."[66] But LeConte was misled by piety towards the memory of his greatest teacher into a serious neglect of chronology, in a matter in which chronology is of the essence of the question at issue. Even if Agassiz could be regarded as the originator of the doctrine of recapitulation, it must be remembered that he announced his evidences for it in his *Poissons du vieux grès rouge*, 1842-44, and repeated them in popular form in his Lowell Lectures of 1848.[67] And in point of fact, the doctrine, and what seemed an important mass of evidence for it, had then long been familiar, so that one finds Lyell in 1830 arguing against the use of it as a proof of evolution. In the *Principles of Geology* he wrote:

> There is yet another department of anatomical discovery to which I must allude, because it has appeared, to some persons to afford a distant analogy, at least, to that progressive development by which some of the inferior animals may have gradually been perfected into those of more complex organization. Tiedemann found, and his discoveries have been most fully confirmed and elucidated by M. Serres, that the brain of the foetus assumes, in succession, forms analogous to those which belong to fishes, birds, and reptiles before it acquires the additions and modifications peculiar to the mammiferous tribe. So that in the passage from the embryo to the perfect mammifer, there is a typical representation, as it were, of all those transformations which the primitive species are supposed to have undergone, during a long series of generations, between the present period and the remotest geological era.

Lyell's reply to this argument was brief and dogmatic: he fully admitted the facts, but denied the inference.

---

followed by thirty pages of "proof," with pictorial illustrations, of embryonic recapitulation. A second edition of LeConte's book was published in 1905.

[66] LeConte, op. cit., ch. II: "The Relation of Louis Agassiz to the Theory of Evolution."

[67] Cf. Agassiz's own words, cited in Marcou, *Louis Agassiz*, I, p. 230.

It will be observed that these curious phenomena disclose, in a highly interesting manner, the unity of plan that runs through the organization of the whole series of vertebrated animals; but they lend no support whatever to the notion of a gradual transmutation of one species into another; least of all, of the passage, in the course of many generations, from an animal of a more simple to one of a more complex structure.[68]

To the mind of Darwin these "curious phenomena" had a very different import. The study of individual "development and embryology" he considered "one of the most important subjects in the whole round of natural history," and in the penultimate chapter of the *Origin of Species* he carefully examined the state of the evidence then available respecting both the resemblances and dissimilarities between the temporary embryonic forms of "higher" species and the adult forms of "lower" species. In some organisms, he pointed out, such resemblances are not found, while in many others they are. In the latter cases, he emphasized, the transitory structures in the embryo "are of no service to it, either at that or at a later period of life." To use an example which Darwin might have used in the *Origin* but did not probably because he was reserving it for *The Descent of Man*—why does the human embryo in its second month sprout a tail which is of no conceivable use to it in that phase of its existence and is sloughed off before birth? A fact of this sort was wholly unintelligible and, indeed, absurd, from the special-creationist's point of view; it seemed to take on meaning when interpreted as indicating that the embryo is a remote descendant of ancestors that had tails—and doubtless had a use for them. The cases in which such detailed morphological resemblances between the embryos of one class or order and the adult or foetal forms of another are not discernible, Darwin explained by the possibility that "in one of [the] two groups the developmental stages may have been suppressed, or may have been so greatly modified through adaptation to new habits of life as to be no longer recognizable." Thus in spite of the existence of some such (so to say) non-conformist groups, the facts of comparative embryology, "second to none in importance," seemed to Darwin to furnish a cogent proof of evolution:

[68] First American edition, 1837, I, p. 526. Serres' work to which Lyell refers had been published as early as 1824.

In two groups of animals, however much they may at present differ from one another in structure and habits in their adult condition, if they pass through closely similar embryonic stages, we may feel assured that they all are descended from one parent form, and are therefore closely related. Thus, community in embryonic structure reveals community of descent; but dissimilarity in embryonic structure does not prove discommunity of descent.... As the embryo often shows, more or less plainly, the structure of the less modified and ancient progenitor of the group, we can see why ancient and extinct forms so often resemble in their adult state the embryos of existing species of the same class.... Embryology rises greatly in interest when we look at the embryo as a picture, more or less obscured, of the progenitor, either in its adult or larval state, of all the members of the same great class.[69]

Yet poor Chambers was reproached for "baseless speculation" because, looking upon facts generally admitted by naturalists of his time, he saw in them precisely the meaning that Darwin saw, but had yet to set forth in print. In the third edition of the *Vestiges*, 1845, Chambers wrote:

First surmised by the illustrious Harvey, afterwards illustrated by Hunter in his wondrous collection at the Royal College of Surgeons, finally advanced to mature conclusions by Tiedemann, St. Hilaire and Serres, embryotic development is now a science. Its primary positions are ... (2) that the embryos of all animals pass through a series of phases of development, each of which is the type or analogue of the permanent configuration of tribes inferior to it in the scale.

And in this Chambers found one of his chief evidences of the transformation of species. Elsewhere in this edition he devoted several pages to the elaboration of the argument from recapitulation in the case of the brain. "Taking as a basis the scale of animated nature as presented in Dr. Fletcher's 'Rudiments of Physiology,'" he pointed out "the wonderful parity observed in the progress of creation, as presented to our observation in the foetal progress of one of the principal human organs."

A few years after Chambers, the Scottish anatomist Robert Knox adduced the same facts of human embryogeny as evidence for the

---

[69] *Origin of Species*, first American edition, N. Y., 1860, p. 391.

same evolutionistic conclusion, of which, in the words of Baden Powell, he was "one of the most sedulous supporters."

> We discover structures in the human embryo, or of any other animal, not persistent but transitory; ... we see that the individual is in fact ... passing through forms which represent the permanent forms of other adult beings belonging to the organic world, not human but bestial, of which some are extinct, some belong to the existing world, while others may represent forms which once existed but are now extinct; or finally forms which are some day to appear, running through their destined course, then to perish as [did] their predecessors. The fully developed, or grown-up, brute forms of birds and fishes, or reptiles and mammals, are represented in the organic structures of the human embryo.[70]

In 1855 Baden Powell included the same phenomena of embryonic recapitulation in his review of "the evidence derived from physiology" for the "probability" of the "idea of the transmutation of species." To quote here his long résumé of the familiar argument would be redundant.[71]

The theory of embryonic recapitulation had by 1850 even found expression in verse—and from no less a writer than the newly appointed Poet Laureate. Tennyson not only knew but was deeply influenced by the conception, and introduced a summary reference to it into the epithalamium with which his long elegy concludes: the child to be born of the coming marriage will first be

> moved through life of lower phase,
> Result in man, be born and think,
> And act and love. . . .

The early Victorians may have "never talked obstetrics when the little stranger came," but their favorite poet did, briefly, talk evolu-

---

[70] *The Races of Men*, 1850, p. 29. In suggesting that the human embryo not only records the forms of its distant forebears but may also prophesy those of its posterity, Knox fell into a characteristic exaggeration; this part of his statement was obviously incapable of present verification and could not provide any support for the theory of descent.

[71] *Essays on the Spirit of the Inductive Philosophy, the Unity of Worlds, and the Philosophy of Creation*, London, 1855, pp. 341-72. Baden Powell also vigorously pressed home (earlier than Huxley) the argument that Lyell's "uniformitarian principle" logically entailed "the theory of the transmutation of species through natural causes."

tionist embryology on the presumable eve of the little stranger's conception.[72]

It is worth while, perhaps, before concluding, to bring the last six arguments together, in a single general view of their logical bearings. No one of them, nor all of them collectively, ever amounted to more than "circumstantial evidence" of the transformation of species; none of them actually exhibited any existing species in *flagrante delicto* of transmutation. These arguments owed their force to the fact that, when taken together, they fitted with striking nicety into the requirements of one of the only two possible hypotheses about the origin of species—a hypothesis already recommended on general grounds of scientific method; while they reduced the rival hypothesis to a grotesque absurdity. "Conceivable" that other hypothesis still remained, as Huxley had once contended. It was, and is, possible, by making a sufficient number of supplementary suppositions, to give to the special-creation doctrine a form in which it is neither explicitly self-contradictory nor explicitly contradictory of any fact established by direct observation. But when thus fitted out with the supplements required by the facts already known to the science of 1840, the doctrine certainly had a singularly odd appearance. It implied that the Creator had produced the different types of organisms by fits and starts, strewing them at irregular intervals along the vast reaches of geological time. Precisely what happened on one of these interesting occasions, the hypothesis left in a bafflling obscurity; after a somewhat extensive

---

[72] *In Memoriam* (Tennyson's Works, 1893 Ed., p. 259). This was certainly not a conventional theme in epithalamia, and it probably shocked some of the contemporary public. Tennyson's use of it is for that reason the more note-worthy—since he was not a writer who sought to gain literary effect by shocking his readers. But the theme was peculiarly suited to his purpose; it led on to, and was of a piece with, the optimistic evolutionism that follows in the poem; when the child's prenatal development, which is a recapitulation of the past ascent of life, "results in man," the new-born human being will be "a closer link Betwixt us and the crowning race Of those that eye to eye, shall look On knowledge, under whose command Is earth and earth's," etc. The poet has now escaped from the terrifying outlook on the future of the human species expressed in Canto VI. Huxley wrote after Tennyson's death that "he was the only modern poet, perhaps the only poet since Lucretius, who has taken the trouble to understand the work and methods of men of science." The poet, in fact, in 1850 saw the biological implications of Lyell's geology better than Huxley did at that time.

reading in the relevant literature of the period, I can not recall that any special-creationist replied to Spencer's request for particulars on this point. Spencer wrote in 1852:

> Let them tell us how a new species is constructed and how it makes its appearance. Is it thrown down from the clouds? or must we hold to the notion that it struggles up out of the ground? Do its limbs and viscera rush together from all points of the compass? Or must we receive the old Hebrew idea that God takes clay and molds a new creature?

On these matters the theory remained judiciously non-committal. But it maintained, at all events, that the majority of species, however created, were destined to be in turn destroyed—and destroyed by the operation of natural forces. The Great Artificer could fashion, but he was either unable or unwilling to protect, the creatures his imagination had devised. When ordinary physical processes were too much for them, sweeping them off by groups, or even, according to one variant of the theory, obliterating them altogether, he was obliged to start afresh; whether this happened four or twelve or twenty-seven or thirty thousand times was a detail about which the partisans of the doctrine could not agree. The forms later produced did not always differ markedly for the better from their unfortunate precursors; many primitive and rather unsuccessful models continued to be repeated. But in general, as time went on, the Creator brought both more diverse and more complicated beings into existence. In doing so, he behaved after the manner of a lazy and incompetent architect, who, instead of "studying" each problem afresh, with reference to the special uses and situation of the edifice to be erected, is content to make a few minor alterations in a single conventionalized plan. The "unity of type" of organisms destined to the most dissimilar modes of existence was generally dilated upon with devout enthusiasm by the special creationists. They seem to have regarded it as an agreeable mannerism of the Creator's personal style. But it was the kind of mannerism which, in a human designer, is commonly ascribed to indolence or limited intelligence. Indeed, the parallel of the lazy architect was inadequate to represent the whole singularity of the Creator's mode of construction. He not only used as few general models as possible, but he also—when, with a cleared field, he created a fresh group of organisms—reproduced in them organs and members which had been functional and useful in their predecessors, but were in the new species useless, meaningless, and even disadvantageous—like the proverbial Chinese tailor, who laboriously imitated all the rents and stains in the discarded European garment given him as a model. Finally, the Creator was supposed to have implanted in all organisms the senseless habit of mimicking, in the early stages of the individual's development, the forms of other and extinct organisms to which that individual bore no relation of kinship.

Such—with the details absolutely required by the accepted scientific knowledge of the time—was the hypothesis tenaciously held by most men of science for at least twenty years before 1859. With the greater number of them the motives for holding it were primarily theological; yet the thing that now impresses us in the theory is its extraordinarily irreligious character. Science might conceivably, after some fashion, have made shift with a hypothesis of this kind; but it is hard to see how any one could suppose it in any degree advantageous to religion. It had not even the poor merit of being anthropomorphic. For no man outside of a madhouse ever behaved in such a manner as that in which, by this hypothesis, the Creator of the universe was supposed to have behaved. Ascribing to him both the ability and the disposition to intervene with absolute freedom in natural—or at least in organic—phenomena, the theory also represented him as incapable of intervening intelligently or effectually.

That men of great abilities were unable to see the true character of the hypothesis which great numbers of them so long embraced, is certainly an interesting, if not an encouraging, fact in the history of the human intellect. But the capacity of theological prepossessions and religious feeling to retard and confuse intellectual processes is an old story. More remarkable, perhaps, is the failure, for an equally long period, of a number of men *not* impeded by theological prepossessions—men who were capable of seeing the absurdities of the special-creation hypothesis—to recognize the methodological superiority and the promise of scientific fruitfulness inhering in the other hypothesis, or even to recognize the logical obligation to choose between the only two hypotheses available. Men of science of the present generation have perhaps little to learn from a consideration of the reasons which prevented a Cuvier, a Miller, a Sedgwick or an Agassiz from accepting the theory of evolution. But there may still be for us profitable matter for reflection in a consideration of the reasons which prevented a Huxley from finding, in

1846, anything of value in facts and reasonings which fifteen years later he was, with unequalled vigor and skill, proclaiming from the housetops.

The only historian of English thought known to me who has quite truly stated what I believe to be the fact about this episode in the history of scientific opinion, is Mr. A. W. Benn. In his *Modern England* (1878) he observed concerning the *Vestiges*:

> Hardly any advance has since been made on Chambers' general arguments, which at the time they appeared would have been accepted as convincing, but for theological truculence and scientific timidity. And Chambers himself only gave unity to thoughts already in wide circulation. . . . Chambers was not a scientific expert, nor altogether an original thinker, but he had studied scientific literature to better purpose than any professor. . . . The considerations that now recommend evolution to popular audiences are no other than those urged in the " Vestiges." [73]

The truth of this is, I think, by no means sufficiently recognized by biologists or by historians of science. I hope that the present study may somewhat contribute to the more general acknowledgment of the correctness of Mr. Benn's statement.

---

[73] Vol. II, 307, I, p. 238. Readers will note that Mr. Benn's statement referred only to the situation in the late 1870's. If applied to the 1950's it would, as will be shown in a subsequent paper, require some qualification.—After the above paper was ready for the press, Professor George Gaylord Simpson, distinguished not only as a biologist but as an analyst of the history of the science, has in a generally excellent review of the new (and unexpurgated) edition of Darwin's autobiography, asserted (*Scientific American*, August, 1958, p. 120) that "to say that Erasmus Darwin or Buffon or Maupertuis anticipated Charles Darwin is not much truer than the absurd claim that science fiction writers invented the atomic bomb." Unfortunately, Simpson here rejects the testimony of Charles Darwin himself that "innumerable facts were in the minds of naturalists . . . ready to take their proper place as soon as any theory which would receive them was proposed." Simpson also ignores the fact, shown in the above paper, that the evolutionist theory had been shown, before Darwin's masterpiece was published, to be the only theory which could be reconciled with those facts. And the evidence of the descent of extant species from other and far more ancient species required to be factually established *before* any verifiable hypothesis about its causes could be invented or appraised. Darwin knew this; he never claimed to have originated the theory of organic evolution, and he found his factual evidences for it almost wholly in the investigations of earlier biologists. But he did claim to have proposed the only adequate general *explanation* of the causes or *modus operandi* of the transformation of species. In this claim he was largely justified; but he was neither the first to present the doctrine of organic evolution through natural causes, nor the first to present the proofs of that doctrine.

FOURTEEN

SCHOPENHAUER AS AN EVOLUTIONIST *

ARTHUR O. LOVEJOY

The Absolute of the philosophy of Schopenhauer is notoriously one of the most complicated of all known products of metaphysical synthesis. Under the single, and in some cases highly inappropriate, name of "the Will" are merged into an ostensible identity conceptions of the most various character and the most diverse historic antecedents. The more important ingredients of the compound may fairly easily be enumerated. The Will is, in the first place, the Kantian " thing-in-itself," the residuum which is left after the object of knowledge has been robbed of all of the " subjective " forms of time and space and relatedness. It is also the Ātman of the Vedantic monism, the entity which is describable solely in negative predicates, though at the same time it is declared to sum up all of the genuine " reality " that there is in this rich and highly colored world of our actual (though illusory) experience. The Will is, again, the " Nature " of Goethe; it is the " vital force " of the late eighteenth and early nineteenth century vitalists in biology; and it is even the physical body of man and animals, in contrast with the mind. It is likewise the absolutely alogical element in reality, the " non-rational residuum," of one period of Schelling's philosophy; and it is an apotheosis of that instinctive, naive, spontaneous, unreflective element in human nature, which had sometimes been glorified by Rousseau and, in certain of his moods, by Herder. It is Spinoza's " striving of each thing *in suo esse perseverare.*" It is the insatiable thirst for continued existence which the Buddhist psychology con-

* First published in *The Monist*, xxi, 1911.

ceives as the ultimate power that keeps the wheel of existence in motion, and it is a hypostasis of the Nirvana in which Buddhism conceives that thirst to be extinguished.

Though thus singularly manifold, these elements are not all necessarily incongruous *inter se*. But, apart from minor discrepancies among them, they all fall into at least two groups, having attributes which obviously cannot be harmonized as characterizations of one and the same entity. The Will, in Schopenhauer, has manifestly a positive and a negative aspect; it is thought of now in concepts to which the term " Will " is truly pertinent, now in concepts to which that term is singularly unsuitable. In so far as the " Will " is a designation for the thing-in-itself, or for the Vedantic Absolute, it is a being which is not only itself alien to time and to space and to all the modes of relation, unknowable, ineffable, but is also *ipso facto* incapable of accounting for, or of being manifested in, a world of manifold, individuated, striving and struggling concrete existences. It is merely the dark background of the world of experience; it is the One which remains while the many change and pass. From the point of view of the world of the many and of change, it is literally, as Schopenhauer said, " nothing." To the understanding it is necessarily as inaccessible, and, indeed, as self-contradictory and meaningless, as is the Unknowable of Herbert Spencer—of which it is, indeed, the twin brother, not to say the identical self. This kind of negative and inexpressible Absolute is a sufficiently familiar figure in the philosophy of all periods. Schopenhauer assuredly did nothing original in reviving it. What was original in his work was that he baptized this Absolute with a new, and startlingly inappropriate, name; and that he gave it this name because, in spite of himself, he was really interested in quite another kind of " ultimate reality " of which the name was genuinely descriptive.

The other aspect of Schopenhauer's " Will " is, of course, that in which it appears, as Spencer's Unknowable intermittently appears, as a real agency or tendency in the temporal world, as a power which is not merely behind phenomena, but also is manifested *in* phenomena; and, more especially, as a blind urge towards activity, towards change, towards individuation, towards the multiplication of separate entities—each of them instinctively affirmative of its own individual existence and also of the character of its kind—towards the diversification of the modes of concrete existence, and towards a struggle for survival between these modes. When Schopenhauer speaks of the Will as a *Wille zum Leben*, it is sufficiently manifest that what he has before his mind is not in the least like the Oriental Brahm, " which is without qualities " and without relations and without change. It is, of course, true that Schopenhauer imagined that he had mitigated the baldness of the incongruity between the two aspects of the Will by calling the one reality and the other mere phenomenon, by insisting that the first sort of characterization tells us, so far as human language can, what the Will is in itself, while the second form refers only to the false appearance which the Will presents when apprehended by the Understanding. But, as a matter of fact, it is quite clear that the characteristics of the world of phenomena, as Schopenhauer habitually thinks of it, are explicable much more largely by the nature of the Will than by the nature of the Understanding. Schopenhauer is fond of reiterating, for example, that space and time constitute the *principium individuationis*; but they are so only in the sense that they provide a means for logically defining individuality. It is very apparent that there is nothing in the abstract notion of either space or time which can explain why that pressure towards individation, that tendency towards the multiplication of concrete conscious individuals, should exist. It is, after all, the cosmic " Will " that must be conceived to be responsible for its own " objectification " in a temporal and spatial universe; for, even from Schopenhauer's own point of view, there is nothing in the conception of the forms under which the Will gets objectified which can account for the necessity of such objectification. It was with the Will in its concrete sense, and in its restless, temporal movement, that Schopenhauer was more characteristically concerned; it was the ubiquity and fundamental significance of this trait of all existence which constituted his personal and relatively novel *aperçu*.

Now the conception of the Will as a force or tendency at work in the world of phenomena is manifestly a conception which might have been expected to lead the author of it into an evolutionistic type of philosophy. Since the Will is characterized as *ein endloses Streben*, as *ein ewiges Werden*, as *ein endloser Fluss*, and since we are told of it that " every goal which it reaches is but the starting point for a new course," its manifestations or products might, it would seem, most naturally be represented as appearing in a gradual, progressive, cumulative order. The phrase " will to live " readily, if not inevitably, suggests a steady movement from less life to more life and fuller, from lower and less adequate to higher and more adequate grades of objectification. But did Schopenhauer in fact construe his own fundamental conception in this way? An examination of his writings with this question in view makes it appear probable that at the beginning of his speculative activity he did not put an evolutionistic construction upon the conception of the Will; but it makes it clear that in his later writings he quite explicitly and emphatically adopted such a construction, connecting with his metaphysical principles a thorough-going scheme of cosmic and organic evolution. Singularly enough, this significant change in Schopenhauer's doctrine upon a very fundamental point has, so far as I know, not hitherto been fully set forth. Not only the most widely read histories of philosophy, but even special treatises on Schopenhauer's system, represent his attitude towards evolutionism wholly in the light of his early utterances; and even where his later expressions upon the subject are not forgotten, their plain import has often been denied, upon the assumption that they must somehow be made to harmonize with the position taken in his early and most famous treatise.

In *Die Welt als Wille und Vorstellung* Schopenhauer is preoccupied chiefly with the negative and "other-worldly" aspect of his philosophy. His emphasis may, upon the whole, be said to be laid upon the consideration that the world of objects is but an illusory presentation of the Will, rather than upon the consideration that the Will is, after all, the kind of entity that presents itself in the guise of a world of objects and of minds. With this preoccupation, Schopenhauer delights to dwell upon the timelessness of the true nature of the Will. Yet, since even in his most mystical and nihilistic moments he is obliged to remember that the Absolute does somehow take upon itself a temporal form, this emphasis upon the eternity of true being did not of itself forbid his representing the temporal side of things as a gradual process of expansion and diversification. The passages in which Schopenhauer speaks of the timelessness of the Will ought not to be quoted, as they sometimes have been quoted, as constituting in themselves any negation of a developmental conception of the world in time; for such passages are not pertinent to the world in time at all. It is rather a subsidiary and somewhat arbitrary detail of his system, which he uncritically took over from Schelling, that leads Schopenhauer in this period to pronounce in favor of the constancy of organic species. Between the Will as a timeless unity and the changing world of manifold phenomena he interpolates a world of Platonic Ideas, or archetypal essences of phenomena. This world, it is true, has only an ideal existence; it has, in a sense, not even the degree of reality that phenomenal objects have. But it has an important functional place in Schopenhauer's scheme of doctrine; since the Ideas, so to say, lay down the limits of diversity within which the phenomena may vary. Each individual being is in some degree different from every other, and the name of them is legion. But the generic forms, the *kinds* of individuals that there may be, are determined by the natures of the Ideas.

Now these Ideas relate primarily to the kinds of natural processes which Schopenhauer regards as the hierarchically ordered grades of the objectification of the Will—mechanism, chemism, organism, etc. But it is evident that Schopenhauer also includes among the Ideas the timeless archetypes of each species of organism. Even from the fact that, upon Schopenhauerian principles, the pure form of each species is eternal, as it behooves a Platonic Idea to be, it could not necessarily be inferred by any cogent logic that the temporal copies of these forms need be changeless. Schopenhauer none the less does appear to draw, in a somewhat arbitrary manner, the inference that species must be everlasting and immutable. He writes, in the Supplement to the third book of *Die Welt als Wille und Vorstellung* (second edition, 1844):

> That which, regarded as pure form, and therefore as lifted out of all time and all relations as the Platonic Idea is, is, when taken empirically and as in time, the species; thus the species is the empirical correlate of the Idea. The Idea is, in the strict sense, eternal, while the species is merely everlasting (*die Idee ist eigentlich ewig, die Art aber von unendlicher Dauer*), although the manifestation of a species may become extinct upon any one planet.

So again (in the chapter on "The Life of the Species," ibid., chapter 42) Schopenhauer writes:

> This desire [of the individuals of a species to maintain and perpetuate the characteristic form of their species], regarded from without and under the form of time, shows itself in the maintenance of that same animal form throughout infinite time (*als solche Tiergestalt eine endlose Zeit hindurch erhalten*) by means

of the continual replacement of each individual of that species by another;—shows itself, in other words, in that alternation of death and birth which, so regarded, seems only the pulse-beat of that form (εἶδος, ἰδέα, *species*) which remains constant throughout all time (*jener durch alle Zeit beharrenden Gestalt*).

These passages seem to be clear in their affirmation of the essential invariability of species.

In *Der Wille in der Natur*, in 1854,[1] we find Schopenhauer passing a partly unfavorable criticism upon Lamarck, which at first sight undeniably reads as if he at that date still retained the non-evolutionistic position of his earlier treatise. He has been asserting that the adaptive characters of organisms are to be explained neither by design on the part of a creative artificer, nor yet by the mere shaping of the organism by its environment, but rather through the will or inner tendency of the organism, which somehow causes it to have the organs which it requires in order to cope with its environment. "The animal's structure has been determined by the mode of life which the animal desired to find its sustenance and not *vice versa*. . . . The huntsman does not aim at the wild boar because he happens to have a rifle: he took the rifle with him, and not a fowling piece, because he intended to hunt boars; and the ox does not butt because it happens to have horns, it has horns because it intends to butt." This, of course, sounds very much like a bit of purely Lamarckian biology; and Schopenhauer is not unmindful of the similarity.

> This truth forces itself upon thoughtful zoologists and anatomists with such cogency that, unless their mind is purified by a deeper philosophy, it may lead them into strange error. Now this actually happened to a very eminent zoologist, the immortal De Lamarck, who has acquired undying fame by his discovery of the classification of animals into vertebrates and invertebrates, so admirable in profundity; for he quite seriously maintains and tries to prove at length that the shape of each animal species, the weapons peculiar to it, and its organs of every sort adapted for outward use, were by no means present at the origin of that species, but have, on the contrary, come into being gradually *in the course of time* and through continued generation, in consequence of the exertions of the animal's will, evoked by the nature of its situation

---

[1] This is the date of the second edition. The first edition appeared in 1836; to it I have not been able to have access.

and environment,—i. e., through its own repeated efforts and the habits to which these gave rise.

Schopenhauer then goes on to urge certain purely biological objections, which may for the moment be passed over, to what he conceives to be the Lamarckian hypothesis. The most serious misconception on Lamarck's part, however, he declares to arise from an incapacity for metaphysical insight, due to the unfortunate circumstance that that naturalist was a Frenchman.

De Lamarck's hypothesis arose out of a very correct and profound view of nature; it is an error of genius, which, in spite of all its absurdity, yet does honor to its originator. The true part of it should be set down to the credit of Lamarck himself, as a scientific inquirer; he saw rightly that the primary element which has determined the animal's organization is the will of the animal itself. The false part of it must be laid to the account of the backward state of metaphysics in France, where the views of Locke and his feeble follower, Condillac, still hold their ground, and where, accordingly, bodies are supposed to be things in themselves, and where the great doctrine of the ideality of space and time and of all that is represented in them . . . has not yet penetrated. De Lamarck, therefore, could not conceive his construction of living beings otherwise than as in time and succession. . . . The thought could not occur to him that the animal's will, as a thing in itself, might lie outside time, and in that sense be prior to the animal itself. Therefore he assumes the animal to have first been without any clearly defined organs, and indeed without any clearly defined tendencies, and to have been equipped only with perceptions. . . . But this primary animal is, in truth, the Will to Live; as such, however, it is metaphysical, not physical. Most certainly the shape and organization of each animal species has been determined by its own will according to the circumstances in which it needed to live; not, however, as a thing physical, in time, but on the contrary as a thing metaphysical, out of time.

As it stands, this passage, apart from its context, is no doubt most naturally interpreted as a rejection, not merely of the details of Lamarck's hypothesis, but also of the general doctrine of a gradual transformation of species in time. Its import has been so understood by a number of expositors of Schopenhauer. Thus Kuno Fischer writes: " Schopenhauer blames De Lamarck for representing animal species as evolved through a genetic and historical process, instead of conceiving of them after the Platonic manner."[2] So Rádl[3]: "Schopenhauer speaks in praise only of the Lamarckian doctrine that the will is the cause of organic forms; Lamarck's genetic philosophy, on the other hand, he rejects." But these writers have neglected to observe that, only a few pages later in the same treatise, Schopenhauer sets down an unequivocal though brief affirmation of the origination of species from one another through descent; and does so on the ground that without such a hypothesis the unity of plan manifest in the skeletal structure of great numbers of diverse species would remain unintelligible. In other words, Schopenhauer argues in favor of transformism by pointing to one of the most familiar evidences of the truth of the theory of descent, viz., the homologies in the inner structure of all the vertebrates. In the neck of the giraffe, for example, (he remarks) we find, prodigiously elongated, the same number of vertebrae as we find in the neck of the mole contracted so as to be scarcely recognizable. This unity of plan, argues Schopenhauer, requires to be accounted for; and it can *not* be accounted for as one of the aspects of the general adaptation of organisms to their environment. For that adaptation might in many cases have been as well, or better, realized by means of a greater diversity in the architectural schemes of species having diverse environments and instincts.

> This *common anatomical feature* (*Element*) which, as has been already mentioned, remains constant and unchangeable, is so far an enigma,—namely, in that it does not come within the teleological explanation, which only begins after that basis is assumed. For in many cases a given organ might have been equally well adapted to its purpose even with a different number and arrangement of bones. . . . We must assume, therefore, that this common anatomical feature is due, partly to the fact that the original forms of the various animals have arisen one out of another (*dass die Urformen der Tiere eine aus der andern hervorgegangen sind*), and that it is for this reason that the fundamental type of the whole line of descent (*Stamm*) has been preserved.[4]

And Schopenhauer himself adds a reference to a passage in the *Parerga und Paralipomena*[5] (to be examined below) in which, at

---

[2] *Arthur Schopenhauer*, 1893, p. 463.
[3] *Geschichte der biologischen Theorien*, II, 465 n.
[4] *Der Wille in der Natur*, 3rd ed., 1878, p. 53.
[5] To #91 of the first edition, 1851 (= #93 of the second edition).

much greater length, his own particular form of organic evolutionism is expounded.

Now, abundant in contradictions though Schopenhauer was, it is difficult to suppose that he can have expressed, within half a dozen pages, diametrically opposed views upon a perfectly definite and concrete question of natural science, in which he manifestly took an especial interest—and that he can, in spite of his habit of carefully revising each edition of his works, have left such a piece of obvious self-contradiction standing in the final version of *Der Wille in der Natur*. If, now, bearing this in mind, we revert to the criticism of Lamarck, which has not unnaturally misled hasty readers of Schopenhauer, we shall see that what is criticized is *not* the doctrine of the derivation of species from earlier species by descent, but only a specific theory of the manner in which " the Will " works in the formation of species. Lamarck, at least as Schopenhauer understood him, placed behind every organ or function of all animals, as its cause and temporal antecedent, a *felt need*, a conscious desire, leading it to the activities by means of which that organ is developed. To this Schopenhauer objects, in the first place, that the hypothesis implies that if we should go back to the beginning of the series of animals we should come to a time in which the ancestor of all the animals existed *without any organs or functions at all*, in the form of a *mere* need, a desire pure and simple;—which implication he regards as reducing the hypothesis to an absurdity. This is an entirely pertinent criticism upon Lamarck's explanation of specific characters as the results of use and disuse of organs, in so far as that explanation is taken as the sole explanation. The criticism applies, not only to the origination of animal organs and functions in general, but also to the origination of any particular class of organs and functions. It is difficult to see how an animal, yearn it ever so strongly, can develop an organ out of its needs merely as such; or how it can modify by use or disuse a type of organ of which it is not yet in possession. Given the rudiments of an eye, with a specific visual sensibility, and it is at least conceivable that the persistent utilization of such a rudimentary organ might somehow lead to its further development; but some sort of eye must necessarily first be given. In other words, Lamarckianism (as apprehended by Schopenhauer) did not sufficiently recognize that the primary thing in species-forming must be the appearance (through obscure embryogenetic processes with which conscious needs and desires can have nothing to do) of suitable congenital variations. The essence of Lamarck's error, as Schopenhauer sees it, is that, according to the French naturalist, "it is the will which arises out of knowledge," i. e., out of the animal's temporally antecedent consciousness of its own need; whereas, in fact, "the will did not proceed from the intellect, nor did the intellect exist, together with the animal, before the will made its appearance." We cannot even say that the will, in the sense of a definite concrete volition, existed before the production of the organ requisite to make the fulfilment of the given kind of volition possible in an animal species. In short, Schopenhauer's doctrine was that the timeless Will, working in time in the form of a blind purposiveness, gives rise to the organs and the potencies of new species by producing new congenital characters *before* any felt need for and endeavor after those characters have arisen; while Lamarck's doctrine, as Schopenhauer believed, was that an actual (though doubtless vague) awareness of need, and a concrete movement of conation, temporally precede the production of each new character or organ. The two doctrines were really distinct; but (as will presently more fully appear) the one was as definitely evolutionistic as the other.

It was, furthermore, an objection in Schopenhauer's eyes to Lamarck's theory (and would have doubtless been urged by him as an objection to the Darwinian theory) that it supposed species to have been formed by the gradual enlargement and accumulation of characters too small and trivial at their first emergence to be functionally significant, or useful in the struggle for survival. He says,

> Lamarck overlooks the obvious objection . . . that, long before the organs necessary for an animal's preservation could have been produced by such endeavors as these carried on through countless generations, the whole species must have died out from the want of them.

Schopenhauer, after his definite adoption of evolutionism, always insisted not only upon the primacy of the fact of variation in the explanation both of species-form and of adaptation, but also upon the doctrine that, though one species descends from another, it descends *ready-made*. In other words—and in the twentieth-century words—Schopenhauer was, in his view concerning species, a mutationist, though one of a somewhat extreme and peculiar sort.

In interpreting the bearing of Schopenhauer's comments on Lamarck in *The Will in Nature* I have, of course, been guided not only by the context of that passage, but also by the passage in the *Parerga und Paralipomena* to which, as has been mentioned, he himself refers his reader for a fuller exposition of his views on the question of species. The latter passage occurs in the small treatise (Chapter VI of *Parerga und Paralipomena*) entitled *Zur Philosophie und Wissenschaft der Natur*, perhaps the most important of its author's later writings, but one which has been amazingly neglected by the historians of philosophy and even by writers of special monographs on Schopenhauer. With the publication of this work (1850)[6] he quite unmistakably announced—what remained his final view— that the philosophy of nature to which his metaphysics of the Will properly led was of a frankly and completely evolutionistic type. Since this part of the *Parerga und Paralipomena* (unlike most of the rest of that collection) has, so far as I know, never been done into English, I shall, in setting forth the teachings of it, for the most part simply give a translation of Schopenhauer's own words.[7]

Organic life originated, Schopenhauer declares, by a *generatio aequivoca* of the organic (under certain definite physical conditions) out of the inorganic; indeed, he believed, with singular scientific naiveté, that spontaneous generation is an everyday occurrence, taking place "before our eyes in the sprouting of fungi from decaying vegetable matter." But only the simplest forms can have been thus produced.

> *Generatio aequivoca* cannot be conceived to occur in the higher grades of the animal kingdom as it does in the lowest. The form of the lion, the wolf, the elephant, the ape, or that of man, cannot have originated as do the infusoria, the entozoa and epizoa,— cannot have arisen directly from the sea-slime coagulated and warmed by the sun, nor from decaying organic substances. The genesis of these higher forms can be conceived of only as a *generatio in utero heterogeneo*,[8]—such that from the womb, or

---

[6] It is evident from the references in *The Will in Nature* that the evolutionistic passages occurred in the first edition of *Zur Philosophie und Wissenschaft der Natur*, though in the text of the second edition from which I shall quote (published posthumously, 1861) they are amplified by additions written by Schopenhauer as late as 1859 or 1860.

[7] What immediately follows is based upon *Parerga und Paralipomena*, II, ##90-94, 74, 87.

[8] Birth from a parent belonging to a different species from that of the offspring; "heterogenesis," in Kölliker's terminology.

rather from the egg, of some especially favored pair of animals, when the life-force of their species was in them raised to an abnormal potency, at a time when the positions of the plants and all the atmospheric, telluric and astral influences were favorable, there arose, exceptionally, no longer a being of the same kind as its parents, but one which, though of a closely allied kind, yet constituted a form standing one degree higher in the scale. In such a case the parent would for once have produced not merely an individual but a species. Processes of this sort naturally can have taken place only after the lowest animals had appeared in the usual manner and had prepared the ground for the coming races of animals.

The reader will observe in the account of the conditions requisite for the production of these exceptional births traces of Schopenhauer's queer weakness for occultism; but the condition which he chiefly insists upon is less remote from the range of conceptions sanctioned by modern natural science. The productive potency of organisms, "which is only a special form of the generative power of nature as a whole," undergoes this "abnormal heightening" when it encounters antagonistic forces, conditions tending to restrict or destroy it; "it grows with opposition." This tendency, for example, manifests itself in the human race in times of war, pestilence, natural catastrophes, and the like; and in such periods of special intensification of the power of reproduction, that power, Schopenhauer seems to conceive, shows also a greater instability and variability, a tendency to the production of new forms which thereafter remain constant. Now, says Schopenhauer—adopting the geological system of Cuvier—a renewal of life through *generatio aequivoca*, followed by an increasing multiplication of diverse descendant species, must have taken place "after each of those great revolutions of the earth, which have at least thrice extinguished all life upon the globe so that it required to be produced anew, each time with more perfect forms, i. e., with forms more nearly approximating our existing fauna. But only in the series of animals that have come into being subsequently to the last of these great catastrophes, did the process rise to the pitch of producing the human race—though the apes had already made their appearance in the preceding epoch."

We have seen Schopenhauer in *The Will in Nature* declaring in favor of the theory of descent on the ground that it affords the only possible explanation of the homologies of the skeletons of the vertebrates. In the present writing he still more emphatically declares in favor of it on the ground of the argument from recapitulation—of the parallelism of the ontogenetic and the phylogenetic series.

> The batrachians visibly go through an existence as fishes before they assume their characteristic final form, and, according to a now fairly generally accepted observation, all embryos pass successively through the forms of lower species before attaining to that of their own. Why, then, should not every new and higher species have originated through the development of some embryo into a form just one degree higher than the form of the mother that conceived it? This is the only reasonable, i. e., the only rationally thinkable, mode of origination of species that can be imagined.

Schopenhauer was thus, as I have already said, not only an evolutionist in his biology but also a mutationist; his speculations are prophetic of the theory of De Vries and Goldschmidt rather than that of Darwin. But the scale on which he supposed these " discontinuous variations " to occur is calculated to make our contemporary mutationists stare and gasp; the changes of form which he assumed are saltatory indeed. He writes:

> We are not to conceive of this ascent as following a single line, but rather as mounting along several lines side by side. At one time, for example, from the egg of a fish an ophidian, and afterwards from the latter a saurian arose; but from some other fish's egg was produced a batrachian, from one of the latter subsequently a chelonian; from a third fish arose a cetacean, possibly a dolphin, some cetacean subsequently giving birth to a seal, and a seal finally to a walrus. Perhaps the duckbill came from the egg of a duck, and from that of an ostrich some one of the larger mammals. In any case, the process must have gone on simultaneously and independently in many different regions, yet everywhere with equally sharp and definite gradations, each giving rise to a persistent and stable species. It cannot have taken place by gradual, imperceptible transitions.

The implication with respect to the simian descent of man Schopenhauer does not shirk:

> We do not wish to conceal from ourselves the fact that, in accordance with the foregoing, we should have to think of the

first men as born in Asia from the pongo (whose young are called orang-outangs) and in Africa from the chimpanzee—though born men, and not apes. . . . The human species probably originated in three places, since we know only three distinct types which point to an original diversity of race—the Caucasion, the Mongolian and the Ethiopian type. The genesis of man can have taken place only in the old world. For in Australia Nature has been unable to produce any apes, and in America she has produced only long-tailed monkeys, not the short-tailed, to say nothing of the highest, i. e., the tailless apes, which represent the next stage before man. *Natura non facit saltus.* Moreover, man can have originated only in the tropics; for in any other zones the newly generated human being would have perished in the first winter. . . . Now in the torrid zones man is black, or at least dark brown. This, therefore, without regard to diversities of race, is the true, natural and distinctive color of the human species; and there has never existed a race white by nature.

In another of the *Parerga und Paralipomena* Schopenhauer speaks of mankind as " this race whose kinship with the ape does not exclude kinship with the tiger."

Schopenhauer does not leave us without a clue as to the writer from whom he learned his evolutionism; though—seldom generous in his acknowledgments, and always prepared to think the worst of the English—he is a good deal more copious in criticism than in appreciation of that writer.

The conception of a *generatio in utero heterogeneo* which has here been expounded was first put forward by the anonymous author of the *Vestiges of the Natural History of Creation* (6th ed., 1847), though by no means with adequate clearness and definiteness. For he has entangled it with untenable assumptions and gross errors, which are due in the last analysis to the fact that to him, as an Englishman, every assumption which rises above the merely physical—everything metaphysical, in short—is forth-with confused with the Hebraic theism, in the effort to escape which, on the other hand, he gives an undue extension to the domain of the physical. Thus an Englishman, in his indifference and complete barbarism with respect to all speculative philosophy or metaphysics, is actually incapable of any spiritual (*geistig*) view of Nature; he knows no middle ground between a conception of it as operating of itself according to rigorous and, so far as possible, mechanical laws, and a conception of it as manufactured according to a preconceived design by that Hebrew God whom he speaks of as its "Maker." The parsons, the English parsons, those slyest of all obscurantists, are responsible for this state of things.

This can scarcely be considered a very clear and coherent criticism of Robert Chambers. But the passage makes it appear highly probable that it was through becoming acquainted, late in the eighteen-forties, with the mutationist evolutionism of Chamber's *Vestiges*, that Schopenhauer was led to adopt and to develop in his own fashion a similar doctrine.

These transformist opinions in biology were, in the treatise *Zur Philosophie und Wissenschaft der Natur*, merely a part of a thoroughgoing scheme of evolutionism, which included a belief in the development of the chemical elements out of an original undifferentiated *Urstoff*, in the gradual formation of the solar system, and in an evolutionary geology.[9] His cosmogony Schopenhauer takes over from Laplace. The general outlines of the history of our planet, as he conceives them in the light of the geology of Cuvier, are set forth in a passage which is interesting enough to be worth quoting at length:

> The relation of the latest results of geology to my metaphysics may be briefly set forth as follows: In the earliest period of the globe, that preceding the formation of the granitic rocks, the objectification of the Will to Live was restricted to its lowest phases—i. e., to the forces of inorganic nature—though in these it manifested itself on the most gigantic scale and with blind impetuosity. For the already differentiated chemical elements broke out in a conflict whose scene was not merely the surface but the entire mass of the planet, a struggle of which the phenomena must have been so colossal as to baffle the imagination. . . . When this war of the Titans had spent its rage, and the granite rocks, like gravestones, had covered the combatants, the Will to Live, after a suitable pause and an interlude in which marine deposits were formed, manifested itself in its next higher stage—a stage in the sharpest contrast with the preceding—namely, in the dumb, and silent life of a purely plant-world. . . . This plant-world gradually absorbed carbon from the atmosphere, which was thus for the first time made capable of sustaining animal life.

---

[9] Op. cit., Sect. 74.

Until this was sufficiently accomplished, the long and profound peace of that world without animals continued. At length a great revolution of Nature put an end to this paradise of plants and engulfed its vast forests. Now that the air had been purified, the third great stage of the objectification of the Will began, with the appearance of the animal world: in the sea, fishes and cetaceans; on the land, only reptilia, though those were of colossal size. Again the curtain fell upon the cosmic stage; and now followed a still higher objectification of the Will in the life of warm-blooded animals—although these were chiefly pachydermata of genera now extinct. After another destruction of the surface of the globe, with all the living things upon it, life flamed up anew, and the Will to Live objectified itself in a world of animals exhibiting a far greater number and diversity of forms, of which the genera, though not the species, are still extant. This more complete objectification of the Will to Live through so great a multiplicity and variety of forms reached as high as the apes. But even this, the world just before ours, must needs perish, in order that the present population of the globe might find place upon fresh ground. And now the objectification of the Will reached the stage of humanity. . . .

An interesting incidental consideration, in view of all this, is that the planets which circle round the countless suns in all space—even though some of them may be still in the merely chemical stage, the scene of that frightful conflict of the crudest forces of Nature, while others may be in the quiet of the peaceful interlude —yet all contain within themselves those secret potencies from which the world of plants and animals must soon or late break forth in all the multiplicity of its forms. . . . But the final stage, that of humanity, once reached, must in my opinion be the last, for this brings with it the possibility of the negation of the Will, whereby there comes about a reversal of the whole inner tendency of existence (*der Umkehr vom ganzen Treiben*). And thus this *Divina Commedia* reaches its end. Consequently, even if there were no physical reasons which made certain a new world-catastrophe, there is, at all events, a moral reason, namely, that the world's continuance would be purposeless after the inmost essence of it has no longer need of any higher stage of objectification in order to make its deliverance (*Erlösung*) possible.

It is thus clear that by 1850 Schopenhauer had formulated his conception of the "objectification of the Will" in thoroughly evolutionistic terms and had incorporated into his philosophy a complete system of cosmogony and phylogeny.[10] It was at about the same time that Herbert Spencer was beginning to formulate the outlines and primary principles of the *Synthetic Philosophy*, which has commonly passed for the first comprehensive attempt by any nineteenth-century philosopher to generalize the conception of evolution and to give to it the principal role in his system. The two doctrines may, in truth, not uninstructively be set side by side. They exhibit, in the first place, a degree of resemblance which is likely to be overlooked by those who cannot discern, beneath diversities of terminology and of emphasis, identities of logical essence. In both systems, for example, the ultimate nature of things is placed beyond the reach of temporal becoming. Spencer's evolutionary process belongs only to the realm of "the knowable," Schopenhauer's to the world of the Will as objectified; behind the one stands, as true reality, the Unconditioned, alien to all the characters of human experience and all the conceptions of human thought; behind the other stands the Will as it is in itself, timeless, indivisible, ineffable. In other words, both systems consist of an evolutionary philosophy of nature projected against the background of an essentially mystical and negative metaphysics. Yet each, as I have already remarked, regards its supratemporal and indeterminate Absolute as the very substance and sum of the world in time; and each is prone to the same inconsistency, that of practically treating this same Absolute as the real ground and explanation of becoming and as a power at work in the temporal movement of things. In the degree of emphasis which they lay upon this negative element in their doctrine, the two philosophers, no doubt, greatly differ. Spencer closes the door upon it after half a dozen chapters, and then forgets it for whole books at a time—reverting to it only at the moments when his logic seems, in the deduction of the laws of "the Knowable," to be on the point of breaking down.

Schopenhauer, too, can forget the obscure background of exist-

[10] It is a singular illustration of the present condition of the historiography of scientific and philosophical ideas, that this fact is ignored, in such reputable histories of philosophy as those of Höffding, Windelband, Kuno Fischer (who devotes a whole volume to Schopenhauer) and in Rádl's *Geschichte der biologischen Theorien* (II, 457). Most of the histories of philosophy which do not contradict the fact, at least fail to mention it. It is, however, correctly though concisely set forth in Frauenstädt's *Neue Briefe über die Schopenhauersche Philosophie*, 1876, p. 103, and in Dacqué's *Der Descendenzgedanke und seine Geschichte*, 1903, p. 82.

ence when he is absorbed in the concrete phenomena of evolution; but he takes it, on the whole, more seriously, and draws the veil from before it more frequently. And the more closely Kantian affinities of his epistemology create for him a difficulty in adjusting his evolutionism to his metaphysics, which Spencer seemingly escapes—though he escapes it only by an evasion. Since, for Schopenhauer, space and time are subjective forms of perception, pre-mental evolution, the formation of planetary systems and of planets themselves before the emergence of consciousness, necessarily has for him an especially equivocal ontological status.

> The geological processes which took place before there was any life on earth were present in no consciousness; ... from lack of a subject, therefore, they had not even objective existence, i. e., they were not at all. But what is meant by speaking of their "having been" (*Dagewesensein*)? The expression is at bottom purely hypothetical; it means that *if* any consciousness had been present in that primeval period, it *would have* then observed those processes. To them the regress of phenomena leads us back; and it therefore lay in the nature of the thing-in-itself to manifest itself in such processes [i. e., if there had been any consciousness for it to manifest itself to].

When Spencer declares that our conceptions of space and time are modes of thought produced *in us* somehow by the Unconditioned, but not ascribable to that entity itself, he involves himself in a similar difficulty about early geological time, and *implies* an identical way of dealing with the difficulty; but so far as I can recall, he does not anywhere directly face the question.

The points of resemblance between the system of Schopenhauer and that of Spencer, however, consist chiefly in the general fact that both were evolutionists, and that their evolutionist cosmology had much the same sort of metaphysical setting. In its spirit, as in its details, Schopenhauer's evolutionism was quite different from Spencer's. He is, but for some analogies in the philosophy of certain of the Romantics, the first representative of a tendency in evolutionistic philosophy that is essentially hostile to the tendency of which Spencer is the representative. Spencer's enterprise is neither more nor less than a resumption of that which Descartes had undertaken in 1633, in his suppressed treatise on "The World"; the nineteenth-century philosopher, like the one of the seventeenth century, conceives it possible to deduce from the laws of the motion of the parts of a conservative material system the necessity for the gradual development of such a world as we now find. Spencer's evolutionism, in short, is, or rather attempts to be, thoroughly mechanistic. And in the course of the whole process, therefore, (though Spencer frequently forgets this) no real novelties can appear except novelties in the spatial arrangement of the particles of matter. Even these novelties are only the completely predetermined consequences of the sum of matter and energy originally present in the universe, and of the laws of relative motion. The whole cosmic history is solely a process of redistribution of matter and change of direction in motion. It is for this reason that Bergson was fond of saying of Spencer that his system contains nothing that really has to do with either becoming or evolution; " he had promised to trace out a genesis, but he has done something quite different; his doctrine is an evolutionism only in name."

Schopenhauer's evolutionism of the ever-expanding, self-diversifying Will, however, is radically anti-mechanistic. For it the universe, even the physical universe, cannot be a changeless closed system, in which no truly new content ever emerges. The primary characteristic of the Will is that it is never satisfied with the attained, and therefore ever goes on to further attainment. Its objectification, in the latest phase of Schopenhauer's thought, becomes necessarily progressive and cumulative. In short, a philosophy which conceives the genesis and movement of the temporal world in terms of the Will necessarily gives a very different account of the biography of the cosmos from that presented by a philosophy which aspires to tell the whole story in terms of mechanics and in accord with the principle that the ultimate content of nature never suffers increase or diminution. This latter program Spencer, it is true, realizes very imperfectly. In the later volumes of the *Synthetic Philosophy*, the *First Principles* seem often pretty completely forgotten. There are not a few strains of what may be called the Romantic type of evolutionism in Spencer. But in him these strains are incongruous with the primary postulate of his system; in Schopenhauer they are the characteristic note of the whole doctrine.

This contrast between the two types of evolutionism found in these two writers is due in part to certain features in their respective doctrines which arose without dependence upon their evolutionism. They had essentially opposed preconceptions about the program and possibilities of science. Spencer was from his youth obsessed with the grandiose idea of a unification of all knowledge. All truths were eventually to be brought under some "highest generalization which is true not of one class of phenomena, but of all classes of phenomena, and which is thus the key to all classes of phenomena." This, of course, meant the theoretical possibility of the reduction of the more complex sciences to the simpler ones—of physiology to chemistry, of chemistry to physics, and of all physics to the mechanics of molecules. This intellectual process of explanation of the more complex by the simpler and more generalized type of phenomena was the counterpart, and in truth a necessary implication, of the objective process of evolution of simple into more complex arrangements of the matter of the universe. Schopenhauer, on the other hand, from the beginning insisted upon the irreducibility of the several sciences to one another, and most emphatically upon the uniqueness and autonomy of biology. When science, he writes, "in the quest for causal explanations (aetiology) declares that it is its goal to eliminate all ultimate forces except one, the most general of all (for example, impenetrability) which science flatters itself upon thoroughly understanding; and when, accordingly, it seeks to reduce (*zurückzuführen*) by violence all other forces to this single force, it then destroys its own foundation and can yield only error instead of truth. If it were actually possible to attain success by following this course, the riddle of the universe would finally find its solution in a mathematical calculation. It is this course that people follow when they endeavor to trace back physiological effects to the form and composition of the organism, this perhaps to electricity, this in turn to chemism, and this finally to mechanism."[11] Just why Schopenhauer adopted this doctrine of the irreducibility and discontinuity of scientific laws at a period when he apparently had not adopted evolutionism, is not wholly clear. He seems to have been partly led to such a view by his conception of the Platonic Ideas. Since for each of the broad divisions of science, which correspond to grades of objectification of the Will, there is a separate Idea, Schopenhauer seems to have felt that the distinctness of the several Ideas forbade the supposition of the complete reducibility of the laws of one science to those of a prior one. But inasmuch as the whole notion of the Platonic Ideas is a logically irrelevant part of the Schopenhauerian system, this

[11] *Die Welt als Wille und Vorstellung*, #27.

explanation does not carry us very far. Whatever his reasons, the fact remains that Schopenhauer attached the utmost importance to his contention that, at the points where one typical phase of the Will's self-manifestation passes over into a higher one, new modes of action, essentially different kinds of being, must be recognized. Consequently, when he eventually arranged the grades of the Will's objectification in a serial, temporal order, thus converting his system into an evolutionism, this contention made his evolutionism one which implied the repeated production of absolute novelties in the universe, and the supervention from time to time of natural laws supplementary to, if not contradictory of, the laws or generalizations pertinent to the phenomena of a lower order.

Another detail of Schopenhauer's body of doctrine which likewise antedates the evolutionistic transformation of his system but yet has an important relation to certain subsequent developments in the philosophy of evolution, was his peculiar form of teleology. He was equally opposed, on the one hand, to the conception of conscious design as an explanation of the adaptive characters of organisms, and on the other hand to the mechanistic elimination of all purposiveness from nature. Between these two extremes he endeavored to find room for a teleology dissociated from anthropomorphism. The Will moves towards ends determined by its own inner nature, though it does not foresee these ends. It triumphs over obstacles in its way, and circumvents obstructions; but it does so blindly and without conscious devices. This notion of a blind purposiveness, which more than any other philosopher Schopenhauer may be said to have introduced into the current of European philosophy, has come in our own day to be a familiar conception in the interpretation of the meaning of evolution, especially in its biological phase. Here again Schopenhauer is the precursor of Bergson. That philosopher too rejected what he called *le finalisme radical* not less than the radical mechanistic doctrine, while insisting upon the indispensability of some notion of "finality" in any attempt to comprehend the development of organisms. From this point of view Bergson has objected, upon grounds altogether similar to those which have been noted in Schopenhauer's references to Lamarck, to the Lamarckian tendency to identify the cause of the production of new characters with "a conscious effort of the individual"; while he at the same time has regarded Lamarckianism as approaching far nearer than does Darwinism, with its essentially mechanistic interpretation of organic evolution, to a correct representation of the developmental process. Like Schopenhauer, Bergson adopted, as the biological theory most congenial to his metaphysics of the *poussée vitale*, a combination of the doctrines of orthogenesis and of mutation. The later writer may or may not have been influenced by the earlier one, but there can be no doubt that in Schopenhauer we find the first emphatic affirmation of the three conceptions most characteristic of the biological philosophy of *L'évolution créatrice*.

It is a somewhat curious circumstance that the trait in Schopenhauer's conception of the action of the " objectified " Will which has hitherto most attracted the notice of writers on the history of biology is closely related to the fundamental conception of precisely that sort of organic evolutionism to which he was most opposed. The universal prevalence of a struggle for existence among organisms was eloquently set forth by Schopenhauer forty years before Darwin published the *Origin of Species*. But it seems never to have occurred to Schopenhauer to regard this struggle as an explanation of the formation of species and the adaptation of organisms to their environments. Why he was unlikely to do so is evident from all that has been already said. The Darwinian hypothesis makes a species and the adaptive characteristics merely the result of a sort of mechanical pressure of external forces. Slight promiscuous variations, due probably to fortuitous displacements in the molecules of the germ-cell, are conserved or eliminated in the course of the jostle for survival, according as they do or do not fit the individuals possessing them to keep a footing in that turmoil. But such a doctrine assigns to the organism itself, and to its inner potencies, an essentially passive role; development is, as it were, extorted from living things by external circumstances, and is not a tendency expressive of all that is most characteristic in the nature of organisms as such. The metaphysician whose ruling conception was that of a cosmic life-force was debarred by the dominent temper of his thought and the deepest tendency of his system from any such account of the causes and the meaning of that progressive diversification of the forms of life, the reality of which he clearly recognized. Thus, though Schopenhauer incidentally shows certain affinities with Darwinism, he is much more truly to be regarded as the protagonist in nineteenth century philosophy—at just the time when Darwin was elaborating a mechanical biology and Spencer a would-be mechanistic cosmogony

—of that other form of evolutionism which a recent French writer has described as "a sort of generalized vitalism." [12] He was thus the first important representative of the tendency which, diversely combined with other philosophical motives, and expressed with varying degrees of logical coherency, has been chiefly represented since his time by such writers as Nietzsche, Bernard Shaw, Guyau, E. D. Fawcett, and Bergson. The Romantic evolutionism of all these writers is, it is true, innocent of the pessimistic coloring of Schopenhauer's philosophy; but the pessimism of Schopenhauer was always connected rather with those preconceptions in his doctrine which were really survivals from older systems, than with that vision of the Will as creatively at work in the temporal universe, which was his real contribution to the modern world's stock of metaphysical ideas.

---

[12] M. René Berthelot, *Evolutionisme et Platonisme*, 1908, p. II.

FIFTEEN

RECENT CRITICISM OF THE DARWINIAN THEORY OF RECAPITULATION: ITS GROUNDS AND ITS INITIATOR

ARTHUR O. LOVEJOY

The title of this paper may lead some readers to infer that it far oversteps the limits of the period to which the present volume is confined. But in fact, the following pages will relate chiefly to a movement in nineteenth-century biology which began more than a decade before Darwin's earliest book appeared, and was initiated by an older contemporary of his. Primarily, the movement was a revolt against an embryological doctrine which, though Darwin himself had not yet publicly given it his support, had already been formulated (as a preceding paper has shown) by several of his precursors, and was, after 1859, to be enthusiastically embraced by all the early converts to his form of evolutionism. But, as we shall see, though the genesis of this movement was pre-Darwinian, it had little effect upon the opinions of most biologists until the last three or four decades. At present, however, it has come to have a wide influence among embryologists. It is one of the fairly numerous examples of the retardation of the impact of an idea upon the thinking even of specialists in the field to which it is pertinent. Yet the fact that its impact was so long delayed is one that seems to need explanation; and the quest of an explanation will lead us back to the years before 1859.

This revolt has, it is true, passed through two phases, but the first phase is wholly post-Darwinian, and is therefore not directly relevant to the theme of the present paper; I shall consequently only cursorily refer to it here. It was preceded by a series of highly important discoveries in genetics, which have been made—or made known—only in the present century: the establishment of the laws of biparental heredity by Mendel, the proof of the occurrence of mutations by de Vries and others after him, the final disproof of the inheritance of acquired characters, and the demonstration of the virtual inaccessibility of the genes—the carriers of the hereditary factors in embryogeny—to any forces external to them, excepting high-energy radiations. Accepting these discoveries, and employing improved techniques of observation, investigators of the phenomena of embryogeny in the last quarter-century came to the conclusion that organisms of the same class do not, as Darwin had assumed, in their larval or foetal stages pass through forms corresponding to the adult forms of their presumable ancestors, but only to their embryonic forms. A brief statement of this emended argument from recapitulation may be quoted from Thomas Hunt Morgan's *The Scientific Basis of Evolution* (1932). "It is a falsification of history," Morgan wrote, to assert that "alterations may take place at any stage in the development of the embryo without changing to the same extent (if at all) the end-stages." But, on the other hand,

> the conclusion that the young stages of those animals are like the young stages of animals lower in the scale is in accord with an idea that has been slowly gaining ground. . . . Darwin accepted the somewhat naive interpretation of his contemporaries. But one of his statements, taken alone, namely, that 'community of embryonic structure reveals community of descent,' is supported by abundant evidence. . . . Hosts of cases [have been] discovered which are contributions to the evolution theory in the sense that they find their simplest interpretation in that theory.

Among such cases Morgan mentions the gill-slits found in the embryos of birds and mammals; the appearance of a notochord in the embryo and its replacement by the vertebral column; and the embryonic history of the reproductive system. He concluded: "These and many similar facts find their simplest and, I think, a satisfactory interpretation in the evolution theory as representing the past history of the embryonic development of the higher groups."[1] This conclusion, from which, I suppose, few, if any, contemporary embryologists would dissent, obviously limits the

---

[1] Morgan, op. cit., pp. 173-174, somewhat abridged in quotation.

scope of the hypothesis of embryonic recapitulation, but does not annul its logical force as an argument for organic evolution.

The second phase of the revolt against the recapitulation-theory—second in the time at which it became influential—seems much more radical, for it literally reverses the older view that the forms through which an embryo of a "higher" and later-evolved species passes correspond to the stages through which its ancestors had passed and are caused by, and can only be explained through, the hereditary transmission of those forms. The spokesmen of the second phase declare that, on the contrary, no such exact recapitulation of ancestral forms can be found, and that "phylogeny does not explain ontogeny at all," though the study of ontogeny can throw light on some—but not the only, nor perhaps the most important—factors in phylogeny. The methods of studying the two processes should be kept distinct: the phenomena of ontogeny should be investigated through the analytical and experimental study of the embryos of extant organisms, the facts of phylogeny must be ascertained from paleontology and from the comparative morphology of adults of existing species. The observations and experiments and the reasonings which in recent years have led a number of embryologists to adopt these conclusions have been succinctly but comprehensively set forth by G. R. de Beer in his little book *Embryos and Ancestors*.[2] But the paternity of this new embryology, as it may be called, is attributed by its adherents to a famous precursor of Darwin—a precursor in date but not in doctrine—whose name has not hitherto been mentioned in this paper though there has been some indication of his influence upon the current phase of the reaction against the pre-Darwinian and Darwinian conception of embryogeny. We must, then, turn back to the writings of the originator, or putative originator, of this movement.

It will be remembered that in the period 1830-1859 Serres[3] was

[2] Oxford, revised edition, 1951. The author is Director of the Department of Natural History in the British Museum and Fellow of the Royal Society.

[3] The Serres to whom the writers mentioned evidently referred was A.-Étienne-Renaud-Augustine Serres (1786-1868), a French anatomist especially well-known for his writings on embryology, a member of the Paris Academy of Sciences, and Director of anatomical studies in the School of Medicine. He has apparently been confused by some biologists and historians of biology with his contemporary, Marcel de Serres, professor of zoology at Montpellier, who did not attain a comparable professional reputation. Étienne Serres, in spite of his emphasis on recapitulation, did not, as late as 1867, regard it as evidence of evolution. In a communication to the

the embryologist whose observations were generally believed to have given the final proof of embryonic recapitulation, and that this was held by the early evolutionists to have furnished absolutely conclusive evidence of the descent of existing from earlier species. In 1824 Serres still reiterated the conclusion that "a man becomes a man only after passing through transitional stages of organization in which he is similar first to a fish, then to a reptile, then to birds and mammals." But in 1828 there was published the first volume of a work by a young German zoologist, Karl Ernst von Baer, *Über Entwicklungsgeschichte der Thiere: Observation und Reflexion*,[4] in which he reported in extraordinary detail *his* observations of the embryonic development of certain vertebrates (especially of barnyard fowl), violently attacked Serres' description of the prenatal stages of the formation of the individual organism, and propounded a new formulation of the "laws" of ontology. Serres' conception and his own, with respect to the sequence of forms which an embryo goes through, were, von Baer wrote, "wholly opposed (*ganz entgegengesetzt*)."[5] The former, which was "the generally accepted conception," was "not based upon accurate observation." What Serres thought he saw in a fertilized hen's-egg was a series of changing shapes of various organs or members, each shape being a more or less modified replica of the corresponding organ or member in the adult stage of one of its ancestors, the temporal order of these shapes corresponding to the relative antiquity of the ancestral class (reptiles before birds, etc.). Von Baer protested that when *he* examined the foetal stages of a chick he could find nothing of the sort. It is true, he admitted, that there are some similarities in embryos to some forms found in more ancient types of organisms; but the resemblances are vague, and the order in which they appear

Academy of Sciences in that year he wrote: "The theory of Lamarck, recently renewed by Mr. Darwin, has once more been found not to be in accord with the known facts." See section 6 of previous article on "The Argument for Organic Evolution Before the *Origin of Species*, 1830-1858."

[4] Königsberg, 1828. The second volume was not published until 1837. *Entwicklungsgeschichte*, in von Baer's use at this time, does not mean "the history of evolution," but simply "ontogeny." Von Baer was born in Esthonia, then under the sovereignty of Russia; but he was of German descent and all his early writings except the first (which was in Latin) were written in German, his professional training was received at Königsberg, and his first academic appointment was in the faculty of that university. He was, however, called to St. Petersburg, and most of his later works, though also written in German, were published there.

[5] Op. cit., I, p. 157.

is not the order in which the supposed ancestors, according to the geological evidence, succeeded one another in past eras. It was, writes de Beer, "in the work of von Baer that the foundations were laid of an accurate descriptive knowledge of the ontogenetic stages through which animals pass. As a result of his extensive researches von Baer came to a number of conclusions which are expressed in four statements which have come to be known as the 'laws of von Baer.'" But if these conclusions, "wholly opposed," as von Baer said, "to the [then] prevailing views," were correct, the argument from embryonic recapitulation, which had seemed to Darwin "second to none" among the evidences of organic evolution, and to Haeckel, LeConte, and many others the most impressive and most cogent of all, had been exploded—and exploded thirty years before the publication of the *Origin of Species*.

Darwin, however, had apparently never read the whole of von Baer's *Entwicklungsgeschichte*, and certainly never gave it serious consideration, in spite of Huxley's efforts to bring it to his attention. It was not at all characteristic of Darwin to neglect the reported observations of earlier specialists in this or any other branch of natural history; he was usually scrupulously careful to examine the observational evidence that could be found in the works of other men on any question which was relevant to his own hypotheses. If he *had* read and pondered von Baer's book, he might possibly have given up the embryonic-recapitulation argument as one of the proofs of organic evolution—or much more probably have merely modified it. If he had done either of these things, his successors would doubtless have done the same. But as he neither abandoned nor emended it, that argument in its original form held its place in biology for almost a century. *Why* Darwin thus neglected von Baer's work I need not here inquire, for that curious enigma has been elucidated in Dr. Jane Oppenheimer's admirably fully documented study elsewhere in this volume. The final result (to date) of Darwin's disregard of the "Observation and Reflection" of his German precursor has been the rejection of the embryological assumptions—or at least some of them—which had been accepted by Darwin and the adoption instead by a number of contemporary embryologists of the slogan "Back to von Baer!"

But this slogan is likely to be misleading. No biologist today (1958), I think, would accept von Baer's embryological doctrine as a whole, or even its most distinctive and essential theses. His rejection of Serres' complete parallelism of ontogeny and phylogeny has been justified by the observations of more recent comparative anatomists; but the embryological conceptions which von Baer sought to substitute for the then "reigning idea" (as he called it) about embryogeny are still more alien to contemporary biology. For those conceptions were not derived simply from his observations of the successive shapes assumed by fertilized *ova*; they were interwoven with, and conceived by him to be necessary deductions from, certain metempirical and essentially metaphysical theorems, from which, *inter alia*, it followed (as he believed) that *any* recapitulation of ancestral forms is, not merely unverified, but impossible *a priori*. Being impossible, recapitulation was, of course, also unverified by empirical evidence, and the critical examination of the "idea that the embryo of higher animals passes through the developed forms of lower animals," to which he devotes a large part of the *Entwicklungsgeschichte*,[6] is based mainly on observational grounds. Its conclusion is, of course, that the "idea" is false: *die individuelle Entwicklung der höheren Thierformen durchläuft nicht die ausgebildeten Formen der niederen Thiere*. This statement, taken alone, might be construed merely as denying that embryos recapitulate the *mature* forms of earlier and "lower" animals. But von Baer at once proceeds to tell his readers what really does take place; he enunciates (in italics) his celebrated "four laws" of embryonic development. The real import of these has apparently been missed by some of the contemporaries who quote them. The fourth law, for example, which is oftenest quoted, might, when taken apart from its context, seem to be only another denial, not of recapitulation as such, but simply of recapitulation of adult forms, for it runs: "the embryo of a higher form of animal never resembles a lower form of animal, but only its embryonic form." The context makes clear, however, that this last clause does not imply that embryos tell us anything whatever as to the class or order from which they are descended. For the fourth law is conceived by von Baer simply as a corollary of the first: "that the form which is the most general (*das Allgemeinste*) in a large group of animals is earlier in the embryo than the more specific form (*das Besondere*)." In other words, the order in which the larval and foetal forms follow one another is not in any degree determined by or correspondent to

[6] Op. cit., I, pp. 199-224.

the temporal order in phylogeny of any other kinds of organisms, and the gradual transition from the first of those forms to the last of them is not a transition from an older to a later type, but a transition from a *logically* more "general" and indefinite type to a logically more determinate and specific type. An expanded statement of the first law is the following:

> When one observes the progress of the formation [of the embryo] what first of all leaps at the eye is that *there is gradually taking place a transition from something homogeneous and general to something heterogeneous and special.* This law of [embryonic] formation it is quite impossible to fail to recognize, and it so completely dominates every single moment of the embryo's metamorphoses that no accurate account of its development can be expressed except in the foregoing terms.[7]

This is the basic and distinctive theorem of von Baer's embryological doctrine of the 1820's. We shall find that, when its meaning is explicated, it is a very curious theorem. But before attempting to analyze it, we must first recall an example given by von Baer himself to illustrate its meaning.

In order to show how very "general" the earliest larval forms of all animals are, von Baer mentions an incident in his own experience: having inadvertently neglected to affix labels (indicating the species of the mother) to two very small embryos which he had preserved in alcohol, he was thereafter never able to tell whether they were embryos of reptiles, birds, or mammals. This passage of von Baer's was quoted by Darwin in the *Origin of Species*, but in the early editions (in the first edition, p. 439) was erroneously attributed to Agassiz; the error was corrected in later editions. To Darwin the fact that "the embryos of distinct animals within the same class are often strikingly similar" seemed an example of the embryonic recapitulation of ancestral characters, and therefore an evidence of the common descent of species now very dissimilar in their adult forms. But, as Darwin failed to note, it is not an example of recapitulation of the *adult* characters of ancestors, but only of the similarities of embryonic characters in organisms of the same class. Von Baer did note, and emphasize, this fact. But to him this was evidence, not that embryos of a given class resemble the forms of their ancestors, but that they are all in the first definite stage of their

[7] Op. cit., I, p. 153; italics mine.

development, fashioned after a single very general and simple model, an *Urtypus* of the class. From this model, however, they immediately began to deviate, and in their subsequent development they depart more and more widely from their initial " general " form, until finally they assume their (relatively) permanent and " most specific " form—i. e., the form of the species to which their parents belong, and into which they would themselves have developed, if permitted to do so.[8] This is von Baer's " second law." It complements the first, and in the two combined he found the universal and fundamental law of the entire world of living things. And when, in the brief final chapter of the first volume of the *Entwicklungsgeschichte*, he sums up the conclusion which the argument of the entire book has, he believes, proved beyond any question, he condenses that universal law into a single sentence: *Die Entwicklungsgeschichte des Individuums ist die Geschichte der wachsende Individualität in jeglicher Beziehung.* (" The story of the development of the individual is the story of the growth of individuality in every respect.") This proposition, he declares, " is so simple that it seems to need no proof through observation, but necessarily to be known *a priori*." However, von Baer judiciously adds, experience has taught us that the results of deductive reasoning are always more trustworthy when they have been corroborated by observation. Such corroboration, he is sure, his book has abundantly provided. And if the " general conclusion expressed above is true, then there is *one* fundamental thought (*Grundgedanke*) which runs through all the forms and stages of animal development, and controls all their various interrelations. It is the same thought which gathers the separate masses of matter in cosmic space and binds them together into a solar system; the same which makes it possible for the scattered dust on the surface of the metallic planet to develop into living forms. This thought, however, is nothing but Life itself, and the words and syllables in which it expresses itself are the diverse forms of the living being." It is with these eloquent and poetic but hardly luminous sentences that the volume ends.[9]

Von Baer's extension of his " one universal *Grundgedanke* " to the formation of solar systems is not relevant to the subject of this paper; it has been cited only to illustrate the boldly speculative

---

[8] In the original: Aus dem Allgemeinster der Formverhältnisse bildet sich das weniger Allgemeine und so fort, bis endlich das Speciellste auftritt.
[9] Op. cit., I, pp. 263-264.

temper of his thinking and the sometimes highly rhetorical, and consequently equivocal character of his style. What is pertinent here is only the question as to what precisely he meant by his fundamental laws of the formation of embryos—which can, as he said, be summarized in a single sentence. We have already seen that those laws implied a total repudiation of the theory of embryonic recapitulation of the characters of ancestors—*both* their embryonic and adult characters—and that he defined the succession of larval and foetal forms in an organism's development as a passage from the most "general" to the less "general" form—*not* as a passage from an earlier-evolved to a later-evolved form. Seeking to make his fundamental theorem fully clear, von Baer again states it, with labored repetitions, as follows:

> The stage which an animal body has reached in its formation consists in the greater or lesser degree of heterogeneity in the parts composing it, . . . *in short, in its greater histological and morphological differentiation. The more similar the entire mass of the body is, the lower is the stage which its formation has reached.* It is a higher stage if nerves and muscles, blood and cell-substance are sharply distinguished. The more different they are, the more developed is animal life in its several directions, or rather, conversely, the more animal life has developed in its various directions, the more heterogeneous are the component parts which will make it possible for that life to manifest itself.[10]

Yet even this, though it leaves the reader in no doubt as to the general import of von Baer's embryological doctrine, still leaves unanswered at least one question that patently requires an answer: it tells us in what "direction" and towards what terminal form an embryo moves in its successive metamorphoses, but it does not tell us about its *terminus a quo*. What degree of "heterogeneity of its parts" must a larva be supposed to start with? But von Baer's answer to this question he has given us in the story of the two little larvae that had no labels. The answer, obviously, is that at the outset an embryo has *no* heterogeneous parts. In short, at the very beginning of its development, it has no definite characters at all. Being a body, it must, of course, have a certain size; but beyond that, von Baer held, it is simply a mass of undifferentiated matter. No portion of it is dissimilar to any other portion, and consequently, all embryos in this incipient stage are exactly alike. Von Baer

[10] Op. cit., I, p. 207; italics in original.

could not determine the class or order to which his unlabeled larvae belonged because they were so completely "general" that they might have been the offspring of parents of any class or any order—or even of any species.

Thus von Baer described the prenatal development of an individual organism in the terms in which Herbert Spencer, many years later, described the evolution of the universe—a movement "from an indefinite, incoherent homogeneity to a definite, coherent heterogeneity." Whether Spencer borrowed the formula from von Baer I do not know. In any case, as used by the author of *The Synthetic Philosophy*, it expressed a conception essentially different from and opposed to the ideas of his German predecessor. For Spencer accepted the biological doctrine of Darwin—the evolution of all now existing species by descent from earlier species. But von Baer was opposed to Darwinism root and branch—to its general principles and to its specific implications. Since his most important book was published long before Darwin's, its polemic, of course, was not explicitly directed against the latter. Shortly after the publication of the *Origin of Species*, and probably before he had read it, von Baer delivered a discourse before the newly founded Russian Entomological Society (of which he was the first president) with the seemingly odd but accurately descriptive title: "What is the right way of interpreting animate nature, and how is this to be applied to Entomology?"[11] It begins with observations about insects and the practical importance of studying them, but ends with a denunciation of "materialism" and a defense of the philosophical doctrine which we now call psychophysical dualism. There is an unbridgeable gulf between man—with his cognitive powers, his moral freedom, his sense of duty, his faith in the aid of "the Universal Source of existence"—and the animals. "It is these feelings through which the human race has been formed and has been ennobled. The animal instincts serve only for the preservation of species and not for their ennoblement." Von Baer laments that "the recent discoveries of the part played by physical and chemical processes in organic life-processes have caused large numbers of cultivated persons, or of persons who fancy themselves cultivated, to think it self-evident that all motions of matter must be subject

---

[11] Delivered in St. Petersburg in May, 1860; published in Berlin, 1862. This is perhaps the most eloquent of von Baer's discourses, but is more notable for its rhetorical than for its logical quality.

to the same laws; . . . they begin to regard even themselves as a product of matter and are unwilling to recognize a moral world-order. . . . Happily, care is being taken that this unworthy and suicidal (*selbstmörderisch*) tendency cannot become universal and permanent." [12] The question how new species originate is not directly discussed; but, as will later appear, von Baer regarded Darwin's answer to that question as a form of materialism.

When, ten years later (1871), Darwin's *Descent of Man* was published, von Baer was at once moved to compose a work which he evidently hoped would demolish the theory of organic evolution as a whole, and not merely one of its principal supports, the hypothesis of embryonic recapitulation. This short but aggressive work appeared in 1873 under the long and mildly interrogative title: *Entwickelt sich die Larve der einfachen Ascidien in der ersten Zeit nach dem Typus der Wirbelthiere?* (Does the larva of the simple ascidian in its first period develop in the manner typical of the vertebrates?) Recent developments in the study of embryogenesis made this question especially timely, and they seemed to von Baer to disclose a fatal inconsistency in Darwin's argument for evolutionism. These developments were due to the work of two Russian zoologists, Kowalevsky and Kupffer, who were pioneers in the investigation of the foetal forms of the ascidians—small marine organisms which have neither a vertebral column nor a notochord (then called *chorda dorsalis*) in their mature forms, and spend their singularly inactive lives firmly attached to rocks, like oysters, though in their foetal stage they are free-swimming. But, wrote Darwin in the *Descent of Man* (1871), Kowalevsky and Kupffer " have lately observed that the larvae of ascidians are closely related to the Vertebrata, in their manner of development, in the relative position of the nervous system, and in possessing a structure closely like the *chorda dorsalis* of vertebrate animals." [13] Accordingly, Darwin continued,

> should M. Kowalevsky's results be well established, the whole will form a discovery of the very greatest value. Thus, if we may rely on embryology, ever the safest guide in classification, it seems that we have at least gained a clue to the source whence

---

[12] Op. cit., p. 56. The last sentence, to a Russian audience, possibly conveyed a threat of governmental action, perhaps of police-action, against the " materialists."

[13] Op. cit., second London edition, p. 158.

the Vertebrata were derived. We should then be justified in believing that at an extremely remote period a group of animals existed resembling in many respects the larvae of our present ascidians, which diverged into two great branches—the one retrograding in development and producing our present class of ascidians, the other rising to the crown and summit of the animal kingdom by giving birth to the Vertebrata.[14]

Darwin, it will be noted, does not here quite assert that Kowalevsky's conclusion about the ascidians *had* been completely established, but he evidently thinks it highly probable that it will be. And its "very great value," if it is true, is that it permits us to extend the argument from embryonic recapitulation beyond the class vertebrates backward to certain very "low" non-vertebrate organisms, which could thus be inferred to be early ancestors of all the vertebrates, including man—whereas in *The Origin of Species* Darwin had usually been careful to limit the argument to animals belonging to the same class.

It was, then, in this passage of *The Descent of Man* that von Baer found, as he believed, the Achilles' heel of Darwin's reasoning. If the reported conclusions of Kowalevsky and Kupffer were well founded, von Baer grants, "the attention given to them by naturalists would be fully justified, for they would provide a powerful support for the bold hypothesis of Darwin that the higher forms of animal organization have developed (*sich entwickelt haben*) in the course of time from the wholly different forms which we usually call 'lower.' It will therefore be very worth-while to test the grounds of this assertion."[15]

The "assertion" which von Baer proceeds to test is, the reader will note, nothing less than the theory of organic evolution. But the "testing" of that theory would have required an examination of *all* the arguments which Darwin and the evolutionists generally had presented as proofs of its validity. Von Baer, however, examined only one of those arguments, that based upon the observations of the foetal form of the ascidians reported by Kowalevsky, which Darwin had recently adduced as evidence that all vertebrates are descended from extremely primitive marine organisms. Thus, even if von Baer's "testing" of that evidence showed it to be

[14] Ibid., p. 159.
[15] The volume of 1873 here cited will hereinafter be briefly called *Ascidiens*. The above quotation is its opening paragraph. It was first presented as a communication to the Academy of Sciences of St. Petersburg.

invalid, this would fall far short of a disproof of the descent of extant species from earlier and very "dissimilar" *vertebrate* organisms—Darwin's embryological thesis in the *Origin*. And the other arguments—from paleontology, from rudimentary organs, from the homologies—obviously could not be refuted by showing that Kowalevsky's observations and inferences concerning the ascidians were erroneous. Von Baer, however, evidently believed that, by his attempted proof that these observations of Kowalevsky were wrong, he was once more—and this time with explicit reference to Darwin—undermining and exploding the whole theory of organic evolution, which by then (1873) had been accepted by nearly all the naturalists who had previously rejected it, including Lyell—with the exception of those few who still took their conception of the origin of species from the Book of Genesis. Thus the scanty remnant of the special-creationists, and von Baer (who did not derive his biological opinions from the Book of Genesis), were, in the 1870's, perhaps the only professional zoologists who still kept up the resistance to evolutionism.

But von Baer was not deterred by the fact—of which he must have been aware—that scarcely any of his fellow-specialists agreed with him. He wrote of them condescendingly, not to say contemptuously:

> It was to be expected that the zealous adherents of the theory of descent would give a very ready acceptance to the results of the investigations of Kowalevsky and Kupffer on the development of the ascidians, for it seemed for the first time to show a formation—even though a wholly transient formation—of an extremely low type of animal through the same processes as in the vertebrates, and thereby to bridge-over the wide gap which separates the headless mollusks from the vertebrates. A longer continuance of this process, it might seem, would convert them permanently into vertebrates. I should regard it as a very unprofitable labor to seek out and quote all such attempts from the side of the theory of descent to turn [those investigations] to the advantage of that theory. But I cannot refrain from mentioning the recent work of the old master, Darwin, *The Descent of Man*.

Von Baer, therefore, summarizes the passage from that work cited above, and goes on to show, as he thinks, that Darwin's reasoning here is simply self-contradictory. Before outlining von Baer's attempt to prove this charge, I will quote one more example of his scorn for all evolutionists. He appends to his brochure a "postscript" apologizing for having refuted at such length the argument from the supposed similarities between the embryos of the ascidians and those of the vertebrates. "Zoologists and anatomists," he writes, "will be likely to assume that I have written for them, but they will perhaps blame me for giving so detailed an exposition, where a mere reminder would have been sufficient." But, he explains, he has not really been writing for biologists: "I have had in view the many dilettantes who believe in complete transmutations and are inclined to think it pure vanity if one is not willing to recognize in the ascidians the ancestors of man. For having, with the dilettantes in mind, presented some repetitions, I ask to be forgiven." [16] Among the "dilettantes" who then "believe in complete transmutations" (i. e. in organic evolution), were Darwin, Wallace, Huxley, Lyell, Haeckel, Asa Gray, LeConte, and virtually the entire generation of younger biologists; the complete list would show the almost incredible lengths to which von Baer's arrogance could lead him.

But it was not enough for von Baer to argue that accurate observation of the ascidians does not justify the theory that they were ancestors of the vertebrates. He was evidently eager to show that the "old master" of the entire evolutionist school was so illogical a thinker that he fell into the most absurd of formal fallacies. Darwin's argument (above quoted from *The Descent of Man*), was, von Baer declared, a simple and obvious self-contradiction.

> According to the customary fashion of reasoning (*Raisonnement*), that [character] which manifests itself very early in the development of an organism must be an inheritance from its earliest ancestors. Consequently it would be necessary to hold that the ascidians are descended from the vertebrates, and not the opposite. But it was also necessary to show the vertebrates to be descended from lower forms. To satisfy this requirement, it is argued that the opposite *is* the case [i. e., that the vertebrates are descended from the ascidians.] [17]

"The hypothesis is flexible!"—such is von Baer's curtly ironic comment on the reasoning which he here attributes to Darwin.

But von Baer, in his zeal to prove Darwin guilty of the crudest sort of self-contradiction, misrepresented, no doubt unconsciously,

[16] *Ascidiens*, 1873, p. 35.
[17] *Ascidiens*, p. 7.

the actual reasoning in the passage in *The Descent of Man* about the ascidians. Darwin had, of course, never said that the *same* ascidians are both the ancestors and the descendants of the vertebrates, nor that *any* ascidians are descended from vertebrates. What he had said (let us recall) was, in substance, this: (a) that the "*present* ascidians," those which we can directly observe, are manifestly invertebrates; but that (b) their larvae, if Kowalevsky's observations were correct, pass through a temporary stage in which, like the larvae of the vertebrates, they possess a *chorda dorsalis*; (c) that this "close resemblance" of the larvae of two organisms of completely dissimilar classes cannot be assumed to be merely a queer accident; and therefore that (d) some intelligible explanation of it not inconsistent with the usual and tested requirements of scientific inference must be sought. Darwin's tentative explanation was—not that the class "ascidians" is descended from the class "vertebrates" —but that the extant ascidians and the vertebrates are both "derived from a group of animals" existing "at an extremely remote period" which had a *chorda dorsalis*, and passed this structural feature on to their posterity. But their descendants have lost it in their various adult forms, though they have retained it in their larval stages, thereby giving evidence of their common remote ancestry. But, in the long course of ages, those descendants became differentiated into two classes. In one of them, the present ascidians, the notochord was simply sloughed off in the adults, because it had no longer any utility in the inactive, almost vegetal, mode of life they had adopted—as the tail of man's more distant simian ancestors was sloughed off when the life of his closer ancestors ceased to be "mainly arboreal." The other class, thanks to some chance variation arising in their less remote ancestors, did not simply slough off the notochord but replaced it with a flexible vertebral column, adapted to maintaining the posture which had become favorable to their success in the struggle for survival—and so were no longer chordates but vertebrates, the most diversely active of all animals— "the crown and summit of the animal kingdom." I do not wish to discuss here whether there is not a somewhat speculative element in Darwin's reasoning here; obviously his postulated primordial group of chordates could not now be verified empirically. But there was nothing self-contradictory in the argument as a whole or in any part of it. Von Baer's charge of logical ineptitude in Darwin's thinking on this subject was wholly without warrant.

Having thus ostensibly disposed, on grounds of formal logic, of Darwin's argument for the ascidian ancestry of the vertebrates, von Baer next seeks to show that the observations of Kowalevsky and Kupffer on which that argument was based were incorrectly reported. He does not deny that in the embryonic stage of the ascidians there is visible a structure which resembles a "chord," later to be discarded; but it is not a *chorda dorsalis*, for it is not just inside the "back," as in a vertebrate, but is adjacent to the belly— is on the *Bauchseite*, not the *Rückenseite* of the creature. Von Baer's attempt to prove this begins with a discussion of the meaning of the word "back," and is presented at much length. Given this dubious discrimination of the front side and the back side of the ascidian larva, von Baer thinks it possible to demonstrate that the question he has propounded must be answered in the negative, and consequently that these organisms are *not* the ancestors of the vertebrates. But the "proof" is found to be wholly unconvincing by contemporary biologists: it demonstrates, rather, von Baer's own deficiencies as an observer.

For the question asked by him in 1873 is no longer an open question; it is a *chose jugée*, and the judgment rendered has been adverse to his contention. The progress of the science of comparative embryology during the eighty-five years since von Baer wrote has led biologists to agree that Kowalevsky's observations, and his and Darwin's tentative inferences from them, were essentially correct. For a single example of the now generally accepted conclusion I quote briefly from a recent (1954) and authoritative book of Professor Alfred S. Romer on the evolution of the vertebrates. Romer first describes the internal structure and habit of life of the tunicates (an alternative name for the ascidians): "In the adult tunicate there is very little nervous system of any sort, not to speak of a brain. Nor is there any trace of a notochord or any skeletal system. Nothing more unlike a typical chordate could, it would seem, be imagined." But observation of the embryo of this organism "opens up a different conception of the tunicate's position in the animal world. The larval form is rather like a tadpole in shape, with a large head region and a slim tail. In the tail is a well-developed notochord, and typical dorsal nerve-chord as well. . . . Presently it becomes attached in the head region to the bottom; active life is abandoned, the gill barrel expands, and the tail—and with it nerve chord and notochord—disappears. Seemingly, the ancestral sea-squirt was a typical primitive chordate, but the modern forms have become highly specialized and degenerate."[18] This, it will be remembered, was precisely the conclusion which Darwin suggested as probable, pending further verification, in *The Descent of Man*.

Von Baer's would-be refutation of the hypothesis of the descent of the vertebrates from invertebrate ancestors was thus based upon erroneous observations of the larvae of certain of the latter class of organisms, supplemented by a fantastic misrepresentation of Darwin's reasoning in favor of that hypothesis. But von Baer's zeal to expose what he considered the errors of Darwinism was, as has already been remarked, not limited to the special question of the ancestry of the vertebrates, which, for him, was important only because of its supposed pertinency to the larger issue concerning Darwinism in general—both the theory of evolution and the hypothesis of natural selection. The energy of his attack upon this was not diminished by age, and his most comprehensive " *Kritik des Darwinismus* " is to be found in the second volume of his posthumously published *Reden*.[19] Two of the discourses in this volume have as their theme *Zweckmässigkeit oder Zielstrebigkeit in der Natur* (*Purposiveness or the Striving towards Ends*): I, *In Nature in General*, and II, *In Living Organisms*. These reveal, even in their titles, the philosophical preconceptions which underlay von Baer's biological doctrines, and it is these preconceptions that inspire his main criticisms of Darwinism in the concluding discourse of the volume *Über Darwins Lehre*—[20] though, characteristically, a number of supposed evidences from " observation "—including, once more, the celebrated ascidians—are introduced to give empirical support to the " reflections." No less characteristically, the discourse begins in a tone of contemptuous irony:

> A resounding rumor is spreading through the countries of Europe: that the secret of creation has at last been definitively revealed. As Newton discovered the laws of the motions of the heavenly

---

[18] *Man and the Vertebrates*, I, pp. 11-14; Pelican Books, London, 1954. The author is Professor of Zoology and Director of the Museum of Comparative Zoology at Harvard University.

[19] St. Petersburg, 1876. This volume has the subtitle *Studien aus dem Gebiete der Naturwissenschaften*.

[20] *Reden*, pp. 235-480. The discourse was evidently a combination and amplification of several journal-articles published earlier.

bodies, so Charles Darwin has laid bare the laws of the formation of living beings, and has thereby brought about a still greater advance in science than did Sir Isaac Newton. One has only to give up the archaic, obsolete prejudice of a controlling purpose in the creation of the world, in order to see that it is all ruled by necessity—that, partly through variations in the forms inherited from parents, partly through the influence of the external world, the great diversity of organisms has been generated. The forms of living beings are inherited from earlier forms, with changes due to the aforesaid modifying influences.

Thus according to Darwin—von Baer continues—life has passed through a vast succession of "transformations and improvements (*Transmutationen und Verbesserungen*), while those creatures which have not improved (*die Unverbesserten*) perish." The ascent of the successive organisms has finally culminated in man (*so steigen die Verbesserungen auf bis zum Menschen*). And those dilettantes who have recently learned of the Darwinian story of the process by which tunicates or fishes or reptiles have been slowly transformed into *Homo sapiens*, "raise loud cries of jubilation and triumph."

But though von Baer could seldom resist the temptation to satirize biologists with whom he disagreed, he of course realized that he was not likely thereby to convict them of error. The discourse *Über Darwins Lehre* is mainly a serious and earnest effort to confute evolutionism, partly by already trite arguments intended to show that the internal structural similarities between organisms of different classes are not evidences of community of descent, but chiefly by a lengthy elaboration of the thesis that the constant ascent of life to higher levels and, above all, the genesis of the mental (or spiritual, *geistig*) powers and the moral consciousness of man, cannot have been produced by the purely material agencies and mechanical processes which Darwin (as interpreted by von Baer) had assumed to be the sufficient explanation of the whole process of evolution. The first type of argument is exemplified by von Baer's renewed attack upon the theory of embryonic recapitulation.[21] This he thought one of the worst of all the errors of the Darwinians because there is no such phenomenon as recapitulation at all. After summarily dismissing several other evidences for the descent-theory, he wrote:

---

[21] Von Baer described this as a post-Darwinian addition to Darwin's own theory—which, of course, it was not, as von Baer in other passages clearly recognized.

> Still less can we admit the validity of the alleged ... law that the development of the individual (Ontogeny) repeats the sequence of forms of its ancestors (Phylogeny), since the former, according to our view, shows only the transition from the more general to the more specialized, but not from one specialized form to another.

In short, the notion of recapitulation was untenable because it was inconsistent with the account of the ontogenetic process which von Baer had given in the *Entwicklungsgeschichte* in 1828—his "one universal law" of the formation of everything, from microorganisms to solar systems.

But von Baer's dominant preoccupation by 1873 was no longer merely the refutation on empirical grounds of the arguments of the Darwinians for transformism. He was now obsessed with a more fundamental issue, which could not be settled by examination of the embryos of various particular types of organisms, but only by reflection upon the total course of the history of animate beings; paleontology was more relevant to it than embryology. That that history showed a long, slow progress from exceedingly "low" forms of life to man, von Baer (as we have seen) asserted not less firmly than Darwin. But Darwin's explanation of this undeniable progress in which new and "improved" classes and orders successively appeared upon the earth was, von Baer declared, wholly inadequate. The constant *Verbesserung* of the forms and powers and modes of action of the types of organisms which have succeeded one another in time cannot be attributed to any antecedent physical causes. The biologist must realize that this great outstanding fact about animate nature must be "explained" teleologically—by final, not by efficient causes. No theory about the "descent" of classes or of species can enable us to account for their ascent. Here von Baer is obviously talking pure metaphysics—and it is the metaphysics of Aristotle, as he points out. But his metaphysical argument is introduced as a weapon against Darwin's fundamental principle that biologists must *not* resort to final causes in order to account for such phenomena as adaptations and the origination of new and more various species, better equipped for survival in the universal struggle for existence. Theories of final causes, Huxley wrote to Lyell in 1858 when seeking to convert him to Darwinism, are "barren virgins." They do not give birth to hypotheses capable of empirical proof or disproof, nor to definite experimental procedures by which the causal relations between various observable characteristics of different organisms can be ascertained. Even Kant, insistent though he had been, for metaphysical and ethical reasons, upon the assertion of an ultimate "purposiveness" in organic processes, had equally insisted that "it is infinitely important for reason in its explanation of nature's processes of production not to pass by or pass beyond the mechanism of nature, because without this no insight into the nature of things can be attained." He had even declared that "apart from causality according to mechanical laws organisms would not be products of nature at all." [22] But von Baer, though he probably had read these *dicta* of Kant's, was determined to regard final causes, and *only* final causes, as explanatory of "nature's processes of production"—that is, of the production of the previously non-existent classes and orders which have arisen in the long "progressive" course of the history of organisms.

But into these controversial issues of metaphysics and epistemology we need not follow von Baer farther, for the present paper is concerned only with the examination of von Baer's role in the history of embryology and, more particularly, with the appraisal of his contribution to the theory of embryonic recapitulation. Of these questions enough has, I think, already been said to permit a brief final summing-up—though perhaps, like von Baer, I should "ask to be forgiven" for repeating what is already evident.

Assuming the correctness of the generally accepted conclusions of twentieth-century biologists, we must, I fear, say that with respect to the question of recapitulation von Baer was *almost* completely wrong. On one point he was, so to say, accidentally right— his denial of the theory of the early Darwinians that embryos of later-evolved organisms recapitulate the adult forms of classes or orders or phyla from which they are descended. The correctness of his contention on this point I have called accidental for two reasons. (1) As a zealous and life-long opponent of Darwinism, he did not admit that later (including extant) classes, orders, or species of organisms *are* descended from any earlier classes or orders. His unsuccessful effort to prove that neither the ascidians nor any other non-vertebrates could have been ancestors of the vertebrates illustrates this. And (2) his correct denial of the recapitulation of adult characters of ancestors were merely part of a single general proposition which, as a whole, was false—viz., that there is no recapitulation in embryos of *any* characters of ancestors other than

[22] *Critique of Judgment*, Sections 71 and 78.

those of forebears of the same species (and perhaps genus) as the offspring. Von Baer's only real reason for asserting the theorem which was true was that he accepted a theorem which was not true; for though he offered observations of his own in support of the latter, the observations were either inaccurate or insufficient. Akin to these errors was his slighting of the evolutionists' evidence from the homologous and the vestigal structures in the adult and/or the embryonic forms of organisms of outwardly extremely dissimilar classes or orders.

Now there *is* a new embryology, admirably expounded in de Beer's little book *Embryos and Ancestors*. It is new in the sense that it differs in important respects from the embryology of Darwin in 1858 and 1871—and still more sharply from the embryological theses of some other early evolutionists, e. g., of Haeckel. And it is in agreement with von Baer's on the one point on which, as I have said, he happened to be right. But since, if the preceding analysis is correct, he was right on this point for the wrong reason, and since none of his distinctive theses are, so far as I know, accepted by any contemporary biologist who has specialized in this field, it is a historical error to credit him with having "laid the foundations of an accurate descriptive knowledge of the ontogenetic stages through which animals pass in their development," [23] or to accept his four laws of ontogeny as valid, in the sense which they had for him. As the most persistent antagonist, among writers eminent in their time, of the doctrine of organic evolution, as an outstanding representative among nineteenth-century naturalists of the philosophical theorem that only final causes can explain the progress of organic forms from "lower" to "higher" types, and as the denier of any inheritance, either in embryos or adults, of internal forms or organs of ancestors belonging to different and earlier classes or orders—in these respects his embryological doctrines were wholly opposed to those now regarded as indisputable.

---

[23] De Beer, op. cit., p. 2.

INDEX

Abiogenesis, 138.
Absolute, the, 415-416, 431.
Accademia del Cimento, 21.
Achillea alpina, 150.
Ackerknecht, Erwin H., 341, 348, 355.
"Acta Eruditiorum," 22.
Activity, 269, 274, 275, 280.
Actualism, 8 (cf. Uniformitarianism).
Adams, C. B., 357, 365 n.
Adams, Frank D., 10, 22, 243, 246.
Adanson, Michel, 151-7, 172.
Adaptation, 124, 128, 134, 139, 142, 325, 337, 339, 342, 353, 422, 424.
Adelmann, 46.
Agassiz, L., 304 f., 307, 319 ff., 329 f., 341, 345, 358, 387, 407, 413, 444.
Agricola (George Bauer), 17.
Akenside, Mark, 91.
Albert of Saxony, 14 f.
Albertus Magnus, 12.
Albino Negro, 63, 71.
Aldrovandi, Ulyssis, 17.
Alessandri, A. degli, 17.
Algae, 339.
Algebra, 275.
Allen, Don C., 18 f.
America, animals of, 108.
Ammonite, 229, 391.
Analogue, 402.
Anatomical feature, 422.
Anatomy, 363 n., 393 n., 398, 402, 407, 443, 451.
Anaximander, 6 ff.
Animalculists vs. ovists, 47, 62.
Ansted, D. T., 357.
Anthropomorphism, 345.
Anti-evolutionism, 367-382, 396.
Anti-mechanism, 433.
Antirrhinum, 147.
Apes, 427 f., 430.
Appetency, 357.

Apuleius of Medaura, 11.
Arduino, G., 243.
Aristotelian mechanics, 287.
Aristotelian vitalism, 268.
Aristotle, 8 ff., 10 f., 31-34, 37, 39, 42, 44, 47, 62, 116 f., 122, 162, 456.
Ark, 18 f.
Arnold, Theodore, 223.
Artificer, 420.
Artificial fertilization, 171.
Ascidian, 448-49, 451-57.
Associationist psychology, 269.
Atheism, 115, 116, 127.
Atkinson, E., 353.
Atom, atomism, 132, 278 ff., 289 f.
Augustine, 11.
Australia, 428.
Autenrieth, 293.
Avicenna, 12.

Bachelard, G., 276.
Bacon, Francis, 36 f.
Baconians, 360.
Baer, K. E. von, 82, 172, 293-314, 316-320, 322, 330 f., 341, 352, 441-457.
Baly, W., 317.
Bärenbach, 207.
Barker, J. E., 222.
Barlow, Nora, 305.
Baron, Walter, 327, 330.
Barry, Martin, 310, 311, 312, 313, 314, 315.
Batrachians, 427.
Bauer, G. (Agricola), 17.
Bauhin, Gaspard, 33, 147.
Baumgärtner, K. H., 337 ff., 343, 345, 352 ff.
Beagle, the, 259 f.
Beaumont, E. de, 22 n.
Beccaria, C. B. Marquis di, 57.
Beer, G. R. de, 440, 442, 458.

459

Bellarmine, Robert, Cardinal, 285
Benn, A. W., 414.
Bentham, Jeremy, 57.
Bergerac, Cyrano de, 123.
Bergman, Tobern, 242.
Bergson, H., 433, 435 ff.
Berkeley, G., 369.
Bernard, Claude, 54.
Berthelot, R., 278, 437 n.
Bertrand, Father, 223, 233.
Berzelius, J. J., Baron, 313.
Bête-machine, 267.
Bible, 19, 329, 330, 335, 352, 354, 364.
Biblical chronology, 3 f., 235, 241, 244.
Biblical story of creation, 276.
Bibliolatry, 4.
Biogenesis, 43, 144.
Biology, 289.
Biology as descriptive and as explanatory, 87 ff., 179 f.
Biology, mechanistic, 325 f., 343, 349, 351, 353.
Birgus latro, 335.
Blanc, Richard le, 17.
Blum, H. F., 287.
Blumenbach, 83, 178.
Boerhaave, H., 136.
Boissier, R., 124.
Bolingbroke, H. St. John, Viscount, 214 f.
Bonnet, C., 39, 51, 65, 78, 88, 116 f., 119, 124, 129, 133 f., 137 f., 144, 164 ff., 169 f., 172, 202 ff.
Borden, Théophile de, 135, 136, 138, 139.
Botany, 277.
Bourguet, L., 232.
Boyle, Robert, 24 f.
Brahm, 417.
Brassica, 36.
Braun, A., 324, 327.
Breeding experiments, 72 ff., 82.
Brongniart, 248.
Brock, J., 175 n., 199 n.
Bronn, 308.
Brown, H. G., 331.
Brücke, E. W. Ritter von, 325.
Brunet, P., 53 n.
Brunswick-Lüneberg family, 22.
Buch, Leopold von, 243.
Buchenau and Classirer, 37 n.
Büchner, Louis, 325, 326, 342, 343, 351, 352, 353, 354.

Buckland, Rev. William, 265 ff.
Buddhism, 415 f.
Buffon, Georges L. L., Comte de, 23, 44, 51, 60 n., 68, 77-82, 116, 122, 124, 126, 128-131, 133, 135, 143, 144, 153, 160, 170, 172, 175, 179, 190, 205 n., 208, 228 f., 232-237, 241, 244, 247, 250, 254, 276, 279, 281, 324, 364, 382 n., 394-398.
Bulfinger, M. de., 71.
Burke, E., 278.
Burmeister, Hermann, 331-332, 341.
Burnet, John, 7.
Burnet, Thomas, 20 f., 228, 232.
Butler, Samuel, 85, 95, 99 n., 104 n., 111, 266, 284 f.

Cabanis, P. J. G., 270, 347.
Caloric, 276.
Camden, William, 17.
Camerarius, E., 145, 158, 223.
Candolle, A. P. de, 327.
Cannon, W. B., 54.
Cardan, Jerome, 17.
Carduus, 148.
Carpenter, William B., 309 ff., 313 f., 317.
Cartesianism, 19 f., 121, 137, 224, 230, 283 f., 289.
Carus, J. V., 327, 328, 329.
Casona, 124.
Cassirer, E., 58, 174 f., 237, 240, 290 f.
Catastrophes, catastrophism, 8, 231, 239 f., 254, 260, 327, 338, 366, 371, 373.
Causation, 283.
Causative agent, 287.
Celestial influences, 10 f.
Cell theory, 68, 331.
Cesalpino, A., 17, 148, 152.
"Chain of being," 35, 37, 43, 115, 118, 119, 133, 134, 135, 137, 153, 265, 290 (v. "Escalator of being," "Scale of being.").
Chalmers, Thomas, 255.
Chambers, Robert, 323, 345, 346, 347, 357-384, 414, 428 f.
Chance, 117, 122, 126 f., 131, 138, 142, 287, 290.
Change of forms, 292.
Characteristics, acquired, 183 ff., 370, 439.

Characters, 395, 400, 420, 423, 424, 435 f., 457.
Charlton, Dr., 25.
Chatelet, Mme du, 52.
Chatelier, H. L. le, 54.
Chelidonia, 146 f.
Chemism, 419.
Chemistry, 269, 271, 273 ff., 277, 281, 286, 289 f.
Chevalier, A., 157 n.
Chorda dorsalis, 452.
Chordates, 452-454.
Christians, 115, 138, 143.
Chromosomes, 140.
Clagett, M., 11, 279.
Clarke, Samuel, 120.
Clematis maritima, 150 .
Climate, influence of, 211.
Clodd, E., 79, 85.
Cohen, I. B., 284.
Cohen, M. R., 7.
Cole, F. J., 46 n., 81.
Cole, William, 28.
Colors, 278.
Columna, F., 17.
Comparative anatomy, 115, 129, 131.
Comparative physiology, 309, 311.
Conation, 424.
Conchae Veneris, 229.
Condamine, C. M. de la, 52.
Condillac, E. G., 421.
Congenital variations, 424.
Conservatism, 278.
Constitutional Party of Germany, 346.
Continuity, 88-91, 153, 237, 280; in Leibniz, 37, 39; in Bonnet, 39.
Continuum, 281, 290.
Corpuscular philosophy, 289.
Cosmic evolution, 133.
cosmogony, 290, 429.
Cosmology, 290.
Cosmos, 121, 143.
Cotta, B., 332, 333, 334, 335, 339, 340, 341, 345, 352, 353.
Creation, 122, 128, 326-330, 354, 359, 364, 368.
Creation, act of, 115, 116, 120, 301.
Creation, date of, 3 f.
Creation, Days of, 235.
Creationism, 365.
Creationists, 8.
Creator, 128, 328, 330, 334, 341, 346 ff. (v. also God).

Crocker, L. G., 240.
Crocodilia, 392.
Croix, Abbe S. de la, 247.
Curtis, W. C., 81.
Cuvier, G., 6, 84, 248, 253 ff., 258, 267, 290, 319, 358, 363, 373, 386, 391 ff., 413, 426, 429.
Cyclical time, 8 (cf. Periodic regeneration).
Czolbe, H. C., 329, 348.

Dacqué, 86, 431 n.
Dahlberg, 148.
D'Alembert, J. Le Rond, 52, 80, 280.
D'Arezzo, R., 12.
D'Argenville, D., 231.
Darwin, Charles, 3 ff., 5, 259 ff., 265 ff., 279, 282, 291, 292-299, 301 f., 304-311, 313, 315-322, 323, 324, 327 ff., 331, 335, 338 f., 341, 349 f., 353, 355, 356 ff., 374-399, 403-408, 424, 435 f., 438-58.
Darwin, Erasmus, 76, 250 f., 414 n.
Darwin, Francis, 4, 259, 292, 307 f., 316.
Darwinism, 131, 142, 143, 297, 299, 324 f., 345, 348, 352 f.
Daubenton, L. J. M., 96, 129, 131, 135, 190, 272, 398
Degeneration in animal species, 106.
Degradation, 271.
Deism, 115.
Delage, Y., 79.
Delphinium, 147.
Deluc, J. A., 252 ff.
Deluge, 12, 33, 222 f., 259, 276.
Democritus, 276.
Depéret, 386 n., 389 n.
Derham, W., 116, 224, 244.
Descartes, R., 19 f., 22 f., 37, 43 f., 122, 137, 237, 290, 432.
Descent, 358, 364, 377, 379, 381, 383, 389, 398 f., 406, 422 ff., 426 ff., 441, 448, 450, 454-457.
Design, 57, 115 ff., 120 f., 126 ff., 131, 139, 142, 435 (v. Teleology).
Desire, 423 f.
Desmarest, N., 243 f., 247.
Determinism, moral, 138.
Deucalion Flood, 9, 11.
Development, 119.
Dianthus, 159-162.

462 INDEX

Diderot, D., 104, 114-143, 237 f., 240, 248, 279 ff., 284, 286.
Dieckmann, H., 136.
Diels, H., 7.
Diluvialism, 243 f., 252, 256.
Diogenes of Apollonia, 8.
Discontinuity, 289.
Divine intelligence, 127.
Divine will, 361.
Divisibility, 280.
Doctor Baumann (Maupertuis), 67.
Dolomieu, G. S. T. de, 243.
Dominance, 82.
D'Orbigny, 386, 389.
Drabkin, I. E., 7.
Drews, 173 n.
Driesch, Hans, 44, 62, 284.
Dualism, 447.
Du Bois-Reymond, E., 325, 343, 353 f.
Duchesne, father and son, 154, 160.
Duhem, P., 8, 282.
Dufrenoy, J. and M. L., 82.
Dynamism, 120, 123, 133, 139, 142.

Earth, 225 f., 371 f.
Earth, age of, 3 f., 364, 368 f.
Echinoderms, 343, 351.
Edwards, Milne, 313.
Ehrenberg, C. G., 313.
Einstein, Albert, 53 f., 290.
Eiseley, L., 81, 82 n., 151 n.
Eliade, M., 8.
Emboîtement, 43, 162, 164, 166, 280 (v. Encasement, Preformation).
Embryo, 127, 130, 139 f., 292, 300, 303 ff., 310, 312 ff., 317 ff., 321 f., 427, 439-446, 451, 457.
Embryogeny, 439 f., 443.
Embryology, 129, 131, 292-322, 406-411, 438 f., 442, 448, 453, 458.
Embryonic development, 352.
Embryonic resemblance, 303, 307, 316-320.
Embryo theory of nature, 240 f., 250.
Emerson, R. W., 362.
Empedocles, 121, 240.
Empiricism, 226, 241 f., 343.
Encasement, 43 ff., 61 f., 82 (v. Emboîtement, Preformation).
Encyclopédie, 136, 222, 231, 238.
Endlicher, S., 330, 331, 339.
Engelhardt, W. V., 22.

English philosophers, Schopenhauer on, 428 f.
Enlightenment, 279, 281, 284.
Entelechy, 44.
Entomology, 447.
Entropy, 141.
Environment, 272, 420, 422, 436.
Eohippus, 176.
Epicurus, 44, 121, 132.
Epigenesis, 42-48, 59, 62, 78, 81, 138, 170, 172, 293.
Epigenesists, Aristotle and Harvey as, 42.
Epigenecist, Maupertuis as, 59.
Epistemology, 457.
Epochs, sequence of (Kant on Bonnet), 200-203.
Eratosthenes, 7 f.
Erosion, 273.
Escalator of being, 271 f.
Essences, 419.
Ethics, 140, 142, 143.
Etiology, 355.
Euler, L., 53.
Evil, metaphysical theory of, 136.
Evolution, 74 ff., 114 f., 122 f., 128, 133, 135, 137-141, 278, 281, 286 f., 293 ff., 301, 323 f., 327, 331, 333, 338, 340 ff., 344, 348, 356-414, 417 f., 423-436, 440, 442, 448 f., 451, 458.
Evolution, cosmic, 417 f., 433, 436, 445.
Evolutionary natural history, 273.
Evolutionary variation, 283.
Evolutionism, 129, 143, 265, 422 ff., 437 f., 451, 458.
Existentialism, 143.
Experimental physics, 130.

Fabre, J. H., 61.
Fabricius, H., 242.
Fallacies, logical, 450 f.
Faust, 122.
Fawcett, E. D., 437.
Fee, J., 53 n., 55.
Fellows, Sir Charles, 123.
Ferdinand II, Grand Duke, 21.
Fertilization, 137 f., 140.
Fiedler, K. G., 7.
Fillon, B., 17.
Final cause, 374, 400, 435, 456 ff.
Fire, 274 f., 290.
Fischer, Kuno, 421, 431 n.

Fish, 33 f., 37, 427 f., 430.
Fiske, 217.
Fitness, 121 f., 127, 283.
Fitzroy, Capt. R., 260.
Fixity, 145 ff., 290.
Fletcher, Dr., 409.
Flood theory, 17.
Flourens, P., 84, 233, 391 n., 392.
Fluid, 276.
Flux, 417.
Foetus, 133, 139, 316.
Fontenelle, B. le B., 230, 233, 283 f.
Formative faculty, 44.
Formey, S., 73.
Fossil enigma, 6.
Forms, 379, 392, 406-411, 412.
Forms, foetal, 439, 443, 448 f.
Forms, larval, 439, 443 f., 448, 452 ff.
Forms, terminal, 446.
Forster, J. R., 195.
Fossils, 10, 123, 131, 319, 363, 368, 381 f., 385 f., 389 ff., 393.
Foster, M., 320.
Frascastoro, G., 17.
Frauenstädt, C. M. J., 431 n.
Frederick II of Prussia, 56, 72.
Frederick William of Prussia, 75.
Freeman, K., 7.
Freiburg Mining Academy, 242.
French Revolution, 277.
Fresne, R. du, 13.
Fungus, 344.
Future transformation, Kant on, 204.

Galápagos Islands, 282.
Galileo, 19, 268, 276, 282, 286 f.
Garner, R. L., 217 n.
Gärtner, J., 163.
Gassendi, Pierre, 117.
Generatio aequivoca, 357, 425 f.
Generation, 425 f.
Generations, alternation of, 343 f.
Generationswechsel, 343.
Geners, 68, 140, 439.
Genesis, 4, 19, 30, 236, 255; (for Kant, 199 ff.).
Genetic dominance, 70.
Genetic particles, 126, 140.
Genetic recombination, 127, 139.
Genetics, 267 f., 280, 285, 289, 394, 398.
Genotype, 168.
Genus, 33, 298, 340.

Geo-chronology, 4.
Geoffroy, 67.
Geographical distribution, 295 ff., 302, 307.
Geographical isolation (in Maupertuis), 60, 76, 82.
Geography, 341.
Geological time, 432.
Geology, 246, 258, 269, 273, 275, 330, 337, 347, 349, 357, 363 f., 371, 398, 426, 429, 442.
Geometric method, 283.
Germ, 166 f., 172.
German Scientists and Physicians, Society of, 325, 329, 349.
Germ plasm, 289.
Gesner, C., 17 f.
Giants, 11.
Giard, A. M., 85.
Gillispie, C. C., 246, 252, 259, 366 n., 367 n.
Glass, Bentley, 125, 127, 133, 324, 398 n.
Glisson, Francis, 132.
Gmelin, L., 147, 154, 242.
Goby-like fish, 11.
God, 115, 117, 120, 127, 131, 136, 224 f., 330, 335, 339, 342, 345 f., 361 n., 364 f., 368, 382, 401, 403, 411 f., 429 (v. also Creator).
Goethe, W. von, 112, 276 ff., 278, 280, 286, 290, 293, 339, 359, 415.
Goldschmidt, R. B. G., 427.
Gomperz, T., 7.
Gould, A. A., 329.
Graaf, Regnier de, 47, 61.
Gråberg, 148.
Gravitation, 283, 289.
Gray, Alonzo, 357, 365 n.
Gray, Asa, 292 f., 306, 358, 451.
Great Years, 8.
Greek philosophies of nature, 277.
Greek view of universe, 143.
Green, E. L., 149 f.
Gronovius, 7.
Günther, R. W. T., 24, 27 f., 178 n.
Guettard, J. E., 243.
Guyau, M. J., 437.
Guyenot, E., 33, 38 f., 41, 43, 82 f., 146, 152, 156, 160 n., 165, 270.

Haartman, 148.

INDEX

Haeckel, E., 173 f., 279, 293, 319, 321, 328, 331, 358, 406, 442, 451, 458.
Haeckelianism, 321.
Hagberg, K., 150 n.
Half-hybrid, 159.
Hall, Sir James, 375.
Hall, T. S., 40 n.
Haller, A. von., 132, 134, 136, 140, 171 f.
Hamilton, Sir W. R., 53.
Hamm, L. D., 47.
Hanna, W., 255.
Hartmann, Eduard von, 323.
Hartmann, Robert, 217 n.
Harvey, William, 40, 42-45, 61 f., 409.
Hatfield, E. J., 284.
Haber, Francis C., 222 ff.
Heat in the earth, 233, 250.
Heer, Oswald, 329.
Hegel, G. W. F., 143, 268, 278, 281.
Heisenberg, W., 54.
Helmholtz, H. L. F. von, 325, 353.
Helmont, J. B. van, 40.
Helvetius, C. A., 57.
Henfrey, A., 300 ff., 304, 309, 335.
Henle, F. G. J., 355.
Henslow, J. S., 259.
Heraclitos, 268, 281, 290.
Herbert, W., 163, 397.
Herder, J. G. von, 112, 136, 189 f., 207, 330, 415.
Heredity, 289, 439 f.
Heritable characters, in Kant, 181 f.
Herodotus, 7.
Hervé, G., 71, 80.
Heterogeneity, 446 f.
Heterogeneous organism, 310, 312 f.
Heterogenesis, 425 n., 428.
Hibiscus, 159.
Hilgendorf, 391.
Higher criticism, 258.
Hippolytus, 6 f.
Histoire de l'Académie des Sciences, 151.
Historicism, 278.
History, geological, 115.
Hitchcock, President E., 365 n., 392.
Hoff, C. E. A. von, 337.
Höffding, H., 431 n.
Hofmeister, W., 329.
Holbach, P. H. D. Baron d', 222, 232, 237 ff., 247f.
Holmes (in Diderot), 120.

Holmes, S. J., 294, 352.
Homeostasis, 54.
Homogeneous organism, 310, 312 f.
Homologies, 96 ff., 210, 401-406, 422, 426 f., 450.
Homo sapiens, 191, 201.
Hooke, R., 24 ff., 121 f.
Hooker, J. D., 293, 307 f., 358.
Horner, L., 260.
Horace, 140.
Human race, 447.
Human races, 428.
Human species, how originated (Kant), 190 ff.
Humboldt, A. von, 7, 22, 243, 332.
Hume, David, 58, 250 f.
Hunter, (probably William), 409.
Hutton, J., 16, 231, 244 ff., 250, 252, 254, 258, 365 f.
Huxley, J., 131, 286.
Huxley, L., 267, 294, 314.
Huxley, T. H., 52, 65, 79, 113, 167, 267, 294-297, 300 ff., 304, 306-309, 311-315, 320 f., 331, 358-363, 374 f., 383, 392 f., 396, 411, 413, 442, 451, 456.
Hybridization, 158-163.
Hybrids, 69, 117, 149, 158, 160, 163, 326, 392, 394-398.
Hybrids, Aristotle on, 31.
Hyoscyamus, 159.

Iceland dogs, 73.
Idealism, 343 f., 352, 354.
Ideas, Platonic, in Schopenhauer, 419, 434.
Idéologues, 269.
Ideoplasm, 289.
Immutability, 290, 350.
Imperator, F., 17.
Individuality, 419, 426, 445.
Individuation, principle of, in Schopenhauer, 416 f.
Infancy, length of human, 213 ff.
Inheritance, 370, 400, 439, 455.
Inheritance of acquired characteristics, 268, 287, 289 (v. Acquired characteristics).
Inorganic nature, 429.
Instincts, 447.
Invertebrates, 452, 454.
Irritability, 141.

INDEX 465

Jaeger, G., 329.
Jameson, R., 6, 253.
Jardin des Plantes, 277.
Jeans, Sir James, 54.
Journal de Physique, 169.
Judd, J. W., 373 n.
Jung, J., 33.
Jussieu, A. L. de, 151 f.
Jussieu, B. de, 152, 154.

Kant, Immanuel, 59, 112, 160, 183, 200 f., 207 f., 395, 457.
Kantianism, 415, 421, 432.
Kelleia family, 65 f., 169.
Kentmann, J., 17.
Kinds, in Schopenhauer, 419.
Ketmia, 159.
Kirwan, R., 246.
Kisch, Bruno, 331.
Knight, 70, 163.
Knox, Robert, 409.
Koelreuter, J. G., 149, 158-163, 172.
Koenig, S., 54 f.
Kofoid, C. A., 369 n.
Kohlbrugge, J. H. F., 293, 324, 355.
Kölliker, R. A. von, 425 n.
Körner, S., 175 n.
Kowalevsky, 448-453.
Koyré, A., 284.
Kupffer, 448-453.
Kurtz, J. H., 258.
Kützing, F. T., 329.

Lagrange, J. J. L., 53, 275.
Lamarck, J. B. P. A. de Monet, 76, 79, 80, 82, 139, 143, 157, 172, 248 ff., 253 ff., 265-291, 324, 335, 353, 357 f., 363, 369-372, 387, 405, 420-425, 435, 441 n.
LaMettrie, J. O. de, 116-119, 124 ff., 132 f., 137 f., 141.
Landrieu, M., 86, 269.
Lanessan, de, 85.
Lankester, E. R., 320.
Laplace, P. S. de, 429.
La Rocque, A., 17.
Lasaulx, E. von, 7 n., 11 n.
Lavoisier, A. L., 187 n., 247, 275 f., 282, 286.
Law of falling bodies, 282.
Law, laws, 79, 341, 361, 366, 394, 403, 441-447, 456, 458.
Least action, 53 ff., 83.

LeConte, J., 357 f., 379, 406, 442, 451.
Lee, H. D. P., 9.
Leeuwenhoek, A. van, 27, 33, 41, 46, 61.
Leibniz, G. W. F., 22 f., 37 ff., 43, 51, 55, 58, 67 f., 88, 91, 117 f., 120, 122, 127, 129, 132, 134, 142, 145, 178, 186, 202 f., 232 f., 237.
Leonardo da Vinci, 12 ff.
Leucojum, 159, 161.
Lhwyd, E., 27 f.
Life, explanation of, 138, 142, 445.
Life, mechanical theory of, 349.
Life, origin of, 132, 137, 141.
Life, problems of, 137, 140.
Life fluids, 271, 273, (v. fluid).
Linaria, 146 ff.
Lingula, 391.
Linnaeus (Carl von Linné), 30, 51, 129, 144 ff., 149 ff., 152, 154, 179, 241 ff., 348, 395.
Linnaean Society, 288.
Linnaean system, 290.
"Lion-marin," 124.
Lister, M., 27 f.
Locke, John, 29, 57, 215 f., 226, 421.
Lotze, R. H. L., 325.
Lovejoy, A. O., 38 n., 39, 52, 55, 57, 79 ff., 91 n., 107 n., 118, 126 ff., 142, 160 n., 220 n., 265, 290 f., 323, 329, 352, 382 n.
Lucretius, 114, 117, 121 ff., 128, 138, 411 n.
Ludolfe, M., 147.
Ludwig, Carl, 325.
Lyceum, 11.
Lyell, Charles, 228, 231, 256 ff., 266, 306, 337, 357-407, 451, 456.
Lyonnet, P., 75.
Lysenko, 156, 286.

Macaulay, T. B., 56.
MacCurdy, E., 12.
Maillet, B. de, 81 f., 123 f., 133, 139, 228-232, 239, 241, 324.
Mairan, J. J. D. de, 233.
Maître-Jan, 46.
Majoli, S., 17.
Malebranche, N., 46.
Malpighi, M., 27, 41 f., 44 ff., 61.
Malthus, Rev. T. R., 260, 282, 288.
Man, 381, 426 ff., 430, 455.
Man, antiquity of, 367 ff.

Marat, J. P., 276, 280.
Marchant, J., 147, 151, 153.
Martyn, J., 233.
Marx, Karl, 143, 268, 281, 285.
Mason, S. L., 13, 243.
Materialism, 116 f., 123, 126, 131, 133, 136, 140, 143, 325 f., 337, 342 f., 345, 351 f., 447 f., 455.
Mathematics of probability, Maupertuis, 72, 82.
Mathematics, 280, 286.
Mather, K. F., 13, 243.
Matter, 115-119, 121 f., 125 ff., 131 f., 137, 142.
Mattioli, A., 17.
Maupertuis, P. L. M. de, 44 n., 51-83, 104, 107, 112, 116, 118 f., 122, 125-133, 138-141, 144, 163 f., 167, 172, 175, 179, 237, 239, 281, 289, 414 n.
Marsupialia, 392.
Matthiola, 159.
Mechanics, 280, 286 f.
Mechanism, 278, 286, 419, 433-436, 455 ff.
Mechanism of sentience, 275.
Mechanism of universe, 274.
Mechanism of variation, 267, 273.
Meckel, J. F., 301, 321.
Medici, Catherine de, 17.
Meles taxus, 393.
Mendel, G., 60, 67 ff., 80, 83, 163, 289, 439.
Mentha, 36.
Mercati, M., 17.
Mercurialis, 147, 151, 154 f.
Mermaids, 124.
Mersenne, R. P., 19.
Merz, J. T., 354.
Metamorphosis, 130, 278, 343.
Metaphysics, 325 f., 339, 341, 351, 415, 421, 425, 428, 431 f., 436 f., 443, 456 f.
Meteorology, 269, 274.
Meyer, E., 327.
Michurin school, 286.
Migration, 76.
Miller, Hugh, 231, 258, 381 n., 386 f., 413.
Milton, John, 20, 259 f.
"Miltonic hypothesis," 258.
Mineralizing force or virtue, 12.
Miracles, 331, 364 f., 376 f.
Missing links, 342, 386, 388 ff., 395.
Mitchell, P. C., 314.

Molecule, 131 f., 140 f.
"Molecules intégrantes," 272.
Moleschott, J. M., 325.
Mollusca, 391.
Monads, 117, 132.
Monboddo, J. B., 124, 220.
Monism, 131.
"Monsters," 56, 120 ff., 125 f., 139, 142.
Monstres par excès and *par défaut*, 74, 155.
Montesquieu, 143.
Moral philosophy, 277.
Morgan, T. H., 439.
Morley, John, 238.
Mornet, D., 226.
Morphology, 306 f., 328, 354.
Mosaic cosmogony, 241, 244, 247 f.
Mosaic geology, 252 f., 258.
Mosaic history, 29, 222 f., 252, 255.
Moscardo, L., 17.
Moscati, 177 f.
Moses, 4, 226, 231, 235, 241 f., 258, 330, 335, 354.
Motion, 116, 119, 122, 126, 130, 133, 137, 142, 287, 289.
Müller, Johannes, 300, 315, 317, 325, 343, 351.
Muséum d'histoire naturelle, 269.
Mutant, 155, 157.
Mutation, 60, 75 ff., 82, 126 f., 146, 164, 172, 339, 343, 353, 427, 439.
Mutation, theory of, in Schopenhauer, 424, 427.
Mutationism, 424, 429, 436.

Nageli, C., 162 f., 289, 324, 331, 343 ff., 352 ff.
Nathorst, A. F., 242.
Natural causation, 376.
Natural history, 271, 277.
Naturalism, 278, 356 f., 360, 365 f., 372 f., 380 f., 396.
Naturalism, social, 281.
Natural law, 344, 346 f., 435.
Natural selection, 60, 82, 136, 142, 266, 282 f., 315, 321, 356, 363, 369 n., 370, 395, 405.
Nature, 115, 117-120, 122, 125, 128-135, 140, 142 f.
Nature, law of, 115, 131, 133, 319.
Naturphilosophie, 279, 293 f., 326 f., 330, 347-353, 355.
Nauck, E. T., 337.

INDEX 467

Nautilus, 229.
Needham, Jain T., 116, 144, 170, 172.
Needham, Joseph, 81.
Negroes, why black, 197.
Neptunism, 243 f., 246 f., 363.
Neumayr, 391.
Newton, I., 28 f., 51 ff., 57 f., 62, 117, 120 f., 226, 232 f., 268, 276 ff., 282 f., 286, 290, 455.
Newtonianism, 224, 280.
Newtonian science, 115, 122, 269, 279, 281.
Nicotiana, 158 f.
Nietzsche, F., 437.
Nieuwentijdt, 116.
Niklaus, Robert, 119.
Noah, 12, 18, 32, 73.
Nollet, J. A., 116.
Nordenskiøld, E., 33 n., 79, 325 f.
Nostradamus, 278.
Notochord, 452 f.
Novelty, 370 f., 433, 435.

Occultism, 426.
Oken, L. (Ockenfuss), 301, 382 n.
Oldenburg, H., 22.
Ontogeny, 406-411, 440-443, 456 ff.
Ontology, 441.
Oppenheimer, Jane, 46 n., 327, 442.
Orang-outang, 124, 134.
Orangutan, 341, 347.
Order, 120 f., 131, 375.
Orestes, 11.
Organic existence, 335 f.
Organic evolution, 122, 132, 134, 135, 141.
"Organic molecules," 116, 124, 126, 128, 131 ff.
Organism, 115, 117, 120, 138 f., 141 f., 278, 280, 284, 286, 299, 310, 316, 318, 325, 331, 333, 338 ff., 345, 351, 419 f., 422, 426, 434. 436.
Organization, 132, 141 f.
Origin of Species, 292-322, 323 f., 331, 338, 349, 351, 355.
Organs, 399 ff., 403, 409, 420, 423 f., 450.
Orthogenesis, 436.
Osborn, H. F., 37 n., 79, 85, 174, 207, 210, 240 n., 326.
Ovid, 25 f.
Ovism, 145, 169.

Owen, Richard, 297, 314, 317, 363 n., 381 n., 393 n., 402.

Packard, A. S., 79, 84.
Paleobotany, 248.
Paleontological indices, 266.
Paleontology, 248, 339, 341, 363 f., 381-394, 443, 456.
Paleozoic Age, 330.
Paley, W., 226, 251 f., 259, 277.
Palingenesis, 200 f.
Palissey, B., 17, 229.
Pallos, P. S., 243.
Poliner, 27
Pangenesis, 68 f., 81.
Panpsychism, 126.
Parallelism, 301, 319, 321.
Parasitism, 343.
Paris Basin, 248, 254.
Parkinson, J., 255.
Parthenogenesis, 51, 169.
Pasteur, Louis, 331, 355.
Patagonia, 260.
Patterson, Louise D., 121.
Paul, Cedar, 355.
Paul, Eden, 355.
Pell, 24.
Peloria, 146 ff., 151, 154.
Perfect Being, 120.
Perfection, 331.
Periodic regeneration, 8 (v. cyclical time).
Perrier, E., 79, 145.
Persistent types, 391-398.
Petrifaction, 12, 24 f., 346.
Phenomena, 416 f., 419, 434.
Philo (or pseudo Philo), 8.
Philosophes, 226 f., 238, 240 f., 244, 248, 250.
Philosophy, 283, 325, 332, 334, 341.
Phlogiston, 276; in Kant, 187.
Phyla, 343.
Phylogeny, 298, 406-411, 431, 440-443, 456.
Physics, 275, 277, 281, 283, 286 f., 289 f.
Physiology, 271, 273, 325, 339, 353.
Picard, 236.
Pictet, F. J., de la Rive, 236.
Pittendrigh, Dr. C. S., 268.
Piveteau, Jean, 128 f., 233.
Planorbis multiformis, 391.
Plato, 34 f., 290.
Platonic year, 8 (v. Great Year).

Platonism, 419 ff., 434.
Playfair, J., 246 f., 366 n.
Pledge, H. T., 55 n.
Plenitude, principle of, 118, 127.
Pliny, 11.
Plot, Robert, 28.
Pluche, Abbe Noel A., 226 f.
Plutonism, 243 f. (v. Vulcanism).
Pneumatic School, 276.
Political economy, 289.
Polydactyly, 62 ff., 67, 70 ff., 163, 169.
Polyp, 116 ff., 129, 132, 134, 317.
Pompeckj, J. F., 324.
Pope, Alexander, 123, 214 f.
Potonié, H., 324, 327, 329.
Poulton, Sir Edward B., 357 n.
Powell, Baden, 410.
Preformation, 31, 43 f., 46 f., 61, 76, 78, 80, 82, 162 f., 166, 169, 172 (v. Emboîtement, Encasement).
Preformationism, in Kant, 186, 188.
Prévost, C., 256.
Priestley, J., 276.
Principles, 276.
Process, 418 f., 421, 423, 426.
Progress, 331 f., 334, 341, 343, 345, 354 f.
Progressionism, 209, 381-391.
Prototype, 128, 130, 134 f.
Providence, 280, 290.
Psychology, 356.
Psychology, 269, 271, 273.
Psycho-physical dualism, 119, 138.
Purposiveness, 284, 435, 454, 457.
Pyrotic theory, 275 f.

Quadruped, 130.
Quatrefages, J. L. A. de, de Breau, 79, 388, 397.
Quantitative mechanism, 116.

Races, 371, 395.
"Races" and "varieties" distinguished from "species," 104, 178, 182.
Rádl, E., 79, 86 n., 93 n., 174, 284, 324, 352, 422, 431 n.
Raspe, R. E., 243.
Rather, L. J., 351.
"Rational philosophy," 130.
Rathles, 313.
Raven, C. E., 23, 28, 35 n.
Ray, John, 27 f., 30, 33-36, 232.

Réaumur, R. A. F. de, 51, 63-66, 116, 138, 150, 170.
Recapitulation, 293 f., 300, 321 f., 345, 406-411, 427, 438-458.
Recessive genes, theory of, 140.
Redi, F., 27, 40 ff., 170.
Reductionism, 434.
Reichenbach, A. P. D., 329.
Reinecke, J. C. M., 324.
Relativism, 143.
Religion, 130, 326, 360, 373, 382, 385, 413.
Remak, R., 331.
Reseda, 154.
Revelation, 128.
Revolutions, 337, 345 f.
Richardson, G. F., 385 n.
Roberts, H. F., 148 n.
Robinet, J. B. R., 81, 91, 134, 138, 324.
Romanes, G. J., 57, 304, 308, 376.
Romantic conservatism, 279.
Romanticism, 276-279, 283, 289, 432 f.
Romanticism, biological, 284, 290.
Romer, A. S., 453.
Rostand, J., 38 n., 150 n., 156, 290.
Rouelle, G.-F., 247.
Rousseau, J. J., 133 f., 216 f., 277, 285, 415.
Roux, W., 308.
Rudberg, D., 146.
Ruhe family, 63 f., 169.
Rütimeyer, L., 391.

Sachs, Julius von, 151 f., 162 f., 329.
Sagnet, Leon, 80.
Sainte-Beuve, C. A., 276.
Saint-Fond, F. de, 243 f.
Saint-Germain, B. de, 22.
Saint-Hilaire, Etienne Geoffroy, 79, 357, 367, 403, 409.
Saint-Hilaire, Isidore Geoffroy, 85.
Saint-Pierre, Bernardin de, 277.
St. Vincent, Bory de, 326.
Sarayna, T., 17.
Saunderson (in Diderot), 120.
Saurians, 385.
Scale of being, 129, 134, 190, 271.
Schaaffhausen, H. S., 339, 341 f., 345, 347, 352 ff.
Schelling, F. W. J. von, 327, 382, 415, 418.
Schindewolf, O. H., 324.
Schleiden, M. J., 330, 335 ff., 339, 352 f.

Schlotheim, Baron von, 248.
Schmidt, O., 391 n.
Schopenhauer, Arthur, 212, 323, 329, 339, 362, 415-437.
Schopenhauer, compared with Spencer, 416, 430 ff., 436.
Schultze, 173 n.
Schwann, Theodor, 325.
Scientific explanation, 266, 275, 280, 283 (Newton and Darwin).
Scientific method, 375-380.
Scientific societies, 23.
Scilla, A., 233.
Scrope, G. P., 3, 256.
"Sea-dogs," 124.
"Sea-girl," 124.
"Sea-men," 124.
"Sea-monkeys," 124.
Secondary age, 330.
Sedgwick, A., 259, 381 ff., 386, 413.
Segregation, Mendel's principle of, 69 f., 82.
Seidlitz, G., 302, 324, 329, 331.
"Semences," 123.
Seneca, 8, 242.
Sensibility, 281.
Sensitivity, 132 f., 136 ff., 141 f.
Serres, A. E. R. A., 407, 409, 440, 443.
Serres, Marcel de, 440 n.
Seward, A. C., 4, 260, 307.
Sexes, the love of, 212.
Shaftesbury, A. A. Cooper, 3d Earl, 121.
Shaw, G. B., 284 f., 437.
Shellfish, 346.
Shryock, R. H., 369 n.
Simpson, G. G., 414 n.
Singer, C., 33 n., 79, 302, 304, 309, 316 f.
Sigerist, H. E., 355.
Sisymbrium, 36.
Smellie, W., 232.
Smith, John Pye, 258.
Smith, William, 247 f., 363.
Snails, 343, 351.
Socialism, 279.
Solovine, M., 117.
Sorbonne, faculty of theology, 232.
Soulavie, Abbe G., 243 f., 247.
Soul, 325 f.
Soviet Union, 285.
Spallanzani, L., 51, 144, 170, 172, 243.
Special creation, 357-411.

Species, 118, 122 ff., 126-131, 134 ff., 138, 140, 142, 271 f., 298, 319, 394 f., 418 f., 421-427, 436.
Species, Buffonian definition of, 93.
Species, constancy of, 327, 329, 332, 334, 341 f., 349 f., 354; in Schopenhauer early, 419 f.
Species, creation of, 325, 331, 335, 343 f.
Species, mutability of, 32, 35, 265.
Species, origin of, 323 f., 327, 330, 333 338 ff., 340, 344 f., 347, 353.
Species, Platonic idea of, 419, 421 f.
Species, transformation of, in Schopenhauer, 421-423.
Species, variations of, 336, 342, 346.
Spencer, Herbert, 357-360, 379, 390, 406, 412, 416, 430 ff., 447.
Spinoza, B. de, 137, 415.
Spontaneous generation, 30-33, 39-43, 61, 116, 122 f., 132, 138, 145, 169 f., 275, 330 ff., 334, 336, 338-341, 344 f., 347 f., 350 f., 354, 360, 425 f.
Sprengel, Konrad, 163.
Sprenger, 146.
Stability, 122 f., 142.
Stahl, G. E., 136.
Steno, Nicolaus (Nils Steenson), 21 f., 27 f., 47, 232.
Sterility of hybrids, 70, 98, 109 f., 176, 394-398; in Kant, 180.
Stieda, L., 294, 296 ff.
Stirring of fluids, 275.
Stoics, 137, 290.
Strabo, 10.
Stratigraphy, 242, 247 f.
Structure, 357, 398-406, 409, 420 ff.
Struggle for existence, 211, 288, 436, 456.
Sub-species, 336, 342.
Substance, 117 f., 133, 137.
Sudhoff, K., 350.
Supreme Being, 117.
Swammerdam, J., 27, 41-45, 61.

Taxonomy, 266, 269, 274, 290.
Teleology, 120, 139, 301, 325, 335, 353, 374 f., 422, 435, 454-458 (v Design).
Teleology and mechanism, Kant, 175, 195-200.
Temple of nature, 226, 258.
Temporal world, 418, 431 f., 437.
Tennyson, Alfred, 382 n., 410.

Tertullian, 12.
Tertiary age, 330.
Thales, 6, 242, 276.
Thalictrum lucidum, 150.
Theism, 428.
Theologians, 283.
Theology, natural, 223 ff., 330.
Theophrastus, 9 ff.
Thermodynamic processes, 287.
Thomson, J. A., 174 n.
Thorndyke, L., 222.
Tiedemann, F., 407, 409.
Time, 4, 281, 287, 418, 420 f.
Toland, John, 116, 137.
Torrey, J., 123.
Tournefort, J. de, 33 f., 145, 153, 179.
Tourneux, J., 129.
Townsend, Rev. J., 248.
Tragopogon, 149, 158 f.
Transcendentalism, 275.
Transformation, 295, 298-301, 307, 362 f., 367, 372, 394 f., 411, 421.
Transformation, law of, 339.
Transformism, 114-143, 324, 329, 332, 334 f., 343, 345, 350, 353 ff., 372, 380-414, 422, 429, 456.
Transmission of acquired characteristics, 124, 127, 134, 138, 141.
Transmutation, 268, 271, 296 f., 299, 327, 329, 333 f., 336, 339 f., 343, 345, 347, 349-355, 357, 359 f., 373 n., 374, 381, 386, 393, 397, 403, 411.
Trautschold, 391.
Trembley, A., 51, 116 ff., 122, 124, 128, 166.
Treviranus, G. R., 313.
Trial and error, 120 f., 126, 133.
Turton, W., 242.
Tyndall, J., 376.
Type, 34 f., 391-398.

Uhlmann, E., 324, 328, 335, 343.
Understanding, 417.
Unger, F., 324, 330, 335, 339-342, 345, 352 ff.
Uniformitarianism, 8 ff., 246 f., 256 ff., 365 ff., 369, 371-380, 384; in Lyell, 266; in Lamarck, 273, 275.
Urpflanze, 339 ff., 354.
Urstoff, 429.
Urtypus, 445.
Ussher, Archbishop James, 4, 364.

Vaillant, 145, 154.
Vallisneri, A., 41 f., 46, 171.
Variation, 144 ff., 283, 370, 392 f., 424, 427, 436.
Vartanian, Aram, 116, 123 f., 128.
Vaughan, T., 134.
"Veau-marin," 124.
Venel, 281.
Verbascum, 159.
Verbena, 148 f.
Verniere, P., 137 f., 280.
Veronica, 148.
Vertebrates, 300, 317 f., 383, 401, 408, 421 f., 427.
Vesicula umbilicalis, 312.
Vestiges of Creation, 323, 334, 345-347, 353, 428.
Vestigiarianism, 361.
Viability, 122, 125, 128 f., 131, 133, 141.
Victorians, 410 f.
Vincent of Beauvais, 12.
Virchow, Rudolf, 329, 331 f., 341 f., 345, 348-353.
Vis lapidificativa, 12.
Vital force, 325, 337, 415, 426, 436 f.
Vitalism, 268, 275, 325, 437.
Vivarais, , 247.
Vogt, Carl, 325 f., 330, 341, 345-348, 351 ff.
Voisins, d'A. de, 243.
Voltaire, F. M. A. de, 52-57, 80, 115, 227 ff., 277.
Voyage of the Beagle, The, 305, 308.
Vries, Hugo de, 60, 67, 75, 80, 83, 157, 427, 439.
Vulcanism, 243 f., 252, 363 (v. Plutonism).

Waagen, G. F., 391.
Wagner, Andreas, 326.
Wagner, Rudolph, 297, 307, 326.
Wallace, William C., 201, 205 n.
Waller, R., 25.
Webster, E. W., 9.
Weismann, A., 69, 289.
Wells, W. C., 369 n.
Weltanschauung, 354.
Werner, A. G., 242 f.
Wernerianism, 243 f.
Whiston, W., 29, 232.
White, R. J., 278.
Whitman, C. O., 164 f.

Will, senses of in Schopenhauer, 415–419.
Willdenow, 242.
Willisell, T., 23.
Will to live, 421, 429 f.
Willughby, F., 33.
Windelband, W., 431 n.
Winter, J. G., 22.
Wolff, C. F., 44, 46, 78, 141, 162, 170, 172.
Wollaston, W., 115.

Woodward, Dr. J., 28 f., 228, 232.
World machine, 224.
Wright, Thomas, 330.

Xanthus of Lydia, 7.
Xenophanes of Colophon, 6 ff.

Zioberg, 146.
Zimmermann, W., 324, 331, 339.
Zirkle, Conway, 286, 369 n.
Zittel, K. A. von, 22.

630